Physicochemical Hydrodynamics

Physicochemical Hydrodynamics

An Introduction
Second Edition

Ronald F. Probstein
Department of Mechanical Engineering
Massachusetts Institute of Technology

WILEY-
INTERSCIENCE

A JOHN WILEY & SONS, INC., PUBLICATION

Published by John Wiley & Sons, Inc., Hoboken, New Jersey.
Published simultaneously in Canada.

To order books or for customer service please, call 1(800)-CALL-WILEY (225-5945).

For general information on our other products and services please contact our Customer Care Department
within the U.S. at 877-762-2974, outside the U.S. at 317-572-3993 or fax 317-572-4002.

Wiley also publishes its books in a variety of electronic formats. Some content that appears in print, however,
may not be available in electronic format.

Library of Congress Cataloging-in-Publication Data is available.

ISBN 0-471-45830-9

10 9 8 7 6 5 4 3 2 1

This book is dedicated with affection to my wife
Irène, whose courage, good humor, and patience
have been an inspiration to me.

Contents

Preface to the Paperback Edition

It has been less than ten years since the Second Edition of this book was published. Despite the many advances that have taken place in this period because of the book's emphasis on rational theory and fundamentals it remains fresh and current to the many fields of application of PCH, including: mechanical, chemical, and environmental engineering, and materials science and biotechnology. In the period since its publication new fields have emerged that rely in whole or in part on the foundations of PCH with the material in the book forming the bases for characterizing the particular application. Two important fields in this category are microfluidics and fluid aspects of nanotechnology. The original edition is unaltered except for the correction of a few typographical errors. Many thanks are due to Bob Esposito and his associates at Wiley who agreed to bring out the book in a less expensive edition.

Ronald F. Probstein

Cambridge, Massachusetts
January, 2003

Preface to the Second Edition

The field of physicochemical hydrodynamics has received much increased attention since the first edition. This has necessitated some revisions and updating. In addition, comments from both students and practitioners suggested that a number of topics not included should be added or topics not treated in sufficient depth should be expanded. The material essentially follows the same outline as the first edition with some topics added through appended sections along with a new chapter on rheology and concentrated suspensions. Among the new topics included are hydrodynamic chromatography, chemical reactions in electrokinetics, and surface tension induced convection. Problems have been added to complement the new material. Suggestions or answers for the problems generally are not included, but a solutions manual is available from the publisher for course instructors to aid in tailoring assigned problems.

The principles followed in the writing were the same as outlined in the original preface except that the field of rheology is now included. This preface is repeated here as given in the first edition.

In the preparation of this volume, I once more acknowledge my gratitude to Mehmet Z. Sengun and R. Edwin Hicks both of whom again provided invaluable comments which have been incorporated in the text. Thanks are also due to Howard Brenner for his discussions and comments. In so far as the book is based on the first edition, the acknowledgments still apply and to that extent are repeated here essentially unchanged.

Preface to the First Edition

Physicochemical hydrodynamics was first set out as a discipline by the late Benjamin Levich in his classic book of the same name. The subject, which deals with the interaction between fluid flow and physical, chemical, and biochemical processes, forms a well-connected body of study, albeit a highly interdisciplinary one. It has applications in many areas of science and technology and is a rapidly expanding field. The aim of this textbook is to provide an introduction to the subject, which I shall refer to here by its acronym PCH.

Emphasis is on rational theory and its consequences, with the purpose of showing the underlying unity of PCH, in which diverse phenomena can be described in physically and mathematically similar ways. The magic of this unity is shown in the similar manner in which solutes concentrate in a flow containing chemically reacting surfaces, reverse osmosis membranes, and electrodialysis membranes or the similarity of particle motions in sedimentation, centrifugation, ultrafiltration, and electrophoresis. Experimental results, numerical solutions, and reference to topics not covered are noted where they serve to illustrate a concept, result, or limitation of what has been presented. Empiricism is not eschewed, but only limited use is made of it and then only when it contributes to a better understanding of an idea or phenomenon.

The book is an outgrowth of a graduate course that I have taught for a number of years at M.I.T. under the joint sponsorship of the mechanical and chemical engineering departments. Like the course, the text is directed toward graduate students in these fields, as well as in materials science, environmental engineering, and biotechnology. An undergraduate course in fluid dynamics and a knowledge of the fundamentals of physical chemistry together with a course in advanced calculus provide sufficient prerequisites for most of the material presented. An effort has been made to include the necessary fundamentals to make the book self-contained. But because of my bias toward the "hydrodynamic" aspect, there undoubtedly has crept in the presumption of a greater knowledge of this area than of the physical-chemical ones.

The subject is a broad one, and since the aim has been to present the fundamentals, it has been necessary to limit the material covered by selecting examples that illustrate the unity of PCH and at the same time put forward its essentials. Consequently, a number of fields, including turbulence, rheology, natural convection, and compressible flows, have been omitted. Numerical methods or formal asymptotic matching procedures are also not included. There is no doubt as to the importance of high-speed computation in PCH, but, consistent with providing an introduction to the fundamentals, the book lets the student first taste the essence of PCH in the form of simple analytical solutions rather than be satiated on a banquet of detailed numerical results.

Problems, which are so important a part of a student's learning experience, are included at the end of each chapter. The problems are ordered following the sequence in which the material is set out. Some of the problems call for numerical answers where it was felt it would be helpful to the student's "feel" for the magnitudes involved. With minor exceptions, SI units are used throughout. The questions range in difficulty, with most requiring an analytic development, but with some asking only for a descriptive answer. All are intended to illustrate the ideas presented, though often the solution goes beyond the explicit discussion in the book, with the answer constituting a generalization or extension of the text material.

Every effort has been made to acknowledge the work of others. However, for pedagogical reasons reference may sometimes be to a recognized text, review, or general reference rather than to the original source, but the person to whom the work is attributed is made clear. On the other hand, the reader is sometimes referred to an early original work where it was felt the examination of the source itself was most illuminating.

Ronald F. Probstein

Acknowledgments for the First Edition

It is with deep appreciation that I gratefully acknowledge the assistance of Mehmet Z. Sengun, who carefully read the manuscript chapter by chapter as it was being prepared for publication, offered important constructive suggestions, and corrected many of the inevitable errors that appear in a textbook of this kind. In addition, I also want to express my gratitude to him for the many excellent problems he contributed. A special note of thanks and appreciation is due R. Edwin Hicks, who read the manuscript and provided numerous valuable comments and corrections which have been incorporated into the text. I wish also to thank C. Ross Ethier for contributing several of the problems and Eric Herbolzheimer, who provided me with an unpublished manuscript.

I acknowledge with gratitude the important contribution of the Bernard M. Gordon Engineering Curriculum Development Fund at M.I.T., which enabled me to be relieved of my teaching duties for a year that I might complete this book more rapidly. Thanks are due David N. Wormley, Chairman of the Mechanical Engineering Department, and Gerald L. Wilson, Dean of Engineering, who were instrumental in bringing this about.

A special note of appreciation is due Virginia Brambilla for her capable handling of many of the secretarial details, and to Robert H. Dano, who expertly prepared all of the figures.

R.F.P.

1 Introduction

1.1 Physicochemical Hydrodynamics

Physicochemical hydrodynamics may be broadly defined as dealing with fluid flow effects on physical, chemical, and biochemical processes and with the converse effects of physical, chemical, and biochemical forces on fluid flows. The interplay between the hydrodynamics and physics or chemistry, including biochemistry, may be *local* or *global*. When it is local, the principal features of the flow may be obtained without a knowledge of the physical or chemical phenomena, and the state of the flow fixes the physical and chemical behavior. When it is global, the physicochemical phenomena control the nature of the entire flow. So far as the fluid mechanics is concerned, the local effects may be considered a class of *weak interactions*, and the global ones a class of *strong interactions*. An explosion is a strong interaction since the energy release associated with the chemical reaction defines the flow. The electroosmotic flow through fine charged capillaries, such as in porous soils, is a strong interaction since the electric field defines the flow. On the other hand, the interaction of a fluid flow with a corrosion reaction at a pipe surface is a weak one since the corrosion will not affect the bulk flow. Similarly, the exothermic chemical reaction in a flow where the reacting components are dilute would only result in a small energy release and would not generally affect the bulk flow.

A distinctive characteristic of physicochemical hydrodynamics is the commonality of behaviors underlying many seemingly diverse phenomena. The commonality is brought about by two factors: (1) the similar character of the continuum, linear constitutive transport relations for mass, heat, and charge; (2) the similarity or, frequently, the identity of boundary conditions for chemical, electrochemical, and biochemical reactions; mass, charge, and heat transfer; and phase change. Throughout the book we shall attempt to exploit the behavioral similarities of these different physical and chemical phenomena.

1.2   Fluid and Flow Approximations

In our treatment of single-component or multicomponent flows of fixed composition, as well as multicomponent flows of species that are reacting, we regard the fluid as a single continuum phase that is continuously and indefinitely divisible. This ensures that all macroscopic physical, chemical, and thermodynamic quantities, such as momentum, energy, density, and temperature, are finite and uniformly distributed over any infinitesimally small volume, and enables a meaning to be attached to the value of the quantity "at a point." The basis of this *continuum approximation* lies in the assumption that the characteristic macroscopic flow scale is large compared with the molecular length scale characterizing the structure of the fluid.

The subject matter will frequently be concerned with situations where the fluid contains a dispersed phase that cannot be considered a component—for example, macromolecules, rigid particles, or droplets. In these cases the continuum approximation is assumed to hold within the suspending fluid and the dispersed phase. The concentration of the rigid or fluid dispersed phase will encompass both dilute and concentrated suspensions.

Although the continuum approximation disregards the molecular nature of the fluid, we shall have recourse to this structure when considering the origin of nonequilibrium, viscous, diffusive, and interfacial effects.

The word *hydrodynamics* is used in the title of this book rather than the more general term *fluid mechanics*, partly because of convention but also to indicate that the fluids we deal with are generally held to be "incompressible" liquids rather than gases. Insofar as the fluids are regarded as continuous, the distinction between liquids and gases is not fundamental with respect to the dynamics, provided compressibility may be neglected. A gas is much less dense and much more compressible than a liquid so long as it is not too close to or above the critical temperature at which it can be liquefied. As a consequence, pressure variations in a gas flow are associated with much larger density changes than in the flow of a liquid. However, the density in a flowing compressible gas can be regarded as essentially constant if the changes in pressure are small. The behavior of a gas flow with small pressure changes is essentially the same as that of an "incompressible" liquid flow. For a single-phase fluid in the absence of temperature gradients, the criterion for *constant-density* flow translates into the *Mach number*, equal to the ratio of the characteristic flow speed to the speed of sound in the fluid, and being small compared to unity.

A word of caution is necessary since the characterization "incompressible" is conventionally interpreted as synonomous with "constant density." However, from our remarks a low-speed flow of, say, air may be regarded a constant-density flow despite the fact that air is a highly compressible fluid. On the other hand, a solution of saltwater subjected to a centrifugal force field in an ultracentrifuge develops a strong density gradient, and the solution, though incompressible, can hardly be considered of constant density.

The fluids we will examine are *real fluids* in that they are characterized by their ability to support shear stresses; as such they are *viscous*. More generally, real fluids support viscous effects, usually termed *transport effects* in the physicochemical literature. These include diffusion of mass, heat, and charge.

Transport effects together with nonequilibrium effects, such as finite-rate chemical reactions and phase changes, have their roots in the molecular behavior of the fluid and are dissipative. *Dissipative phenomena* are associated with thermodynamic irreversibility and an increase in global entropy.

Viscous flows may be classified into the limiting regimes of laminar and turbulent flows. In *laminar flow* the motion is regular and the fluid moves as if it were layered, with each layer having a different velocity. On the other hand, *turbulent flow* exhibits an irregular and chaotic behavior, though there may be some persistence of order present. In forced convection, the motion is laminar or turbulent, depending on whether the Reynolds number is respectively small or large compared with a critical value (the *Reynolds number* is defined as the characteristic flow speed multiplied by the characteristic flow length divided by the fluid kinematic viscosity). The numerator in the dimensionless ratio characterizes the flow and measures momentum transport by convection, whereas the denominator characterizes the fluid and measures momentum transport by diffusion.

The important practical features of turbulence are the sharply increased rates of transfer and mixing compared with molecular diffusion. Although a fundamental understanding of turbulence remains elusive, there are nevertheless many technical and engineering problems that can be handled by empirical and phenomenological modeling, examples of which include the mixing length and eddy diffusivity concepts. Because our aim is to display phenomena that arise from the interplay between physical chemistry and flow, we shall restrict our considerations to laminar flow, for which there is a rational and well-defined theory at least for Newtonian fluids. Another reason for this choice is that many of the important problems in physicochemical hydrodynamics are concerned with flow systems of small scale, so the Reynolds numbers are low and the motions are laminar. Examples are fluid systems involving macromolecules and particles, porous media and capillaries, and significant interfacial forces.

We shall also examine laminar flows for non-Newtonian fluids, where the stress is not linear in the rate of strain. Such flows, which fall under the science of rheology, encompass a spectrum of materials from elastic fluids at one end to Newtonian fluids at the other. Included are polymeric fluids and suspensions, both of which play an important role in physicochemical hydrodynamics.

In the spirit of our restriction to the laminar regime, we shall only briefly touch on *natural convection*—that is, flows produced by buoyancy forces acting on fluids in which there are density differences. A common example is buoyant motion in a gravitational field where the density difference arises from heat exchange. Even in weakly buoyant motions generated by small density differences, turbulence is ubiquitous. We shall, however, consider convection induced by surface tension gradients.

1.3   Particle and Pore Geometry

Throughout the book we shall frequently deal with suspensions of small "particles," including macromolecules, colloids, cells, and flocs. The geometry of these particles is important for defining their interactions with the fluid

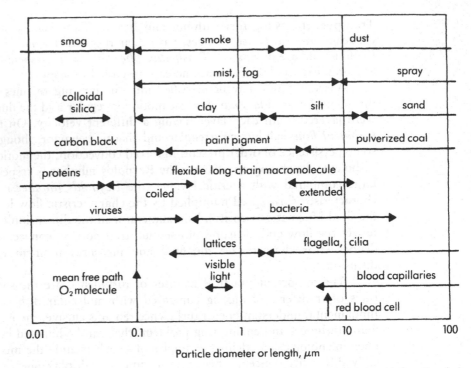

Figure 1.3.1 Some particles of interest and their characteristic sizes (after Batchelor 1976).

system. Often their shapes are complex, and nonrigid particles may differ in configuration under static and dynamic conditions and in different environments. Figure 1.3.1 shows some of the particles of interest and their characteristic sizes.

We will generally consider macromolecules to represent the smallest dispersed phase not considered a component. A *macromolecule* is a large molecule composed of many small, simple chemical units called *structural units*. It may be either biological or synthetic. Biological macromolecules contain numerous structural units, in contrast to synthetic macromolecules. Sometimes all macromolecules are referred to as *polymers*, although a polymer may be distinguished as a macromolecule made up of repeating units. Polyethylene, for example, is a synthetic polymer built up from a single repeating unit, the ethylene group. It has a simple linear chain structure in which each structural unit is connected to two other structural units.

A *protein* is a biological macromolecule composed of amino acid residues of the 20 common amino acids, joined consecutively by peptide bonds. Hemoglobin, the oxygen-carrying protein in red blood cells, is nearly spherical, with a diameter of about 5 nm (Stryer 1988). A model of the hemoglobin molecule as deduced by Perutz (1964) from x-ray diffraction studies is shown in Fig. 1.3.2. The model is built up from blocks representing the electron density patterns at various levels in the molecule. A larger protein, one that is fundamental to the blood-clotting process, is fibrinogen, a long slender molecule with a length of

5 nm

Figure 1.3.2 Model of hemoglobin deduced from x-ray diffraction studies. [After Perutz, M.F. 1964. The hemoglobin molecule. *Sci. Amer.* 211(5), 64–76. Copyright © 1964 by Scientific American, Inc. All rights reserved. With permission.]

about 50 nm (Stryer 1988). On a scale often an order of magnitude larger are viruses, which are very symmetric rigid macromolecules consisting of infectious nucleic acids surrounded by coats made up of protein subunits. Figure 1.3.3 is an electron micrograph of a tobacco mosaic virus of length about 300 nm.

Given the variety of particles and their diverse shapes, the question arises as to how they are represented or "modeled" in a rational treatment of their interactions in fluid systems. In our treatments we shall consider the particles to be regular geometrical shapes in the Euclidean sense. Thus particles will, for example, be represented by spheres (the model used most often), prolate and oblate ellipsoids of revolution, rods, and disks. Many protein macromolecules can be regarded as spherical, as, for example, hemoglobin in Fig. 1.3.2. Synthetic polymers dispersed in suspension, like the polystyrene latex particles shown in the micrograph of Fig. 1.3.4, are spherical or very nearly spherical, as are the particles of numerous colloidal systems. Many proteins can be regarded as ellipsoids of revolution. Clays and many crystalline materials are platelike and can be modeled as thin disks, and proteins such as fibrous collagen and the tobacco mosaic virus of Fig. 1.3.3 can be regarded as cylindrical rods.

Synthetic polymers and biological macromolecules are often modeled as a cluster of spheres or as a string of rods and spherical beads. The rod-and-bead configuration may be rigid, as a dumbell, or flexible, where a bead connects to two rods as in a ball-and-socket joint or jointed chain. The protein fibrinogen has the character of a linear, rod-and-bead configuration with two rods and three beads. Most synthetic polymers and many biological macromolecules are flexible because of rotations about the chemical bonds.

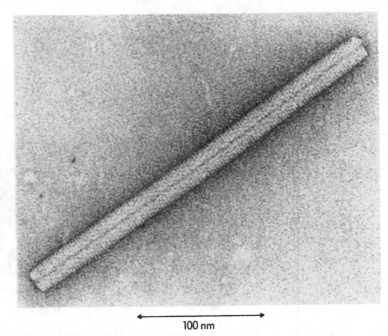

100 nm

Figure 1.3.3 Electron micrograph of tobacco mosaic virus particle. [Courtesy of Prof. Emeritus Robley C. Williams, Virus Laboratory and Dept. of Molecular Biology, Univ. of California, Berkeley.]

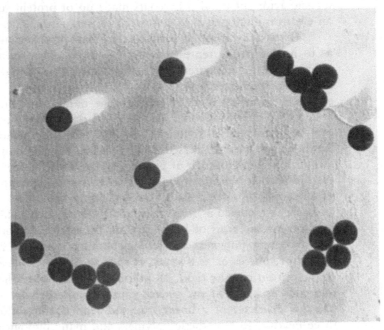

Figure 1.3.4 Electron micrograph of shadowed 300 nm diameter polystyrene latex particles. [Courtesy of Olga Shaffer, Emulsion Polymers Institute, Lehigh University.]

The individual particles of which we have spoken seem, in many cases, amenable to a relatively simple geometric description. In solution, however, particles may floc or aggregate due to random particle-particle and particle-floc collisions, and generally complex shapes arise that belie the much simpler shape of the original particle. Figure 1.3.5 from Weitz & Oliveria (1984) shows in two-dimensional projection an irreversible aggregate of uniform-size, spherical gold particles with diameter 15 nm.

The treelike cluster in Fig. 1.3.5 is one of a general class of shapes named *fractals* by Mandelbrot (1982). A fractal is a shape whose regularities and irregularities are statistical and whose statistical properties are identical at all scales, that is, scale-invariant under a change of length scale. From Fig. 1.3.5 it may be seen that associated with the cluster are open spaces with scales extending down from that of the cluster to that of a single particle. There is no characteristic length scale for the open spaces or other geometrical features between their extremes, indicating dilation symmetry. If a sphere of radius r is

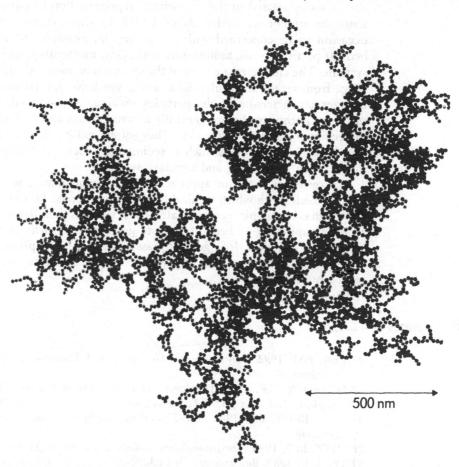

500 nm

Figure 1.3.5 Electron micrograph of irreversible aggregate formed in suspension of spherical gold particles with diameter 15 nm. [Courtesy of Dr. David A. Weitz. From Weitz, D.A. & Oliveria, M. 1984. Fractal structures formed by kinetic aggregation of aqueous gold colloids. *Phys. Rev. Letters* 52, 1433–1436. With permission.]

drawn around an arbitrary point on the cluster and the number of particles N is counted, it is found that $N(r) \sim (r)^{1.75}$ for r ranging from about the particle size to the cluster size (Weitz & Oliveria 1984). It is argued theoretically that the hydrodynamic interactions of the cluster in a low-speed, inertia free flow (low Reynolds number flow) are as if the cluster were a hard sphere of radius a spanning the cluster (Witten & Cates 1986). Such behavior provides some justification for the approach of representing even irregularly shaped particles by regular geometrical shapes.

An additional point regarding particle suspensions is that we shall generally assume the dispersion to be *monodisperse*; that is, the particles are all the same. Most suspensions are *polydisperse*, with the particles characterized by a distribution of sizes and shapes. When we do not account for the polydispersity specifically, we will assume that the particular property being determined can be defined by an appropriate average over the particle system as, for example, number, mass, or volume average.

Closely related to the geometrical representation of particles is the representation of porous media (Adler 1992). In the text we will have frequent occasion to be concerned with porous media, examples of which are packed beds of particles, soils, sedimentary rock, gels, membranes, and many biological systems. The characteristic sizes of the channels or interstices in such media may range from several molecular diameters in synthetic, reverse osmosis membranes to sizes characteristic of the particles making up a packed bed or a natural medium. Porous media are generally heterogeneous and characterized by three-dimensional random networks. They often exhibit a fractal nature, as in geophysical environments, such as sedimentary rocks, and in biological environments, such as the lung and capillary systems.

Consistent with our approach to particle geometry, we shall assume the porous media with which we deal to be homogeneous. Moreover, we will model the media by simple geometrical means such as bundles or assemblages of straight capillaries or beds of discrete geometrically defined particles, such as spheres or cylinders. It is generally assumed that appropriate averages can be defined for a real porous media that enable the simplified models to be representative.

References

ADLER, P.M. 1992. *Porous Media: Geometry and Transports*. Boston: Butterworth-Heinemann.

BATCHELOR, G.K. 1976. Developments in microhydrodynamics. In *Theoretical and Applied Mechanics* (ed. W.T. Koiter), pp. 33–55. Amsterdam: North Holland.

MANDELBROT, B.B. 1982. *The Fractal Geometry of Nature*. San Francisco: W.H. Freeman.

PERUTZ, M.F. 1964. The hemoglobin molecule. *Sci. Amer.* 211(5), 64–76.

STRYER, L. 1988. *Biochemistry*, 3rd edn. New York: W.H. Freeman.

WEITZ, D.A. & OLIVERIA, M. 1984. Fractal structures formed by kinetic aggregation of aqueous gold colloids. *Phys. Rev. Letters* 52, 1433–1436.

WITTEN, T.A. & CATES, M.E. 1986. Tenuous structures from disorderly growth processes. *Science* 232, 1607–1612.

2   Transport in Fluids

2.1   Phenomenological Models

The principles of conservation of momentum, energy, mass, and charge are used to define the state of a real-fluid system quantitatively. The conservation laws are applied, with the assumption that the fluid is a continuum. The conservation equations expressing these laws are, by themselves, insufficient to uniquely define the system, and statements on the material behavior are also required. Such statements are termed *constitutive relations*, examples of which are Newton's law that the stress in a fluid is proportional to the rate of strain, Fourier's law that the heat transfer rate is proportional to the temperature gradient, Fick's law that mass transfer is proportional to the concentration gradient, and Ohm's law that the current in a conducting medium is proportional to the applied electric field.

The constitutive equations to be adopted are defined empirically, though the coefficients in these equations (viscosity coefficient, heat conduction coefficient, etc.) may be determined at the molecular level. Often, however, these coefficients are determined empirically from the phenomena themselves, though the molecular picture may provide a basis for the interpretation of the data. It is for this reason that the description of the fluid state based on a continuum model and concepts is termed a *phenomenological description or model*.

Sometimes when dealing with a fluid that contains a dispersed particle phase that cannot be considered a component, we treat the suspension fluid as a continuum with a constitutive relation that is modified because of the presence of the particles. An example to be discussed in Chapter 5 is Einstein's modification of the Newtonian viscosity coefficient in dilute colloidal suspensions due to hydrodynamic interactions from the suspended particles. As with molecular motions, the modified coefficient may be determined from measurements of the phenomenon itself by using results from analyses of the particle behavior in the fluid as a guide. These ideas are further expanded upon in Chapter 9 where the behaviors of concentrated suspensions of colloidal and non-colloidal particles are examined.

From physical experience we know that a flow of energy or matter can be set up in a conducting system whenever there is a spatial gradient of a state variable, for example, temperature, pressure, or voltage. All fluxes will vanish under conditions of *spatial homogeneity* where the spatial gradients of all state variables are zero. Irreversible thermodynamics provides a more specific statement. Recall that an *intensive thermodynamic property* is independent of the mass or size of the system, the converse holding for an *extensive property*. Thus, the transport of fluxes (that is, the rates of flow per unit cross-sectional area of energy and matter in a conducting or transporting medium) are determined by the nature of the transporting medium and its local intensive thermodynamic state and by the local gradients in the natural intensive properties.

The phenomenological or constitutive equations describe the manner in which the fluxes depend upon the spatial gradients of the intensive properties. In what follows we shall discuss the momentum flux, which is related to velocity gradient by fluid viscosity; heat flux, which is related to temperature gradient by fluid thermal conductivity; mass flux, which is related to concentration gradient by the fluid diffusivity; and current density, which is related to electrostatic potential gradient by specific conductivity. The fluxes of mass, heat, and charge are all vector quantities, and their transport characteristics are quite similar and often analogous. Momentum flux or stress, however, is a second-order tensor, and except when there is only a single component of stress, similarities in transport behavior to the other quantities are often limited. This may be interpreted as a consequence of the fact that the components of a vector transform like the coordinates themselves, whereas the components of a second-order tensor transform like the squares of the coordinates.

2.2   Viscosity and Momentum Transport

The real fluids we consider can support shear stresses and are viscous. We emphasize this again because real-fluid effects per se are identified with dissipation. However, dissipation may arise in a fluid not only because of viscous and other transport effects but also because of nonequilibrium effects associated with finite-rate reactions. Nonequilibrium effects can be present in an inviscid fluid, that is, a fluid unable to support shear stresses. The distinction between nonequilibrium effects, on the one hand, and transport effects, on the other, is based on the concept of spatial homogeneity. Transport effects, like viscosity, are defined as dissipative effects that depend inherently upon spatial gradients. A nonequilibrium effect is defined as a dissipative effect that is present with spatial homogeneity, with the proviso that the fluid density can change with time (Hayes & Probstein 1966).

Newton's law of viscosity states that there is a linear relation between the shear stresses and rates of strain. Let us first examine this law for the case of simple shear where there is only one strain component. For explicitness consider the planar Couette problem of a steady shear flow generated by the parallel motion of one infinite plate at a constant speed U with respect to a second fixed infinite plate, the plates being separated by a small distance h with the pressure p constant throughout the fluid (Fig. 2.2.1). The role of boundary conditions in

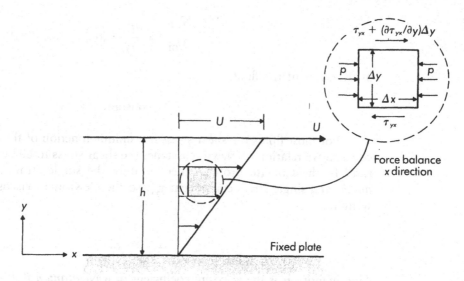

Figure 2.2.1   Shear flow between two parallel plates.

a viscous flow is critical, and we assume the *no-slip condition*—that is, the fluid "sticks" to both plates. A tangential force is required to maintain the motion of the moving plate, and this force must be in equilibrium with the frictional forces in the fluid.

A force balance for the shaded fluid element in Fig. 2.2.1 gives for the net force acting on the element in the x direction

$$\sum F_x = \left(\frac{\partial \tau_{yx}}{\partial y}\right) \Delta y \, \Delta x \, 1 \qquad (2.2.1)$$

Here, τ_{yx} is the shear stress exerted in the x direction on a fluid surface of constant y. By convention, on a positive y face the shear is positive in the positive x direction, and on a negative y face the shear is positive in the negative x direction.

From Newton's second law

$$\frac{\sum F_x}{\text{Vol}} = \rho \frac{Du}{Dt} \qquad (2.2.2)$$

where ρ is the mass density and D/Dt is the *material derivative* or rate of change of a physical quantity following a fluid element:

$$\frac{D}{Dt} = \frac{\partial}{\partial t} + u_i \frac{\partial}{\partial x_i} \qquad (2.2.3)$$

The notation D/Dt is used to distinguish from the usual total time derivative.

For steady flow with the plates infinite in the x direction, $\partial/\partial t = 0$ and $\partial/\partial x = 0$. As a consequence, the velocity component parallel to the plates is $u = u(y)$ while the normal component $v = 0$. Hence, $Du/Dt = 0$, and

$$\frac{\sum F_x}{\text{Vol}} = \frac{\partial \tau_{yx}}{\partial y} = 0 \qquad (2.2.4)$$

or, throughout the fluid,

$$\tau_{yx} = \text{constant} \qquad (2.2.5)$$

For most fluids the shear stress is a unique function of the strain rate. The constitutive relation of Newton assumes the shear stress to be linear in the strain rate. In the Couette problem there is only the single strain-rate component du/dy and single stress component τ_{yx}, so the Newtonian viscosity law may be written

$$\tau_{yx} = \mu \frac{du}{dy} \qquad (2.2.6)$$

The quantity μ is the viscosity coefficient of a *Newtonian fluid*—that is, a fluid that follows the Newtonian viscosity law. It is an intensive property and is generally a function of temperature and pressure, although under most conditions for simple fluids it is a function of temperature alone. All gases and most simple liquids closely approximate Newtonian fluids. Polymeric fluids and suspensions may not follow the Newtonian law, and when they do not they are termed *non-Newtonian fluids*. Non-Newtonian behavior falls under the science of rheology which will be discussed in Chapter 9.

In writing Newton's law of viscosity with a positive sign, we have followed the convention of applied mechanics that all stresses are positive. Chemical engineers use a negative sign in relating stress and strain rate in parallel with heat and mass transport where, for example, heat flux is proportional to the negative of the temperature gradient. The difference in convention is unimportant.

For Couette flow the shear force per unit area τ_{yx} is constant, and since it is proportional to the local velocity gradient, it follows that the velocity profile is linear. If the no-slip boundary condition is satisfied at both plates,

$$\frac{u}{U} = \frac{y}{h} \qquad (2.2.7)$$

from which

$$\tau_{yx} = \mu \frac{U}{h} \qquad (2.2.8)$$

Bird et al. (1960) point out that τ_{yx} has another interpretation. With reference to the Couette problem, in the neighborhood of the moving surface at $y = h$ the fluid acquires a certain amount of x momentum. This fluid, in turn, imparts some of its momentum to the adjacent layer of fluid, causing it to remain in motion in the x direction. Hence x momentum is transmitted to the fluid in the negative y direction. Consequently, τ_{yx} may be thought of as the viscous flux of x momentum in the negative y direction. Bird et al. argue that this interpretation ties in better with other heat and mass transport behavior, although, as we have already observed, this analogy is not necessarily appropriate in multidimensional flows.

The units of viscosity are defined by Newton's law of viscosity and are shown for some of the common systems of units in Table 2.2.1. The ratio

$$\nu = \frac{\mu}{\rho} \qquad (2.2.9)$$

occurs frequently in viscous flows and is termed the *kinematic viscosity*. It has dimensions $[L^2][T^{-1}]$, and its units are given in Table 2.2.1. Of importance is that ν has the same dimensions as the coefficient of diffusion in a mass transfer problem and may be interpreted as a diffusion coefficient for momentum.

Tables 2.2.2 and 2.2.3 give the viscosity of water, air, and some other common gases and liquids. An important point to observe is that for gases the viscosity increases with temperature, whereas for liquids the viscosity usually decreases. The dependence on pressure is not strong.

Table 2.2.1
Units of Viscosity

Name	Symbol	SI	cgs	British Grav.
Viscosity	μ	$N\,s\,m^{-2}$ $Pa\,s$	$dyne\,s\,cm^{-2}$ poise	$lb_f\,s\,ft^{-2}$ $slug\,ft^{-1}\,s^{-1}$
Kinematic viscosity	ν	$m^2\,s^{-1}$	$cm^2\,s^{-1}$	$ft^2\,s^{-1}$

Table 2.2.2
Viscosity of Water and Air at Atmospheric Pressure[a]

| Temperature T | | Water | | Air | |
| | | Viscosity $\mu \times 10^3$ | Kinematic Viscosity $\nu \times 10^6$ | Viscosity $\mu \times 10^5$ | Kinematic Viscosity $\nu \times 10^5$ |
K	°C	Pa s	$m^2\,s^{-1}$	Pa s	$m^2\,s^{-1}$
273	0	1.787	1.787	1.716	1.327
293	20	1.002	1.004	1.813	1.505
300[b]	27	0.823	0.826	1.853	1.566
313	40	0.653	0.658	1.908	1.692
333	60	0.467	0.474	1.999	1.886
353	80	0.355	0.365	2.087	2.088
373	100	0.282	0.294	2.173	2.298

[a]After Bird et al. (1960).
[b]After Lienhard (1987).

Table 2.2.3
Viscosity of Some Gases at Atmospheric Pressure and Some Saturated Liquids[a]

	Temperature T		Viscosity $\mu \times 10^5$	Kinematic Viscosity $\nu \times 10^5$
	K	°C	Pa s	$m^2 s^{-1}$
Gas				
Hydrogen	300	27	0.896	10.95
Carbon dioxide	250	−23	1.259	0.581
	300	27	1.496	0.832
	350	77	1.721	1.119
Carbon monoxide	300	27	1.784	1.567
Nitrogen	300	27	1.784	1.588
Oxygen	300	27	2.063	1.586
Liquid			$\mu \times 10^3$	$\nu \times 10^6$
Carbon dioxide	260	−13	0.115	0.115
	280	7	0.092	0.104
	300	27	0.059	0.088
Freon 12	300	27	0.254	0.195
Ethyl alcohol[b]	293	20	1.194	1.513
Mercury	300	27	1.633	0.12
Glycerin	293	20	1412.	1120.

[a]After Leinhard (1987).
[b]Viscosity from Bird et al. (1960); kinematic viscosity calculated.

The reason for the different behaviors of viscosity with temperature lies in the different mechanisms of momentum transport in gases, where the molecules are on average relatively far apart, and in liquids, where they are close together. The origin of shear stress arises from molecular motions in which molecules that move from a region of higher average transverse velocity toward a region of lower average transverse velocity carry more momentum than those moving in the opposite direction. This transfer of excess molecular momentum manifests itself as a macroscopic shear. In a gas the momentum transport of the molecules from a region of lower to higher velocity, or vice versa, is proportional to the random thermal motion or mean molecular speed. Calculation leads to a coefficient of "momentum diffusivity" or kinematic viscosity

$$\nu \sim \bar{c}l \qquad (2.2.10)$$

where l is the mean free path between collisions and $\bar{c}$ is the mean molecular speed, a quantity that increases as the square root of the absolute temperature.

In a liquid the situation is considerably different. Due to the close molecular packing, the molecules have a preferred motion because they acquire sufficient activation energy to "jump" to a neighboring vacant lattice site. The velocity gradient normal to the main direction of motion, du/dy, is proportional to the shear stress τ_{yx} multiplied by $\exp(-\Delta G/RT)$, where ΔG is the activation

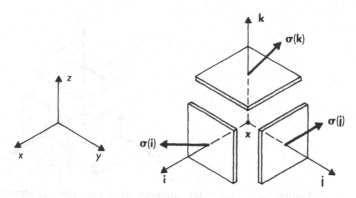

Figure 2.2.2   Reference stresses at point x in fluid.

energy for the molecule to escape to a vacant site in the fluid, R is the gas constant, and T is the absolute temperature. The exponential term characterizes the probability of a molecule in a fluid at rest escaping into an adjoining "hole." For a fluid flowing in the direction of the molecular jump, this probability is increased in proportion to the shear stress because of the additional work done on the molecules by the fluid motion. If we replace τ_{yx} by $\mu(du/dy)$, the velocity gradient terms cancel, and it follows that the viscosity of a liquid is proportional to $\exp(\Delta G/RT)$. The exponential decrease of viscosity with temperature agrees with the observed behavior of most liquids.

We next generalize the Newtonian viscosity law to three dimensions. To do so, we must recognize that the description of stress at a point in a fluid depends on the orientation of the surface element on which it acts. In rectangular Cartesian coordinates we choose as the *reference stresses* at point x and time t the values of the stresses exerted on surfaces in the positive x, y, and z directions, respectively (Fig. 2.2.2). Each of these reference stresses are vectors $\sigma(\mathbf{i})$, $\sigma(\mathbf{j})$, $\sigma(\mathbf{k})$. More generally, the stress on a surface having any orientation $\mathbf{n}$ at point x can be expressed in terms of the reference stresses. These stresses may be written in terms of their components as

$$\sigma_x = \tau_{xx} n_x + \tau_{yx} n_y + \tau_{zx} n_z$$
$$\sigma_y = \tau_{xy} n_x + \tau_{yy} n_y + \tau_{zy} n_z \qquad (2.2.11)$$
$$\sigma_z = \tau_{xz} n_x + \tau_{yz} n_y + \tau_{zz} n_z$$

The first subscript on τ indicates the axis to which the face is perpendicular (i.e., the surface orientation), and the second the direction to which the shear stress is parallel (i.e., the force component; see Fig. 2.2.3).

The nine reference stress components, each of which depends on position x and time t, and referred to as the *stress tensor components*. In Cartesian tensor notation we may write

$$\sigma_i = \tau_{ji} n_j \qquad (2.2.12)$$

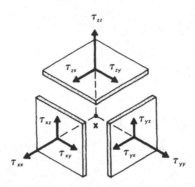

Figure 2.2.3 Cartesian components of the reference stresses (stress tensor components).

The *stress tensor* thus allows us to completely describe the state of stress in a continuum in terms of quantities that depend on position and time only, not on the orientation of the surface on which the stress acts. More precisely, the stress tensor should be referred to as a "tensor of second order" or "tensor of second rank" because its components transform as squares of the coordinates. We shall, however, simply use the term *tensor*, since tensors of order higher than second generally are not dealt with in fluid mechanics. We note in passing that a vector is a "tensor of first order," its components transforming like the coordinates themselves, and a scalar is a "tensor of zeroth order," a scalar being invariant under coordinate transformation.

The off-diagonal or shear terms in the stress tensor are symmetric; that is,

$$\tau_{ji} = \tau_{ij} \qquad i \neq j \tag{2.2.13}$$

This means that the stress tensor has only six independent components.

The assumptions of a Newtonian fluid generalized to three dimensions are as follows:

1. The fluid is isotropic; that is, the properties are independent of direction.
2. In a static or inviscid fluid the stress tensor must reduce to the hydrostatic pressure condition; that is,

$$\tau_{ij} = -p\delta_{ij} \qquad \text{where } \delta_{ij} = \begin{cases} 1 & \text{if } i = j \\ 0 & \text{if } i \neq j \end{cases} \tag{2.2.14}$$

3. The stress tensor τ_{ij} is at most a linear function of the *rate-of-strain tensor* ε_{ij}, where

$$\varepsilon_{ij} = \frac{1}{2}\left(\frac{\partial u_i}{\partial x_j} + \frac{\partial u_j}{\partial x_i}\right) \tag{2.2.15}$$

With these assumptions it can be shown (see, for example, Batchelor 1967) that the expression for the stress tensor in a Newtonian fluid becomes

$$\tau_{ij} = -p\delta_{ij} + 2\mu(\varepsilon_{ij} - \tfrac{1}{3}\varepsilon_{kk}\delta_{ij}) \tag{2.2.16}$$

where

$$\varepsilon_{kk} = \frac{\partial u_k}{\partial x_k} = \nabla \cdot \mathbf{u} \qquad (2.2.17)$$

is the *dilatation*.

For constant-density flows where $\nabla \cdot \mathbf{u} = 0$, the stress tensor components in rectangular Cartesian coordinates x, y, z with velocity components u, v, w are

$$\tau_{xx} = -p + 2\mu \frac{\partial u}{\partial x}$$

$$\tau_{yy} = -p + 2\mu \frac{\partial v}{\partial y}$$

$$\tau_{zz} = -p + 2\mu \frac{\partial w}{\partial z}$$

$$\tau_{xy} = \tau_{yx} = \mu \left(\frac{\partial u}{\partial y} + \frac{\partial v}{\partial x} \right) \qquad (2.2.18)$$

$$\tau_{xz} = \tau_{zx} = \mu \left(\frac{\partial u}{\partial z} + \frac{\partial w}{\partial x} \right)$$

$$\tau_{yz} = \tau_{zy} = \mu \left(\frac{\partial v}{\partial z} + \frac{\partial w}{\partial y} \right)$$

2.3   Thermal Conductivity and Heat Transport

The transfer of heat in a fluid may be brought about by conduction, convection, diffusion, and radiation. In this section we shall consider the transfer of heat in fluids by conduction alone. The transfer of heat by convection does not give rise to any new transport property. It is discussed in Section 3.2 in connection with the equations of change and, in particular, in connection with the energy transport in a system resulting from work and heat added to the fluid system. Heat transfer can also take place because of the interdiffusion of various species. As with convection this phenomenon does not introduce any new transport property. It is present only in mixtures of fluids and is therefore properly discussed in connection with mass diffusion in multicomponent mixtures. The transport of heat by radiation may be ascribed to a photon gas, and a close analogy exists between such radiative transfer processes and molecular transport of heat, particularly in optically dense media. However, our primary concern is with liquid flows, so we do not consider radiative transfer because of its limited role in such systems.

If a temperature gradient is maintained in a fluid between two points, there will be a flow of heat from the region of higher to lower temperature. Fourier's law of heat conduction states that there is a linear relation between this heat flux and the temperature gradient. By analogy with the Couette flow problem, consider two fixed plates separated by a distance h between which is contained a heat-conducting fluid. The upper plate is maintained at a temperature T_1 that is slightly greater than the temperature T_2 at which the lower plate is maintained (Fig. 2.3.1). The fluid in contact with each plate is assumed to be at the temperature of the plate with which it is in contact (the no-temperature jump condition). Heat must be supplied at the upper plate and the same amount removed at the lower plate to maintain an equilibrium temperature distribution. For this one-dimensional problem where $T = T(y)$, we may write Fourier's law as

$$q_y = -k \frac{T_1 - T_2}{h} = -k \frac{dT}{dy} \qquad (2.3.1)$$

where the limiting differential form supposes the plate separation distance h to approach zero. Here, q_y is the heat flux in the positive y direction, and the transport coefficient k is the *coefficient of thermal conductivity*. The analogy with momentum transport is evident at least in one dimension.

In a fluid in which the temperature varies in all three directions, Fourier's law may be written

$$\mathbf{q} = -k\nabla T \qquad (2.3.2)$$

where $\mathbf{q}$ is the vector rate of heat flow per unit area. Note that because heat flux is a vector, it has three components, whereas the shear stress or momentum flux is a tensor having nine components.

Since most fluids are isotropic, the coefficient k has no directional characteristics. This assumption is not valid for some solids, such as crystalline solids and laminated materials.

The units of thermal conductivity are defined from Fourier's law and are shown for some of the common systems of units in Table 2.3.1. The *thermal diffusivity* defined by

$$\alpha = \frac{k}{\rho c_p} \qquad (2.3.3)$$

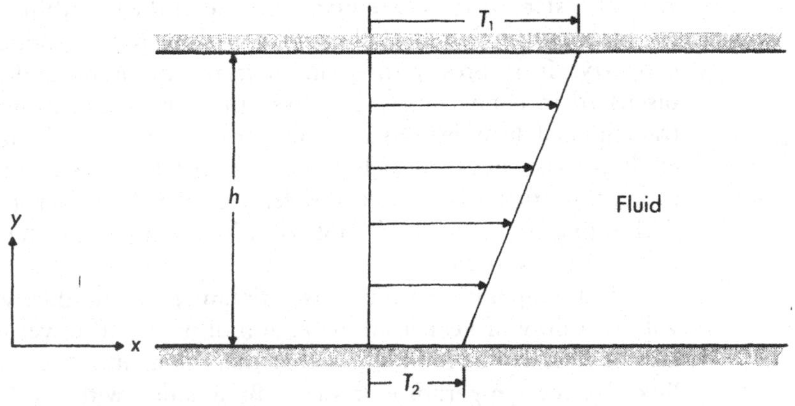

Figure 2.3.1 Temperature distribution between parallel plates.

Table 2.3.1
Units of Thermal Conductivity and Diffusivity

Name	Symbol	SI	cgs	British
Thermal conductivity	k	$W\,m^{-1}\,K^{-1}$ $J\,m^{-1}\,K^{-1}\,s^{-1}$	$cal\,s^{-1}\,cm^{-1}\,°C^{-1}$	$Btu\,hr^{-1}\,ft^{-1}\,°F^{-1}$
Thermal diffusivity	α	$m^2\,s^{-1}$	$cm^2\,s^{-1}$	$ft^2\,s^{-1}$

Table 2.3.2
Thermal Conductivity and Diffusivity of Some Gases at Atmospheric Pressure and Some Saturated Liquids[a]

	Temperature T		Thermal Conductivity $k \times 10^2$	Thermal Diffusivity $\alpha \times 10^5$
	K	°C	$W\,m^{-1}\,K^{-1}$	$m^2\,s^{-1}$
Gas				
Hydrogen	300	27	18.2	15.54
Carbon dioxide	250	−23	1.288	0.740
	300	27	1.657	1.059
	350	77	2.047	1.481
Carbon monoxide	300	27	2.525	2.128
Nitrogen	300	27	2.59	2.22
Oxygen	300	27	2.676	2.235
Liquid			k	$\alpha \times 10^7$
Carbon dioxide	260	−13	0.123	0.584
	280	7	0.102	0.419
	300	27	0.076	0.146
Freon 12	300	27	0.069	0.539
Ethyl alcohol[b]	293	20	0.167	0.891
Mercury	300	27	8.34	44.1
Glycerin	293	20	0.285	0.962
Water	273	0	0.575	1.368
	300	27	0.608	1.462
	373	100	0.681	1.683
	400	127	0.686	1.726
	500	227	0.635	1.463
	600	327	0.481	1.108

[a]Lienhard (1987).
[b]Conductivity from Bird et al. (1960); diffusivity calculated.

occurs in transient heat conduction problems. Here, c_p is the specific heat at constant pressure. For an incompressible material $c_p = c_v = c$, where c_v is the specific heat at constant volume. The quantity ρc_p is just the volumetric heat capacity. Like kinematic viscosity, α has dimensions $[L^2][T^{-1}]$ and is a measure of the rate at which heat is transported through a material.

Table 2.3.2 gives the thermal conductivity and diffusivity of some common liquids. Table 2.3.3 gives the same quantities for air and liquid sodium. An interesting observation from the data of Table 2.3.3 is that the thermal diffusivity for liquid sodium and air, which is a measure of the rate at which heat is transported, is of the same order, although the thermal conductivity of sodium is some thousand times larger.

Brenner & Edwards (1993) have examined mixed systems of materials of the same thermal diffusivity but differing thermal conductivity, emphasizing that the resultant effective diffusivity of the mixed system on the macroscale is not a constant scalar diffusivity but rather an anisotropic tensor diffusivity. This point is illustrated by the simple example of a laminated material, composed of alternating layers of equal diffusivity air a and liquid metal m of thicknesses l_a and l_m, which extends to infinity in all three spatial directions. Applying the formulas for parallel and series conductivities, the effective diffusivity in the direction parallel to the strata is just $\bar{\alpha}_{\parallel} = \alpha_a = \alpha_m$. In the limit where $k_m \gg k_a$, the relation for the effective diffusivity in the direction perpendicular to the strata reduces to (Brenner & Edwards 1993)

$$\bar{\alpha}_{\perp} = \frac{1}{\phi_a \phi_m} \frac{k_a}{\rho_m c_m} \qquad (2.3.4)$$

where $\phi_i = l_i / (l_a + l_m)$ $(i = a, m)$ is the volume fraction of the phase i. The effective diffusivity perpendicular to the strata is seen to be small because of the

Table 2.3.3
Thermal Conductivity and Diffusivity of Air at Atmospheric Pressure and of Liquid Sodium

Substance	Temperature T		Thermal Conductivity k	Thermal Diffusivity $\alpha \times 10^5$
	K	°C	W m^{-1} K^{-1}	m^2 s^{-1}
Air[a]	300	27	2.61×10^{-2}	2.20
	500	227	3.95	5.44
	700	427	5.13	9.46
	900	627	6.25	14.22
Sodium[b]	473	200	8.15×10	4.78
	673	400	7.12	6.58
	873	600	6.27	5.25

[a]Lienhard (1987).
[b]Calculated from data in Weast (1986).

low thermal conductivity of the air coupled with the high density of the metal. Thus, although each of the materials has the same isotropic thermal diffusivity, the effective macroscale diffusivity is highly anisotropic.

As with viscosity, the thermal conductivity of gases increases with increasing temperature, whereas for most liquids it decreases with increasing temperature, although the reason for this decrease is somewhat different than for viscosity. Polar liquids, like water, may be exceptions, exhibiting a maximum in the curve of thermal conductivity versus temperature. Also, as with viscosity, the effect of pressure is small.

To explain the temperature behavior of thermal conductivity, we note that by analogy with $\nu \sim \bar{c}l$ the thermal diffusivity for gases is given by

$$\alpha \sim \bar{c}l \tag{2.3.5a}$$

or

$$k \sim \rho c_p \bar{c}l \tag{2.3.5b}$$

Thus, the thermal conductivity, like viscosity, increases as the square root of the temperature. Bridgman (1923) argued that for a liquid the mean free path may be identified with the lattice spacing δ. The volumetric heat capacity will be inversely proportional to the cube of this spacing, so $\rho c_p l \sim \delta^{-2}$, with the dimensional factor in the proportionality equal to the Boltzmann constant. Bridgman also argued that the mean speed of the energy transfer due to molecular collisions is proportional to the sound speed in the liquid c_{liq}, just as in a low-density gas. It follows that

$$k \sim c_{\text{liq}} \delta^{-2} \tag{2.3.6}$$

The sound speed in a liquid decreases with increasing temperature for most liquids, as does δ^{-2}, so the conductivity decreases. Polar liquids are exceptions, and, for example, in water the sound speed increases with temperature. The decrease in δ^{-2} is insufficient to compensate this increase, so the conductivity of water increases with temperature, as observed earlier.

2.4 Diffusivity and Mass Transport

A transport of mass or "diffusion" of mass will take place in a fluid mixture of two or more species whenever there is a spatial gradient in the proportions of the mixture, that is, a "concentration gradient." Mass diffusion is a consequence of molecular motion and is closely analogous to the transport of heat and momentum in a fluid.

Consider a two-component or binary system, for example, a glass of water into which a drop of colored dye is injected. As is known from experience, the dye will diffuse outward from the point of injection where the concentration is highest to the other portions of the water where there is no dye. The transport of the dye molecules is equal and opposite to the transport of the water molecules, and after a sufficient time an equilibrium state is achieved of a uniform mixture of dye and water. This is "ordinary" binary diffusion, which will be the principal subject of our later analyses.

A diffusional flux may also be induced by imposing a pressure gradient on the system. This is the basis of centrifugal separations of mixtures, discussion of which is reserved for Section 5.5. In solution mixtures not subjected to high pressure gradients, the pressure diffusion effect is, in general, small. Diffusion can also be brought about by a temperature gradient, and this diffusional effect is known as the Soret effect. It too is usually small, provided the temperature gradients are not large. We shall not consider thermal diffusion further. Finally, external forces such as an electric field or gradient in a magnetic field can bring about diffusive effects if there are molecules or particles in the mixture that are, respectively, ionic or magnetic. The transport of particles by the application of magnetic fields falls under the subject of magnetohydrodynamics and will not be covered in this book. Ionic diffusion set up under the action of an electrical field is discussed separately in the following section, where electrical conductivity and charge transport are examined.

In a solution or mixture there are a variety of ways of defining concentration. Before introducing some of these, let us recall a few fundamental definitions. Using the example of a water molecule, we note that its mass is made up of the mass of an oxygen atom plus the mass of two hydrogen atoms: $16 + (2 \times 1) = 18$. This relative mass expressed in grams represents an amount of water called a *mole*. The mass of a mole of substance is called the *molar mass* M and in units kg kmol^{-1} has the same magnitude as the molecular weight. The molar mass of water is thus 18 kg kmol^{-1} or $18 \times 10^{-3} \text{ kg mol}^{-1}$. In general, m grams of a substance is equal to $n \equiv m/M$ moles of that substance.

Consider now a solution made up of i species. The two basic concentration units used are the *mass concentration (density)* and *molar concentration*, defined, respectively, by

$$\rho_i = \frac{m_i}{V} = \frac{\text{mass of species } i}{\text{volume of solution}} \quad (\text{kg m}^{-3}) \tag{2.4.1}$$

$$c_i = \frac{n_i}{V} = \frac{\text{no. of moles of species } i}{\text{volume of solution}} \quad (\text{mol m}^{-3}) \tag{2.4.2}$$

Two related dimensionless concentrations are the *mass fraction* and *molar (mole) fraction*, defined, respectively, by

$$\omega_i = \frac{\rho_i}{\rho} = \frac{\text{mass concentration of species } i}{\text{mass density of solution}} \tag{2.4.3}$$

$$x_i = \frac{c_i}{c} = \frac{\text{molar concentration of species } i}{\text{molar density of solution}} \tag{2.4.4}$$

The *mean molar mass* of the mixture in kg mol^{-1} is

$$\bar{M} = \frac{\rho}{c} \tag{2.4.5}$$

Some useful formulas for concentrations are given in Table 2.4.1.

Table 2.4.1
Formulas for Concentrations

Mass				Molar			
Basic		Useful		Basic		Useful	
$m_i = n_i M_i$	$\rho_i = c_i M_i$	$\sum \omega_i = 1$		$n_i = \dfrac{m_i}{M_i}$	$c_i = \dfrac{\rho_i}{M_i}$	$\sum x_i = 1$	
$m = \sum m_i$	$\rho = \sum \rho_i$	$\left(\sum \dfrac{\omega_i}{M_i} \right)^{-1} = \bar{M}$		$n = \sum n_i$	$c = \sum c_i$	$\sum x_i M_i = \bar{M}$	
$\omega_i = \dfrac{m_i}{m}$	$\omega_i = \dfrac{\rho_i}{\rho}$	$\omega_i = \dfrac{x_i M_i}{\sum x_i M_i}$		$x_i = \dfrac{n_i}{n}$	$x_i = \dfrac{c_i}{c}$	$x_i = \dfrac{\omega_i / M_i}{\sum \omega_i / M_i}$	

There are a number of definitions of the "average" velocity characterizing the bulk motion of a multicomponent system, where each species is moving at a different speed, just as there are for concentration. Often the solvent velocity is used as the reference velocity. Another reference velocity is the *mass average velocity* familiar in fluid mechanics and defined by

$$\mathbf{u} = \frac{1}{\rho} \sum \rho_i \mathbf{u}_i \tag{2.4.6}$$

The quantity $\rho \mathbf{u}$ is the mass flux through a unit area normal to $\mathbf{u}$. A *molar average velocity* $\mathbf{u}^*$ may be correspondingly defined by

$$\mathbf{u}^* = \frac{1}{c} \sum c_i \mathbf{u}_i \tag{2.4.7}$$

where $c \mathbf{u}^*$ is the molar flux through a unit area normal to $\mathbf{u}^*$. In flow systems one may also be interested in the species velocity $\mathbf{u}_i$ with respect to the averaged velocity $\mathbf{u}$ or $\mathbf{u}^*$ (Bird et al. 1960).

The actual choice of reference velocity is arbitrary, and in sufficiently dilute solutions the distinction is unimportant since they all become approximately the same. Thus at infinite dilution $\mathbf{u} = \mathbf{u}^*$, and also in the special cases $M_1 = M_2 = \cdots = \bar{M}$ for any concentration. It follows that the greater the differences in the molar masses of the species present, the more dilute must the solution be for the approximation $\mathbf{u} = \mathbf{u}^*$ to be valid.

The terms in the sums of Eqs. (2.4.6) and (2.4.7) represent the individual mass and molar fluxes of each species i with respect to fixed coordinates and are written, respectively,

$$\mathbf{j}_i = \rho_i \mathbf{u}_i \quad (\text{kg m}^{-2} \text{ s}^{-1}) \tag{2.4.8}$$

$$\mathbf{j}_i^* = c_i \mathbf{u}_i \quad (\text{mol m}^{-2} \text{ s}^{-1}) \tag{2.4.9}$$

In flow systems the mass and molar fluxes, with each species motion referred to the mass average and molar average velocities are, respectively,

$$\mathbf{J}_i = \rho_i(\mathbf{u}_i - \mathbf{u}) \qquad (2.4.10)$$

$$\mathbf{J}_i^* = c_i(\mathbf{u}_i - \mathbf{u}^*) \qquad (2.4.11)$$

The fluxes $\rho_i\mathbf{u}$ and $c_i\mathbf{u}^*$ simply represent the bulk convective fluxes. In examining systems containing reacting chemical species it is generally useful to use molar fluxes.

At this point let us restrict our considerations to a binary system in which there is a spatial concentration gradient. *Fick's first law of diffusion* states that there is a linear relation between the species flux and the concentration gradient:

$$\mathbf{J}_1 = \mathbf{j}_1 - \rho_1\mathbf{u} = -\rho D_{12}\nabla\omega_1 \qquad (2.4.12)$$

$$\mathbf{J}_1^* = \mathbf{j}_1^* - c_1\mathbf{u}^* = -cD_{12}\nabla x_1 \qquad (2.4.13)$$

where $D_{12} = D_{21}$ is the *mass diffusivity* or *mass diffusion coefficient* in a binary system, with units of $m^2 s^{-1}$. These equations show that the species 1 diffuses *relative to the mixture* in the direction of decreasing mass or mole fraction in direct analogy with the transfer of heat in the direction of decreasing temperature. In a binary mixture $J_1 = -J_2$ and $J_1^* = -J_2^*$.

A particular case of interest for liquids is constant ρ where

$$\mathbf{J}_1 = \mathbf{j}_1 - \rho_1\mathbf{u} = -D_{12}\nabla\rho_1 \qquad (2.4.14)$$

The diffusion coefficient or diffusivity D_{12} is frequently referred to as the *binary diffusion coefficient*, though often the term *binary* together with the subscripts is dropped.

In gases the diffusivity is almost independent of composition, increases with temperature, and varies inversely with pressure. In liquids, on the other hand, diffusivity is strongly dependent upon concentration and generally increases with temperature. This contrasts with the thermal diffusivity and kinematic viscosity, which for most liquids decrease with increasing temperature. Tables 2.4.2 and 2.4.3 give some typical gaseous and liquid diffusivities for dilute systems.

So far we have made no statement regarding the solution concentration, except when discussing a reference velocity and other than to observe that in liquids the diffusivity is a strong function of concentration. For example, with a ternary, instead of a binary, system there would be two concentration gradients, and the diffusive flux of each species could be affected by both concentration gradients. One instance where this is not so is that of infinitely dilute solutions for which each component is unaffected by the presence of the other. Here, Fick's law for the diffusive fluxes is simply

$$\mathbf{J}_i = \mathbf{j}_i - \rho_i\mathbf{u} = -D_i\nabla\rho_i \qquad (2.4.15)$$

Table 2.4.2

Diffusivities of Some Dilute Gas Pairs at Atmospheric Pressure[a]

Gas Pair	Temperature T		Diffusivity $D_{12} \times 10^5$ m^2 s^{-1}
	K	°C	
CO_2–N_2O	273	0	0.96
CO_2–N_2 [b]	273	0	1.44
	288	15	1.58
	298	25	1.65
O_2–N_2	273	0	1.81
H_2–CO_2	273	0	5.50
H_2–O_2	273	0	6.97

[a]After Roberts (1972).
[b]After Bird et al. (1960).

Table 2.4.3

Diffusivities of Some Electrolytes and Nonelectrolytes at Infinite Dilution in Water[a]

	Temperature T		Diffusivity $D \times 10^9$ m^2 s^{-1}
	K	°C	
Electrolyte			
$MgSO_4$	298	25	0.849
$CaCl_2$	298	25	1.335
KCl	298	25	1.994
NaCl	278	5	0.919
	288	15	1.241
	298	25	1.612
	308	35	2.031
Nonelectrolyte			
Sucrose	274	1	0.242
	298	25	0.523
Glycine	298	25	1.064
Urea	298	25	1.382

[a]After Longsworth (1972).

$$\mathbf{J}_i^* = \mathbf{j}_i^* - c_i \mathbf{u}^* = -D_i \nabla c_i \qquad (2.4.16)$$

Keep in mind that the formulation is valid only for dilute solutions.

The diffusivity for dilute liquid solutions may be estimated theoretically from simple hydrodynamic considerations. Estimates for concentrated solutions are far more difficult. The dilute solution analysis will be carried out in greater detail when we examine Brownian motion, but for the moment we content ourselves with an order-of-magnitude estimate with concentration effects neglected. It is assumed that the solute particle diffusion through the liquid solvent is a consequence of its translational kinetic energy, which is about kT (per particle), where k is the Boltzmann constant. On the other hand, viscous forces exert a drag on the particle that resists its thermal motion. From Stokes' drag law for a low Reynolds number flow, the force is proportional to $\mu_2 \bar{u}_1 d_1$, where d_1 is the mean particle diameter and $\bar{u}_1$ is its mean speed. The work done by the drag over the mean distance l between collisions is therefore $\mu_2 \bar{u}_1 d_1 l$. Equating the work and energy and setting

$$D_{12} \sim \bar{u}_1 l \tag{2.4.17}$$

we see that

$$\frac{D_{12} \mu_2}{kT} \approx \frac{1}{d_1} \tag{2.4.18}$$

This result is applicable to dilute solutions of buoyant particles as well as molecules. For molecules, assuming a cubic lattice with the molecules touching, $d_1 \sim \delta \sim (\bar{V}_1 / N_A)^{1/3}$, where δ is the lattice spacing, $\bar{V}_1$ is the molar volume of the solute particle 1, and N_A is Avogadro's number. Since $\mu_2 \sim \exp(\Delta G / RT)$, the diffusivity increases exponentially with temperature, and this is generally the observed behavior in liquids.

2.5   Electrical Conductivity and Charge Transport

In this section we consider the transfer of mass that takes place in a mixture of species when, under the action of an applied electric field, unequal electrical forces act on the different species as a result of differences in the species charge. This mass transfer is termed *migration in an electric field* or simply *electromigration*. It is really a diffusion in a "preferred" direction and follows from the fact that the electric field accelerates a charged particle, which subsequently collides with other solute or solvent particles. The result is a migration rather than directed movement in the direction of the field. This is quite analogous to the particle diffusion that takes place in a concentration gradient.

For simplicity we discuss mainly solutions sufficiently dilute that the solute species and their gradients do not interact. The solution might be an un-ionized solvent containing ionized electrolytes. If a gradient in electrostatic potential is applied to the solution, there will be an electric force exerted on the ion species, which is proportional to the potential gradient. The electric field $\mathbf{E}$ is the negative gradient of the electrostatic potential:

$$\mathbf{E} = -\nabla \phi \tag{2.5.1}$$

The force exerted on a particle is the magnitude of the particle charge multiplied by the sign of the charge and the electric field. The force per mole may therefore be written as

$$-z_i F \nabla \phi$$

Here, F is Faraday's constant equal to the charge of 1 mole of singly ionized molecules, and z_i is the charge number of the species i. Note that

$$F = N_A e$$

$$= 6.022 \times 10^{23} \text{ mole}^{-1} \times 1.602 \times 10^{-19} \text{ coulombs} \qquad (2.5.2)$$

$$= 9.65 \times 10^4 \text{ C mol}^{-1} \qquad \text{(singly ionized)}$$

where e is the elementary charge.

A statement of the constitutive relation analogous to those for mass, heat, and momentum is that the flux due to migration in an electric field is proportional to the force acting on the particle multiplied by the particle concentration. The molar flux in stationary coordinates is then

$$\mathbf{j}_i^* = -v_i z_i F c_i \nabla \phi = v_i z_i F c_i \mathbf{E} \quad (\text{mol m}^{-2} \text{ s}^{-1}) \qquad (2.5.3)$$

The proportionality factor v_i is a transport property, like thermal conductivity or diffusivity, called the *mobility* because it measures how "mobile" the charged particles are in an electric field. The mobility may be interpreted as the average velocity of a charged particle in solution when acted upon by a force of 1 N mol^{-1}. The units of mobility are therefore $\text{mol N}^{-1} \text{ m s}^{-1}$ or mol s kg^{-1}. The concept of mobility is quite a general one, since it can be used for any force that determines the drift velocity of a particle (a magnetic force, centrifugal force, etc.). The flux relation can also be expressed in terms of mass by

$$\mathbf{j}_i = -v_i z_i F \rho_i \nabla \phi = v_i z_i F \rho_i \mathbf{E} \quad (\text{kg m}^{-2} \text{ s}^{-1}) \qquad (2.5.4)$$

In electrochemical studies the molar formulation is almost always used, and we shall follow that practice.

Diffusivity and mobility are directly related. This is readily shown by considering an infinitely dilute solution and applying the simple kinetic arguments used to derive the temperature dependence of diffusivity. For a particle undergoing diffusion as a consequence of its translational kinetic energy, $D \sim \bar{u} l$ (Eq. 2.4.17), with $\bar{u}$ the mean particle speed and l the mean distance between collisions. Multiplying and dividing the right-hand side of this relation by Avogadro's number and the viscous drag force exerted on a particle, we obtain

$$D \sim v(\text{Force} \times l) N_A \qquad (2.5.5)$$

Here, v is the mobility defined with a unit force per mole. The term in parentheses is the work done per particle by the viscous drag, and it is of the order of the kinetic energy per particle kT. From the definition of the gas constant,

$$R = kN_A \quad (\text{J mol}^{-1}\,\text{K}^{-1}) \tag{2.5.6}$$

we arrive at the relation

$$D_i = RTv_i \tag{2.5.7}$$

This important equation is known as the *Nernst-Einstein equation*, which we shall discuss in more detail later.

The motion of charged species gives rise to a current, which, expressed as current density **i**, is

$$\mathbf{i} = F \sum z_i \mathbf{j}_i^* \quad (\text{A m}^{-2}) \tag{2.5.8}$$

If there are no concentration gradients and no flow, the motion of the charge is due only to the applied electric field and

$$\mathbf{i} = \sigma \mathbf{E} = -\sigma \nabla \phi \tag{2.5.9}$$

where

$$\sigma = F^2 \sum z_i^2 v_i c_i \quad (\text{S m}^{-1}) \tag{2.5.10}$$

is the electrical conductivity of the solution. Equation (2.5.9) is readily recognized as an expression of *Ohm's law*. When ordinary diffusion is present and there are concentration gradients, Ohm's law does not hold because there is a contribution to the current from diffusion.

To compare electrolyte conductivities with concentration normalized, we define a *molar conductivity* by

$$\Lambda_i = \frac{\sigma_i}{c_i} = F^2 z_i^2 v_i \quad (\text{S m}^2\,\text{mol}^{-1}) \tag{2.5.11}$$

This is the conductivity a solution would have if there were 1 mole of the substance in $1\,\text{m}^3$ of the solution. Now the molar conductivities of two electrolyte solutions can often most usefully be compared if the charges of the charge carriers in the solutions are the same. If there are singly charged ions in one solution and doubly charged ions in another, the same quantities of electrolytes would contain different amounts of charge. To get around this, we define a conductivity in which a mole of charge is compared, that is, 1 mole of ions divided by z_i, which is termed one *equivalent* of the substance. The molar conductivity so defined is termed the *equivalent conductivity* and is given by $\Lambda_i |z_i|^{-1}$ with units of S m^2 equiv^{-1}. Equivalents are not recognized in SI units. We have discussed them here because conductivities frequently appear in the literature as equivalent conductivities.

Table 2.5.1.
Molar Conductivity of Some Ions at Infinite Dilution in Water[a]

Ion	Temperature T K	°C	Molar Conductivity $\Lambda \times 10^3$ S m^2 mol^{-1}	Ion	Temperature T K	°C	Molar Conductivity $\Lambda \times 10^3$ S m^2 mol^{-1}
H$^+$	288	15	30.1	OH$^-$	298	25	19.8
	298	25	35.0	Cl$^-$	288	15	6.14
	308	35	39.7		298	25	7.63
Li$^+$	298	25	3.87		308	35	9.22
Na$^+$	288	15	3.98	Br$^-$	288	15	6.33
	298	25	5.01		298	25	7.83
	308	35	6.15		308	35	9.42
Ca^{2+}	298	25	11.9	NO$_3^-$	298	25	7.14
Cu^{2+}	298	25	10.8	HCO$_3^-$	298	25	4.45
La^{3+}	298	25	20.0	SO$_4^{2-}$	298	25	16.0

[a]After Atkinson (1972).

Table 2.5.1 lists the molar conductivities of some ions in dilute solution. There is a strong temperature dependence and, as with mass diffusivity, the conductivity increases exponentially with absolute temperature. The molar conductivity also depends on the electrolyte concentration falling off with increasing concentration, the drop being more rapid for weak electrolytes than for strong ones (see, e.g., Castellan 1983).

References

ATKINSON, G. 1972. Electrochemical information. In *American Institute of Physics Handbook*, 3rd edn. (ed. D.E. Gray), pp. 5-249–5-263. New York: McGraw-Hill.

BATCHELOR, G.K. 1967. *An Introduction to Fluid Dynamics*. Cambridge: Cambridge Univ. Press.

BIRD, R.B., STEWART, W.E. & LIGHTFOOT, E.N. 1960. *Transport Phenomena*. New York: Wiley.

BRENNER, H. & EDWARDS, D.A. 1993. *Macrotransport Processes*. Boston: Butterworth-Heinemann.

BRIDGMAN, P.W. 1923. The thermal conductivity of liquids under pressure. *Proc. Am. Acad. Arts. Sci. 59*, 141–169.

CASTELLAN, G.W. 1983. *Physical Chemistry*, 3rd edn. Reading, Mass.: Addison-Wesley.

HAYES, W.D. & PROBSTEIN, R.F. 1966. *Hypersonic Flow Theory. vol. I. Inviscid Flows*, 2nd edn. New York: Academic.

LIENHARD, J.H. 1987. *A Heat Transfer Textbook*, 2nd edn. Englewood Cliffs, N.J.: Prentice-Hall.

LONGSWORTH, L.G. 1972. Diffusion in liquids. In *American Institute of Physics Handbook*, 3rd edn. (ed. D.E. Gray), pp. 2-221–2-229. New York: McGraw-Hill.

ROBERTS, R.C. 1972. Molecular diffusion of gases. In *American Institute of Physics Handbook*, 3rd edn. (ed. D.E. Gray), pp. 2-249–2-252. New York: McGraw-Hill.
WEAST, R.C. (ed.) 1986. *CRC Handbook of Chemistry and Physics*, 67th edn., p. F-58. Boca Raton, Fla.: CRC Press.

Problems

2.1   Comparison of the tabulated momentum, thermal, and mass diffusivities (ν, α, D) shows them to be of the same order for dilute gases but markedly different for liquids. Using molecular arguments, briefly explain why they should be the same for gases and different for liquids. Explain also the differences among ν, α, and D for liquids.

2.2   Show that for a dilute binary solution, where the solute mole fraction is x_1 and the molar masses of solute and solvent are, respectively, M_1 and M_2, that the mass average velocity and molar average velocity are approximately the same if the diluteness criterion $x_1|(M_1/M_2) - 1| \ll 1$ is satisfied.

2.3   Show that in a binary system the mass flux with respect to the mass average velocity is $J_1 = -J_2$ and that the molar flux with respect to the molar average velocity is $J_1^* = -J_2^*$.

2.4   The annular gap between two infinitely long concentric cylinders of opposite charge is filled with water. Charged colloidal particles are to be moved from the inner cylinder (the source) to the outer cylinder (the collector) as a result of the voltage drop across the gap. The particle volume concentration is sufficiently small that each particle may be assumed to behave independently of the others.

The width of the annular gap is 0.02 m, and there is an applied voltage drop of 2 V across the gap. The particles are spherical, have a radius a of 1 μm, a density ρ twice that of water, and carry a charge q of 10^{-14} C. The temperature is constant at 300 K, and gravitational effects may be neglected.

The drag force on a spherical particle moving in water at low speeds (low Reynolds number) is $F = -6\pi\mu aU$, where μ is the viscosity of the water and U is the particle speed.

a.   Carry out an order-of-magnitude analysis to show that the particle inertia can be neglected.

b.   Define the particle diffusivity.

c.   Give a criterion to measure the effect of diffusive forces with respect to electrical forces. Is diffusion important?

d.   Estimate the particle velocity at the collector surface. Assuming the particles start from rest at the source, estimate their time to traverse the gap.

3   Equations of Change

3.1   Isothermal

Here we set out the equations of conservation of mass and momentum for a viscous Newtonian fluid of uniform and homogeneous composition. These two equations together with the appropriate boundary conditions are sufficient to describe the changes in velocity with respect to position and time for a viscous, isothermal flow of a uniform and homogeneous fluid. In the following sections we shall write the energy conservation, species conservation, and charge conservation equations, as well as the appropriate momentum equation together with the constitutive relations needed to describe changes in temperature, concentration, and other variables.

The equation of conservation of mass may be derived by considering the fixed control surface shown in Fig. 3.1.1 through which the fluid is flowing. Now the flux of mass through the surface must equal the decrease of mass within the volume due to unsteadiness, or

$$\int_S \rho \mathbf{u} \cdot d\mathbf{A} = -\int_V \frac{\partial \rho}{\partial t} \, dV \qquad (3.1.1)$$

Applying Gauss's theorem, we get

$$\int_S \rho \mathbf{u} \cdot d\mathbf{A} = \int_V \boldsymbol{\nabla} \cdot \rho \mathbf{u} \, dV \qquad (3.1.2)$$

It follows that

$$\int_V \left\{ \frac{\partial \rho}{\partial t} + \boldsymbol{\nabla} \cdot \rho \mathbf{u} \right\} dV = 0 \qquad (3.1.3)$$

Since this is true for every control volume,

31

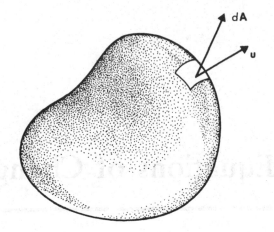

Figure 3.1.1   Fixed control surface through which fluid flows.

$$\frac{\partial \rho}{\partial t} + \boldsymbol{\nabla} \cdot \rho \mathbf{u} = 0 \tag{3.1.4}$$

or

$$\frac{\partial \rho}{\partial t} + \rho \boldsymbol{\nabla} \cdot \mathbf{u} + (\mathbf{u} \cdot \boldsymbol{\nabla})\rho = 0 \tag{3.1.5}$$

In terms of the material derivative defining a change in a quantity moving with a fluid element (Eq. 2.2.3), we have

$$\frac{D\rho}{Dt} + \rho \boldsymbol{\nabla} \cdot \mathbf{u} = 0 \tag{3.1.6}$$

For the special case where ρ is constant following a fluid particle, $D\rho/Dt = 0$ and

$$\boldsymbol{\nabla} \cdot \mathbf{u} = 0 \tag{3.1.7}$$

This is the equation of continuity for "incompressible" flow.

Conservation of momentum as expressed through Newton's second law applied to a fluid particle may be written as

$$\rho \frac{D\mathbf{u}}{Dt} = \mathbf{f}_{\text{body}} + \mathbf{f}_{\text{surf}} \tag{3.1.8}$$

where the applied force per unit volume on the fluid particle is divided into surface and body forces. The body forces are proportional to the total volume or mass of the fluid element, examples of which are the gravitational body force, electrical body force, and electromagnetic body force. Here, only the gravitational body force is considered, which per unit volume is

$$\mathbf{f}_{body} = \rho \mathbf{g} \tag{3.1.9}$$

where $\mathbf{g}$ is the local gravitational acceleration.

The surface forces are those applied by the external stresses defined through Eq. (2.2.12), where

$$\mathbf{f}_{surf} = \nabla \cdot \mathfrak{T} = \frac{\partial \tau_{ij}}{\partial x_j} \tag{3.1.10}$$

Here the divergence of the stress tensor is the net force per unit volume acting on a fluid element. Note that $\nabla \cdot \mathfrak{T}$ is not a simple divergence because $\mathfrak{T}$ is a dyadic and not a vector, although the term is interpretable physically as a rate of momentum change.

From Eqs. (3.1.8) to (3.1.10) we may write

$$\rho \frac{D\mathbf{u}}{Dt} = \nabla \cdot \mathfrak{T} + \rho \mathbf{g} \tag{3.1.11}$$

In Cartesian tensor notation, substituting for τ_{ij} from Eq. (2.2.16), we have

$$\rho \frac{Du_i}{Dt} = -\frac{\partial p}{\partial x_i} + \frac{\partial}{\partial x_j} \left\{ \mu \left(\frac{\partial u_i}{\partial x_j} + \frac{\partial u_j}{\partial x_i} - \frac{2}{3} \delta_{ij} \frac{\partial u_k}{\partial x_k} \right) \right\} + \rho g_i \tag{3.1.12}$$

This equation is usually referred to as the *Navier-Stokes equation*.

In the special case of incompressible (constant-density) flow, $\partial u_k / \partial x_k = \nabla \cdot \mathbf{u} = 0$ and the dilatation term vanishes. Further, if the viscosity μ is constant, then

$$\frac{\partial}{\partial x_j} \left\{ \mu \left(\frac{\partial u_i}{\partial x_j} + \frac{\partial u_j}{\partial x_i} \right) \right\} = \mu \frac{\partial^2 u_i}{\partial x_j^2} + \mu \frac{\partial}{\partial x_i} \left(\frac{\partial u_k}{\partial x_k} \right)$$

$$= \mu \nabla^2 \mathbf{u} \tag{3.1.13}$$

In this important special case the Navier-Stokes equation reduces to

$$\rho \frac{D\mathbf{u}}{Dt} = -\nabla p + \mu \nabla^2 \mathbf{u} + \rho \mathbf{g} \tag{3.1.14}$$

For incompressible, isothermal, single-phase flows where μ may be treated as constant, the above equation, together with the continuity equation, is sufficient to describe the flow under appropriately specified boundary conditions. For example, in the case of a fluid-solid interface the no-slip condition will generally be applied, wherein

$$\mathbf{u}_{fluid} = \mathbf{u}_{solid} \tag{3.1.15}$$

At a stationary boundary this implies zero tangential velocity of the fluid particle at the surface, and, where the boundary is also impermeable, it implies zero normal velocity of the fluid particle. This latter condition will not apply if there is blowing, suction, evaporation, or condensation at the surface.

At a fluid-fluid interface, with interfacial tension neglected, we shall generally apply the conditions of constancy of normal and tangential velocity components, pressure, and shear across the interface. Discontinuities in some or all of these quantities are, however, not precluded by the conservation conditions.

3.2   Nonisothermal

Nonisothermal conditions will prevail in a fluid flow system when there is an appreciable temperature change in the fluid resulting from heating (or cooling) either by heat addition or generation. Examples are the release of heat within the fluid by chemical reaction, electrical heat generation, the transfer of heat into the fluid by a heated surface, heat generated within the fluid by viscous dissipation as in high-speed gas flows, rapid polymer extrusion, or high-speed lubrication. In such cases the fluid system is no longer isothermal, and it is necessary to supplement the equations of conservation of mass and momentum with a conservation of energy equation.

Heat energy, which is associated with temperature, is a form of energy distinct from either the kinetic energy associated with the mean fluid motion (that is, the $\rho|\mathbf{u}|^2$ energy) or the potential energy associated with position. Heat energy flows from one neighboring element to another when the temperatures are different. The heat energy of a fluid particle is defined by the internal energy, which depends on the local thermodynamic state. With e the internal energy per unit mass, we formulate the principal of conservation of energy mathematically as the first law of thermodynamics, which may be written in terms of specific properties (per unit mass) as

$$\delta \tilde{q} = de - \delta w \tag{3.2.1}$$

Here, $\tilde{q}$ is the heat supplied, e is internal energy, and w is the work done on the fluid. We shall only consider heat conduction and shall not discuss heat generation from chemical or electrical sources.

Consider a fixed control surface enclosing a fluid element. It follows from the first law that the sum of the rate of increase of the internal and kinetic energies of the element must equal the sum of the rate at which heat is being conducted across the surface into the fluid plus the rate at which work is being done on the fluid by the stresses at the surface and by the body forces. With the aid of the continuity equation (3.1.5),

$$\int_S [\boldsymbol{\tau} \cdot \mathbf{u}] \cdot d\mathbf{A} - \int_S \mathbf{q} \cdot d\mathbf{A} + \int_V \rho \mathbf{g} \cdot \mathbf{u} \, dV$$

$$= \int_V \rho \left\{ \frac{D}{Dt} \left(\frac{1}{2} |\mathbf{u}|^2 + e \right) \right\} dV \tag{3.2.2}$$

Applying Gauss's theorem and the condition that the resulting equation must be true for every control volume, we get

$$\rho \frac{D}{Dt}\left(e + \frac{1}{2}|\mathbf{u}|^2\right) = -\nabla \cdot \mathbf{q} + \rho \mathbf{g} \cdot \mathbf{u} + \nabla \cdot [\boldsymbol{\tau} \cdot \mathbf{u}] \qquad (3.2.3)$$

or, in Cartesian tensor notation,

$$\rho \frac{D}{Dt}\left(e + \frac{1}{2}u_i^2\right) = -\frac{\partial q_i}{\partial x_i} + \rho g_i u_i + \frac{\partial}{\partial x_i}\tau_{ij}u_i \qquad (3.2.4)$$

By decomposing the work term associated with the stress tensor, and then using the momentum equation, the Newtonian constitutive relation, and Fourier heat law, along with some manipulation, we obtain the following equation for the rate of change of internal energy (see, for example, Howarth 1953):

$$\rho \frac{De}{Dt} = -p\frac{\partial u_i}{\partial x_i} + \frac{\partial}{\partial x_i}\left(k\frac{\partial T}{\partial x_i}\right) + \Phi \qquad (3.2.5)$$

The *dissipation function* Φ is the rate of dissipation of energy per unit time per unit volume, and is defined by

$$\Phi = 2\mu(\varepsilon_{ij}\varepsilon_{ij} - \tfrac{1}{3}\varepsilon_{ii}^2) \qquad (3.2.6)$$

or, in rectangular Cartesian coordinates,

$$\Phi = 2\mu\left[\left(\frac{\partial u}{\partial x}\right)^2 + \left(\frac{\partial v}{\partial y}\right)^2 + \left(\frac{\partial w}{\partial z}\right)^2\right]$$

$$+ \mu\left[\left(\frac{\partial v}{\partial x} + \frac{\partial u}{\partial y}\right)^2 + \left(\frac{\partial w}{\partial y} + \frac{\partial v}{\partial z}\right)^2 + \left(\frac{\partial u}{\partial z} + \frac{\partial w}{\partial x}\right)^2\right] \qquad (3.2.7)$$

$$- \frac{2}{3}\mu\left(\frac{\partial u}{\partial x} + \frac{\partial v}{\partial y} + \frac{\partial w}{\partial z}\right)^2$$

The dissipation is a positive-definite quantity and represents the irreversible conversion of mechanical energy to thermal energy due to the action of fluid stresses. The remaining part of the stress work that goes into internal energy, $-p\partial u_i/\partial x_i$, represents reversible strain energy and can carry either sign.

Two other useful forms of the energy equation give, respectively, the dependence of the rate of change of specific enthalpy h and specific entropy s on the dissipation and heat conduction. From Eq. (3.2.5) and the thermodynamic relations

$$dh = de + d\left(\frac{p}{\rho}\right) \qquad T\,ds = dh - \frac{dp}{\rho} \qquad (3.2.8)$$

these equations are shown to be expressible as

$$\rho \frac{Dh}{Dt} = \frac{Dp}{Dt} + \frac{\partial}{\partial x_i}\left(k\,\frac{\partial T}{\partial x_i}\right) + \Phi \tag{3.2.9}$$

$$\rho T \frac{Ds}{Dt} = \frac{\partial}{\partial x_i}\left(k\,\frac{\partial T}{\partial x_i}\right) + \Phi \tag{3.2.10}$$

Heat addition due to chemical reaction or electrical heating has not been taken into account but is readily done so by adding an appropriate source term to the right-hand sides of the energy equations as written.

An important special case is that of incompressible flow. As discussed in Section 1.2, the term *incompressible* is something of a misnomer, since what is generally meant in fluid mechanics is constant density. However, a flow in which there are temperature gradients is not quite one of constant density since the density varies with temperature. But the criterion for a constant-density flow is that the flow velocity be small compared with the sound speed in the fluid; that is, the Mach number must be small. For a small Mach number the pressure changes are small. Therefore when evaluating the derivatives of thermodynamic quantities for an incompressible flow with an imposed spatial variation in temperature, we must hold the pressure, not the density, constant (Landau & Lifshitz 1987), whence

$$\frac{\partial s}{\partial t} = \left(\frac{\partial s}{\partial T}\right)_p \frac{\partial T}{\partial t} \qquad \nabla s = \left(\frac{\partial s}{\partial T}\right)_p \nabla T \tag{3.2.11}$$

where the specific heat at constant pressure $c_p = T(\partial s/\partial T)_p$.

Provided the mean absolute temperature differences are small, the approximation of constant density is still viable so long as the Mach number is small. That the temperature differences are small does not mean that the temperature gradients are necessarily so. For small Mach number and small temperature differences, the flow can be considered incompressible, that is, of constant density, with $\nabla \cdot \mathbf{u} = 0$. With small temperature differences the fluid properties will be constant, and the energy equation reduces to

$$\frac{DT}{Dt} = \alpha \nabla^2 T \tag{3.2.12}$$

where α is the thermal diffusivity and where the viscous dissipation Φ has been neglected. Neglect of the dissipation term is satisfactory so long as the viscous heating is small compared with the conductive heat flow that arises from the temperature differences. This will generally hold except in those instances where there are large velocity gradients and the heat generated by viscous dissipation cannot be conducted away rapidly enough. An example is high-speed lubrication.

To complete the thermodynamic description of a homogeneous fluid system, we need to specify an equation of state relating three thermodynamic properties. We shall not generally be concerned with gases, but we note that away from the critical point for moderate and low densities the perfect or ideal gas equation is normally a satisfactory description. Written in terms of density, pressure, and temperature the perfect gas relation is

$$p = \rho \, \frac{R}{M} \, T \tag{3.2.13}$$

where M is the molar mass and R is the gas constant $8.314 \, \mathrm{J \, K^{-1} \, mol^{-1}}$. An alternative form, with volume in place of density, is

$$pV = nRT \tag{3.2.14}$$

since $\rho = nM/V$.

As a first approximation we may write for liquids

$$\rho = \rho_0[1 - \alpha(T - T_0) + \beta(p - p_0)] \tag{3.2.15}$$

or, in terms of specific volume v,

$$v = v_0[1 + \alpha(T - T_0) - \beta(p - p_0)] \tag{3.2.16}$$

where the subscript 0 denotes the reference value. Here, α is the *thermal expansion coefficient* or *volume expansivity* defined by

$$\alpha = \frac{1}{v} \left(\frac{\partial v}{\partial T} \right)_p = -\frac{1}{\rho} \left(\frac{\partial \rho}{\partial T} \right)_p \tag{3.2.17}$$

and β is the *compressibility coefficient* or *isothermal compressibility* defined by

$$\beta = -\frac{1}{v} \left(\frac{\partial v}{\partial p} \right)_T = \frac{1}{\rho} \left(\frac{\partial \rho}{\partial p} \right)_T \tag{3.2.18}$$

Values of these coefficients are for water at 20°C and atmospheric pressure $\alpha = 2.1 \times 10^{-4} \, \mathrm{K^{-1}}$ and $\beta = 4.6 \times 10^{-10} \, \mathrm{Pa^{-1}}$.

3.3   Multicomponent

In setting down the conservation equations, we considered only fluids of uniform and homogeneous composition. Here we examine how these conservation equations change when two or more species are present and when chemical reactions may also take place. In a multicomponent mixture a transfer of mass takes place whenever there is a spatial gradient in the mixture proportions, even in the absence of body forces that act differently upon different species. In fluid flows the mass transfer will generally be accompanied by a transport of momentum and may further be combined with a transport of heat.

For a multicomponent fluid, conservation relations can be written for the individual species. Let $\mathbf{u}_i$ be the species velocity and ρ_i the species density, where the index is used to represent the ith species rather than the component of a vector. The overall continuity equation developed earlier is valid for each species, provided no species are produced (or consumed). If, however, species are produced by chemical reaction, say at a mass rate r_i kg m^{-3} s^{-1}, one must take this production into account when balancing the mass flux through a control surface. In this case the integral form of the continuity equation should be modified to

$$\int_S \rho \mathbf{u}_i \cdot d\mathbf{A} = -\int_V \frac{\partial \rho_i}{\partial t} \, dV + \int_V r_i \, dV \qquad (3.3.1)$$

giving

$$\frac{\partial \rho_i}{\partial t} + \nabla \cdot \rho_i \mathbf{u}_i = r_i \qquad (3.3.2)$$

The species production rate r_i must be obtained from chemical kinetics considerations and is dependent on the local thermodynamic state and the stoichiometric coefficients associated with the chemical reactions.

In terms of the mass average velocity $\mathbf{u}$, defined by $\rho \mathbf{u} = \Sigma \, \rho_i \mathbf{u}_i$, and mass flux with respect to the mass average velocity $\mathbf{J}_i$, defined by $\mathbf{J}_i = \rho_i(\mathbf{u}_i - \mathbf{u})$, the continuity equation for individual species, Eq. (3.3.2), can be rewritten in the form

$$\frac{\partial \rho_i}{\partial t} + \nabla \cdot \rho_i \mathbf{u} = -\nabla \cdot \mathbf{J}_i + r_i \qquad (3.3.3)$$

In terms of the material derivative,

$$\frac{D\rho_i}{Dt} + \rho_i \nabla \cdot \mathbf{u} = -\nabla \cdot \mathbf{J}_i + r_i \qquad (3.3.4)$$

Adding all the continuity equations for the i species, Eq. (3.3.2), gives

$$\frac{\partial \sum \rho_i}{\partial t} + \nabla \cdot \sum \rho_i \mathbf{u}_i = \sum r_i \qquad (3.3.5)$$

But $\rho = \Sigma \, \rho_i$, $\rho \mathbf{u} = \Sigma \, \rho_i \mathbf{u}_i$, and, since mass is conserved in a chemical reaction, $\Sigma \, r_i = 0$, whence the continuity equation for the mixture is identical to that for a pure fluid.

The momentum equation, as represented by the Navier-Stokes equation, is not restricted to a single-component fluid but is valid for a multicomponent solution or mixture so long as the external body force is such that each species is acted upon by the same external force (per unit mass), as in the case with gravity. In the following section we consider external forces associated with an applied external field, which differ for different species. The reason for there being no distinction between the various contributions to the stress tensor associated with diffusive transport is that the phenomenological relation for the stress is unaltered by the presence of concentration gradients. This is seen from the fact that the stress tensor must be related to the spatial variations in fluid properties in a tensorially appropriate manner, for example, second-order tensor to second-order tensor. In the Newtonian constitutive relation the stress tensor is linearly related to the rate-of-strain tensor. If a dependence of the stress on the spatial variation in concentration is assumed, then it must be related to a second-order tensor associated with this variation, for example, $\nabla \rho_i \nabla \rho_i$. But such a quantity is of higher order than the rate-of-strain tensor and, for consistency, should not be retained.

The energy equation is also unchanged for a multicomponent mixture so long as the external force is, for example, that due to gravity, where each species is acted upon by the same external force. However, when concentration gradients are present, the phenomenological relation for the heat flux vector $\mathbf{q}$ does change from that for a pure fluid. For fluid mixtures the energy flux must be modified to incorporate an added flux arising from interdiffusion of the species. There is also an energy flux, termed the *Dufour effect*, arising from diffusive thermal conduction; however, it is usually small and will be neglected here. The modified phenomenological relation for the energy flux can be shown to be (Bird et al. 1960)

$$\mathbf{q} = \sum h_i \mathbf{J}_i - k \nabla T \tag{3.3.6}$$

where h_i is the partial specific enthalpy of the ith species, and $\mathbf{J}_i$ is the mass flux relative to the mass average velocity. The terms in h_i represent the transport of energy caused by interdiffusion, and include the transport of chemical potential energy. Equation (3.3.6) is the phenomenological relation normally employed in double diffusive problems of heat and mass transfer.

In both the multicomponent continuity equation and in the energy flux relation given above, it is necessary to specify the multicomponent mass flux $\mathbf{J}_i$ relative to the mass average velocity. The phenomenological relation for the mass flux in a general multicomponent system can be expressed as a sum of the diffusive flux due to concentration gradients (Section 2.4) plus a number of other diffusive fluxes. In particular, there is diffusion due to applied external forces, where the same force per unit mass does not act on all the species. An example is the diffusion that takes place when an applied electric field acts on a mixture of charged species (Section 2.5). This phenomenon will be treated separately in the following section. A diffusive flux termed *pressure diffusion* also arises from an imposed pressure gradient. For most molecular species this flux is generally small unless the pressure gradients are very large, as in centrifugal separations. For this reason pressure diffusion is neglected here and considered later in the context of centrifugal separations. The last diffusive flux is that due to temperature gradients, and it is termed *thermal diffusion* or the *Soret effect*. Like the Dufour effect, with which it is interrelated, thermal diffusion is generally small unless the temperature gradients are very large. It will not be discussed further.

The derivation of these various flux contributions may be found in the classic text by Hirschfelder et al. (1954). A summary of the resulting expressions is given in Bird et al. (1960). The basis of the derivation for the ordinary diffusive flux is that the driving force is the chemical potential gradient, not the concentration gradient. The ordinary diffusive flux, measured relative to the mass average velocity, can be shown from principles of nonequilibrium thermodynamics to be expressible in the form

$$\mathbf{J}_i = \frac{c^2}{\rho} \sum_j M_i M_j \mathscr{D}_{ij} \left[\frac{x_j}{RT} \, \boldsymbol{\nabla}\mu_j \right] \tag{3.3.7}$$

Here, μ_j is the chemical potential, with the term in brackets representing the dimensionless driving force. The $\mathscr{D}_{ij}$ are generalized diffusion coefficients for the pair ij in the multicomponent mixture, defined such that they are consistent with Onsager's reciprocity relation (Van de Ree 1967).

The chemical potential may be written in terms of the *activity a* as

$$\mu_j = \mu_j^\circ(T, p) + RT \ln a_j \tag{3.3.8}$$

where μ_j° is independent of the solution composition. From this relation, Eq. (3.3.7) can be put into the more familiar form

$$\mathbf{J}_i = \frac{c^2}{\rho} \sum_j M_i M_j \mathscr{D}_{ij} \left(\frac{\boldsymbol{\nabla} \ln a_j}{\boldsymbol{\nabla} \ln x_j} \right)_{T,p} \boldsymbol{\nabla}x_j \tag{3.3.9}$$

In the special case of an ideal solution, which generally implies dilute solutions, the activity can be replaced by the mole fraction, and Eq. (3.3.9) reduces to

$$\mathbf{J}_i = \frac{c^2}{\rho} \sum_j M_i M_j \mathscr{D}_{ij} \boldsymbol{\nabla}x_j \tag{3.3.10}$$

For a binary mixture

$$\mathbf{J}_1 = -\frac{c^2}{\rho} M_1 M_2 D_{12} \boldsymbol{\nabla}x_1 \tag{3.3.11}$$

which is one form of Fick's first law. We have replaced $\mathscr{D}_{12}$ by the diffusivity D_{12} of a pair in a binary solution since the two diffusivities are identical here.

That Eq. (3.3.11) is the same as the previously given form of Fick's law follows from the definition

$$x_1 = \frac{\omega_1/M_1}{\omega_1/M_1 + \omega_2/M_2} \tag{3.3.12}$$

or

$$\boldsymbol{\nabla}x_1 = \frac{\boldsymbol{\nabla}\omega_1}{M_1 M_2 (\omega_1/M_1 + \omega_2/M_2)^2} = \frac{\bar{M}^2}{M_1 M_2} \boldsymbol{\nabla}\omega_1 \tag{3.3.13}$$

With $\bar{M} = \rho/c$, on substituting into Eq. (3.3.11), we get

$$\mathbf{J}_1 = -\rho D_{12} \boldsymbol{\nabla}\omega_1 \tag{3.3.14}$$

Restricting our considerations to a binary system, we can write the continuity equation (Eq. 3.3.3) in the form

$$\frac{\partial \rho_1}{\partial t} + \nabla \cdot \rho_1 \mathbf{u} = \nabla \cdot (\rho D_{12} \nabla \omega_1) + r_1 \qquad (3.3.15)$$

where r_1 is the mass rate of production of species 1 per unit volume $(\text{kg m}^{-3}\,\text{s}^{-1})$. Equivalently, in terms of molar fluxes,

$$\frac{\partial c_1}{\partial t} + \nabla \cdot c_1 \mathbf{u}^* = \nabla \cdot (c D_{12} \nabla x_1) + R_1 \qquad (3.3.16)$$

where R_1 is the molar rate of production of species 1 per unit volume $(\text{mol m}^{-3}\,\text{s}^{-1})$. Because moles are not conserved, this equation is not a molar continuity equation.

An important special case is that of diffusion in dilute liquid solutions. In that situation $\nabla \cdot \mathbf{u} = 0$, and Eq. (3.3.15) reduces to

$$\frac{\partial \rho_1}{\partial t} + \mathbf{u} \cdot \nabla \rho_1 = D_{12} \nabla^2 \rho_1 + r_1 \qquad (3.3.17)$$

The corresponding relation with molar concentration as the dependent quantity is obtained by dividing through by M_1 to give

$$\frac{\partial c_1}{\partial t} + \mathbf{u} \cdot \nabla c_1 = D_{12} \nabla^2 c_1 + R_1 \qquad (3.3.18)$$

With $R_1 = 0$ Eq. (3.3.18) is termed the *convective diffusion equation*. When, in addition, $\mathbf{u} = 0$, the equation reduces to the ordinary diffusion equation, which is also referred to as *Fick's second law of diffusion*. It is applicable to diffusion in solids or stationary liquids and has the same form as the heat conduction equation in stationary media with constant thermal conductivity.

3.4   Charged Species

We saw in Section 2.5 that when a mixture of charged species is subjected to an applied electric field a mass transfer (migration) takes place. In a hydrodynamic system there are also contributions to the species flux due to ordinary convection and diffusion, as discussed in the last section. Here, we wish to set down the appropriate equations of change governing the motion and behavior of species acted upon by an electric field. In doing so, we shall assume that the phenomenological relation defining the flux of the ith species due to diffusion plus electromigration will be that for a dilute solution. This is dictated not only by the general applicability of this approximation in electrochemical flow systems, even though it is only strictly applicable for dilute solutions, but also because of the complexity of dealing with a general flux relation of the form of Eq. (3.3.7). In this regard we observe that Eq. (3.3.7) is still valid with charged species and an applied electric field, provided the chemical potential is understood to be the electrochemical potential, which is a function of pressure, temperature, chemical composition, and the electrical state of the phase. The characterization of the electrical state of phases of different composition itself introduces a number of subtle questions (Newman 1991), Inevitably the phenomenological relations for concentrated solutions rest largely on empirical determinations of the generalized diffusion coefficients.

For dilute solutions the flux contributions from diffusion, electromigration, and convection can be linearly superposed, whence from Eqs. (2.4.16) and (2.5.3) we have for the molar flux of the ith species:

$$\mathbf{j}_i^* = -v_i z_i F c_i \nabla\phi - D_i \nabla c_i + c_i \mathbf{u} \tag{3.4.1}$$

where the molar average velocity $\mathbf{u}^*$ has been replaced by the mass average velocity $\mathbf{u}$, since in sufficiently dilute solutions $\mathbf{u}^* = \mathbf{u}$. In electrochemical systems it is convenient to use molar flux but not molar average velocity, since fluid dynamic solutions are expressed through the mass average velocity, for which a continuity equation can be written. From Eqs. (2.4.15) and (2.5.4) the corresponding mass flux is

$$\mathbf{j}_i = -v_i z_i F \rho_i \nabla\phi - D_i \nabla\rho_i + \rho_i \mathbf{u} \tag{3.4.2}$$

Recall that v_i is the mobility, which is related to the diffusivity by the relation $D_i = RTv_i$. Equations (3.4.1) and (3.4.2) are called the *Nernst-Planck equations*.

Because of the motion of the charged species, there will be a current. Specifically, the current density is given by $\mathbf{i} = F \sum z_i \mathbf{j}_i^*$. Employing the dilute solution relation for the molar flux, we obtain

$$\mathbf{i} = -F^2 \nabla\phi \sum z_i^2 v_i c_i - F \sum z_i D_i \nabla c_i + F\mathbf{u} \sum z_i c_i \tag{3.4.3}$$

or

$$\mathbf{i} = -\sigma\nabla\phi - F \sum z_i D_i \nabla c_i + F\mathbf{u} \sum z_i c_i \tag{3.4.4}$$

where $\sigma = F^2 \sum z_i^2 v_i c_i$ is the scalar electrical conductivity of the solution. The current is seen to be made up of contributions from the electric field, the concentration gradients, and the convection of charge.

With current and electric fields present the laws of electrodynamics, as specified by Maxwell's equations, must be employed in addition to those of fluid mechanics. If there are no applied magnetic fields and if we neglect as small the magnetic field induced by the current, as well as any magnetic fields induced from time-varying electric fields, then the electrodynamic problem reduces to an electrostatic one. This problem can be specified through *Poisson's equation* relating the spatial variation in the electric field to the charge distribution, which for a medium of uniform dielectric constant is

$$\nabla^2\phi = -\frac{\rho_E}{\epsilon} \tag{3.4.5}$$

Here, ρ_E is the electric charge density $(C\,m^{-3})$. The permittivity of the medium ϵ is equal to the permittivity of a vacuum, $\epsilon_0 = 8.854 \times 10^{-12}\,C^2\,N^{-1}\,m^{-2}$ $(C\,V^{-1}\,m^{-1})$ multiplied by the dielectric constant also known as the relative permittivity ϵ_r, which is dimensionless. For water $\epsilon_r = 78.3$ at 25°C.

The electric field is coupled with the fluid mechanics through the Lorentz relation for the force on a charged particle:

$$\mathbf{f}_E = \rho_E \mathbf{E} \tag{3.4.6}$$

where $\mathbf{f}_E$ is the electric body force per unit volume and the charge density is

$$\rho_E = F \sum z_i c_i \tag{3.4.7}$$

In the important case where the solution is electrically neutral,

$$\sum z_i c_i = 0 \tag{3.4.8}$$

This condition is only an approximation in fluids, and the $=$ should really be replaced by $\approx$. The more correct statement would be that $\sum z_i c_i$ is not zero but that its absolute value is small compared with the maximum (absolute) value of $z_i c_i$. Equation (3.4.8) states that there is no local accumulation of charge, which corresponds to $\nabla \cdot \mathbf{i} = 0$. This is the assumption employed in current theory at low frequencies.

Electroneutrality or the absence of charge separation holds closely in aqueous electrochemical solutions, although not necessarily organic solutions, everywhere except in thin regions near charged boundaries. These regions are termed double layers or Debye sheaths and have thicknesses on the order of 1 to 10 nm. The double layer is important when we consider very small charged particle interactions and charged surface phenomena, but is generally unimportant with respect to bulk flow characteristics. Clearly, when the solution is electrically neutral, the electrical body force is identically zero and the momentum and energy equations will be unaltered. In addition, the convective contribution $Fu \sum z_i c_i$ to the current density in Eq. (3.4.4) vanishes. However, even in this case it is only when there are no concentration gradients that the current density reduces to Ohm's law.

From Poisson's equation it would appear that electroneutrality implies that the potential distribution is governed by Laplace's equation. For exact electroneutrality this is true, which seems to be inconsistent with Eq. (3.4.4), for on using $\nabla \cdot \mathbf{i} = 0$ (current continuity or conservation of charge) we get

$$\nabla \cdot (\sigma \nabla \phi) + F \sum z_i \nabla \cdot (D_i \nabla c_i) = 0 \tag{3.4.9}$$

for the governing potential distribution. For constant properties this equation says that $\nabla^2 \phi = 0$ only when $\sum z_i D_i \nabla^2 c_i = 0$. The resolution of the seeming paradox lies in the observation that one cannot use both Poisson's equation and the electroneutrality condition, since this overspecifies the problem. Equation (3.4.9) is only an approximation because of the approximation introduced by the use of the electroneutrality condition. Poisson's equation always holds, and in principle one can measure the error in the electroneutrality approximation by obtaining the solution of ϕ from Eq. (3.4.9), plugging it into Poisson's equation, and in turn determining the concentration distribution.

In practice, the electroneutrality condition will never be applicable very close to a boundary where charge separation takes place, even in aqueous media. In these thin layers the gradients in the electric field become very large and balance the right-hand side of Poisson's equation, which is also large, though $\Sigma z_i c_i$ may be very small, because the coefficient F/ϵ is very large ($\sim 10^{14}$ V m mol^{-1} for water).

If electroneutrality may be assumed, the fluid mechanical conservation equations of mass, momentum, and energy remain unchanged from those discussed in the last section for multicomponent systems. If electroneutrality is not assumed, the mass conservation equation remains unchanged but the Lorentz body force must be added to the right-hand side of the Navier-Stokes equation. In addition, to the right-hand side of the energy equation is added the corresponding work term $\rho_E \mathbf{E} \cdot \mathbf{u}$.

As an important special case it is of interest to compare the equations governing the concentration distribution in a dilute binary electrolyte with those for a neutral binary system. By dilute binary electrolyte is meant an un-ionized solvent and dilute fully ionized salt—that is, one composed of one negative charged species and one positive charged species that do not enter into any reactions in the bulk of the fluid ($R_i = 0$). If the positive ions are denoted by the subscript $+$ and the negative ions by the subscript $-$, the condition of electroneutrality is expressible by

$$z_+ c_+ + z_- c_- = 0 \tag{3.4.10}$$

With ν_+ and ν_- the number of positive and negative ions produced by the dissociation of one molecule of electrolyte, we can introduce the reduced ion concentration c defined by

$$c = \frac{c_+}{\nu_+} = \frac{c_-}{\nu_-} \tag{3.4.11}$$

The above relation automatically satisfies the electroneutrality condition.

From species conservation (Eq. 3.3.4), assuming an incompressible fluid and no reactions, and the expression for mass flux Eq. (3.4.2), we have

$$\frac{\partial c}{\partial t} + \mathbf{u} \cdot \nabla c = z_\pm \nu_\pm F \nabla \cdot (c \nabla \phi) + D_\pm \nabla^2 c \tag{3.4.12}$$

where the mobility and diffusion coefficients have been assumed constant. Subtracting the equation for the negative ions from that for the positive ions gives

$$(z_+ \nu_+ - z_- \nu_-) F \nabla \cdot (c \nabla \phi) + (D_+ - D_-) \nabla^2 c = 0 \tag{3.4.13}$$

This relation may be used to eliminate the potential from either the positive ion or negative ion convective diffusive equation (Eq. 3.4.12) to give

$$\frac{\partial c}{\partial t} + \mathbf{u} \cdot \nabla c = D \nabla^2 c \qquad (3.4.14)$$

Here, D is an effective diffusion coefficient defined by

$$D = \frac{z_+ v_+ D_- - z_- v_- D_+}{z_+ v_+ - z_- v_-} \qquad (3.4.15a)$$

or equivalently, since $D_\pm = RT v_\pm$,

$$D = D_+ \frac{1 - z_+/z_-}{1 - z_+ D_+/z_- D_-} \qquad (3.4.15b)$$

where we note that for $D_+ = D_-$ the effective diffusion coefficient $D = D_+ = D_-$.

The interesting result shown by Eq. (3.4.14) is that the concentration distribution in a dilute binary electrolyte is governed by the same convective diffusion equation as for a neutral species even though there is a current flow. The potential distribution is given from Eq. (3.4.13), which is simply an expression of current continuity. We can see this from Eq. (3.4.4) for the current, which we may write, using $z_+ v_+ = -z_- v_-$,

$$-\frac{\mathbf{i}}{z_+ v_+ F} = (z_+ v_+ - z_- v_-)Fc\nabla\phi + (D_+ - D_-)\nabla c \qquad (3.4.16)$$

But $\nabla \cdot \mathbf{i} = 0$, giving Eq. (3.4.13).

Note that the concentration distribution for each of the ith species with constant diffusion coefficients will also be governed by the convective diffusion equation (Eq. 3.4.14) when electromigration (and reactions) can be neglected. This will be a good approximation when the electric field is reduced because the solution is highly conducting. Hence, the mass transfer is due principally to diffusion and convection. Practically, one may achieve this in an electrochemical system by increasing the conductivity by adding an *indifferent electrolyte*, which is present in large excess over the others and which does not react with them (Newman 1991, Koryta & Dvorak 1987).

3.5   Characteristic Parameters

To this point the equations of change have been set down for pure fluids under both isothermal and nonisothermal conditions, and for multicomponent fluids and charged species. The boundary and initial conditions have, however, been considered only to a limited extent. They will be discussed in the context of the specific subject areas: for example, diffusion, chemical reaction, surface tension, and heat transfer. Here, the form of the equations of change will be analyzed so that some of the more important characteristic similarity parameters can be brought out and the stage set for subsequent analyses over restricted ranges of these parameters.

We shall not examine the equations of change in their most general form but will instead limit our attention to dilute or binary solutions in incompressible flow, without species production. For electrolyte solutions we will restrict ourselves to electrically neutral binary solutions or highly conducting solutions in which the electric field is small so that the convective diffusion equation for neutral species is applicable and the momentum and energy equations are unaltered. For simplicity the transport and physical properties are also taken to be constant.

Below are summarized the overall continuity, momentum, convective diffusion, and energy equations under the restrictions noted.

$$\nabla \cdot \mathbf{u} = 0 \tag{3.5.1}$$

$$\frac{D\mathbf{u}}{Dt} = -\frac{1}{\rho} \nabla p + \nu \nabla^2 \mathbf{u} + \mathbf{g} \tag{3.5.2}$$

$$\frac{Dc}{Dt} = D\nabla^2 c \tag{3.5.3}$$

$$\frac{DT}{Dt} = \alpha \nabla^2 T \tag{3.5.4}$$

These equations repeat those previously set down. Here, ν is the kinematic viscosity, and α is the thermal diffusivity. The subscripts have been dropped in the convective diffusion equation, and D can be the binary diffusion coefficient, the effective electrolytic diffusion coefficient, or the diffusion coefficient of the ith species. The molar concentration is to be interpreted in the same context. In the energy equation, sometimes referred to as the *heat conduction equation* in the form written, heat flux due to interdiffusion and due to viscous dissipation have been neglected as small. Heat sources are also absent.

If the problem is an electrochemical one, there is still an additional equation to define the potential distribution given by Eq. (3.4.9). This equation will introduce a dimensionless similarity parameter characterizing the potential drop. For a characteristic potential drop $\Delta\phi$, one form this parameter takes will later be shown to be $zF\Delta\phi/RT$.

Let us rewrite the system, Eqs. (3.5.1) to (3.5.4), in terms of reduced dimensionless variables, recognizing that the choice of characteristic scales is to some extent an arbitrary one. With L, U, and τ, respectively, the characteristic length, characteristic speed, and characteristic time of the problem, and with asterisks denoting a reduced dimensionless variable, we have

$$x = L\mathbf{x}^* \qquad u = U\mathbf{u}^* \qquad t = \tau t^* \tag{3.5.5}$$

In many problems the characteristic time is not independently imposed and is given by $\tau = L/U$.

There are several choices for the pressure difference, but for forced convection problems where the pressure difference is due mainly to the dynamic force, we set

$$p - p_0 = \rho U^2 p^* \qquad (3.5.6)$$

where the subscript 0 denotes a convenient reference state, for example, an initial or free stream value. Selecting a characteristic driving temperature difference $(T_0 - T_w)$ and a characteristic driving concentration difference $(c_0 - c_w)$, we write

$$T - T_0 = (T_0 - T_w)T^* ; \qquad c - c_0 = (c_0 - c_w)c^* \qquad (3.5.7)$$

The subscript w may, for example, represent a known value of the temperature or concentration at a wall or surface in the flow.

In terms of the above reduced variables, Eqs. (3.5.1) to (3.5.4) become

$$\nabla^* \cdot \mathbf{u}^* = 0 \qquad (3.5.8)$$

$$St \frac{\partial \mathbf{u}^*}{\partial t^*} + \mathbf{u}^* \cdot \nabla^* \mathbf{u}^* = \frac{1}{Re} \nabla^{*2} \mathbf{u}^* - \nabla^* p^* + \frac{1}{Fr} \frac{\mathbf{g}}{g} \qquad (3.5.9)$$

$$St \frac{\partial c^*}{\partial t^*} + \mathbf{u}^* \cdot \nabla^* c^* = \frac{1}{Pe_D} \nabla^{*2} c^* \qquad (3.5.10)$$

$$St \frac{\partial T^*}{\partial t^*} + \mathbf{u}^* \cdot \nabla^* T^* = \frac{1}{Pe_T} \nabla^{*2} T^* \qquad (3.5.11)$$

The equations as nondimensionalized are appropriate only under a suitable assumption of order of magnitude. A term may be, but is not necessarily, dominant or negligible if the coefficient, for example Re^{-1}, is correspondingly large or small. Such a situation is known to occur in boundary layers where the gradients become large. Moreover, the various phenomena have been assumed to scale similarly in space and time, and a forced convection flow is implicitly assumed in which inertial and viscous forces are important. In buoyancy or free-convection flows a different scaling would be required since there is no imposed characteristic velocity. Finally, despite the similarity of the equations of mass and heat transfer, the solutions may be quite different, depending on the boundary and initial conditions.

Two of the dimensionless parameters appearing in the foregoing equations are dynamical in character and do not relate to molecular transport. They are

$$St = \frac{L}{\tau U} = \text{Strouhal number}$$

$$= \frac{L/U}{\tau} = \frac{\text{flow time scale}}{\text{unsteady time scale}} \qquad (3.5.12)$$

$$Fr = \frac{U^2}{gL} = \text{Froude number}$$

$$= \frac{\rho U^2 / L}{\rho g} = \frac{\text{inertial force}}{\text{gravitational force}} \qquad (3.5.13)$$

The *Strouhal number* is a measure of the unsteadiness of the motion. The *Froude number* is important in free surface flows, for example. The last parameter appearing in the dimensionless momentum equation is

$$\text{Re} = \frac{UL\rho}{\mu} = \text{Reynolds number}$$

$$= \frac{\rho U^2/L}{\mu U/L^2} = \frac{\text{inertial force}}{\text{viscous force}} \tag{3.5.14}$$

An alternative interpretation may be ascribed to the Reynolds number, consistent with our earlier analogy of the similarity of momentum, heat, and mass transport. We may then interpret the dimensionless parameters appearing in the energy and diffusion equations in an analogous manner; that is,

$$\text{Re} = \frac{UL}{\nu} = \text{Reynolds number}$$

$$= \frac{\rho U^2/L}{\mu U/L^2} = \frac{\text{momentum transported by convection}}{\text{momentum transported by viscous diffusion}} \tag{3.5.15}$$

$$\text{Pe}_T = \frac{UL}{\alpha} = \text{Peclet number (thermal)}$$

$$= \frac{\rho c_p U(T_0 - T_w)/L}{k(T_0 - T_w)/L^2} = \frac{\text{heat transported by convection}}{\text{heat transported by conduction}} \tag{3.5.16}$$

$$\text{Pe}_D = \frac{UL}{D} = \text{Peclet number (diffusion)}$$

$$= \frac{U(c_0 - c_w)/L}{D(c_0 - c_w)/L^2} = \frac{\text{mass transported by convection}}{\text{mass transported by diffusion}} \tag{3.5.17}$$

Both UL/α and UL/D are termed the Peclet number and usually given the same symbol. When there is no reason for confusion, we shall do the same; otherwise we distinguish between the two by reference to the thermal Peclet number or the diffusion Peclet number. The Peclet number plays a similar role in heat and mass transport as the Reynolds number in momentum transport. The thermal and diffusion Peclet numbers may be written somewhat differently to bring out their relation to the Reynolds number; in particular,

$$\text{Pe}_T = \frac{UL}{\nu} \frac{\nu}{\alpha} = \text{Re Pr} \tag{3.5.18}$$

$$\text{Pe}_D = \frac{UL}{\nu} \frac{\nu}{D} = \text{Re Sc} \tag{3.5.19}$$

Here, we have introduced two new dimensionless parameters, the *Prandtl number* and the *Schmidt number*, defined by

$$\text{Pr} = \text{Prandtl number} = \frac{\nu}{\alpha} \tag{3.5.20}$$

$$Sc = \text{Schmidt number} = \frac{\nu}{D} \qquad (3.5.21)$$

The importance of writing the parameters this way is that the Prandtl and Schmidt numbers are properties of the fluid, whereas the Reynolds number is a property of the flow.

Table 3.5.1 shows the Prandtl numbers for some common fluids, and Table 3.5.2 lists Schmidt numbers for dilute mixtures. The values in these tables are obtained from the data given in Sections 2.2 to 2.4. In some instances these data have been interpolated.

The consequence of a large Schmidt number, common in liquids, is that convection dominates over diffusion at moderate and even relatively low Reynolds numbers (assuming consistent order of magnitude in the terms). In gases these effects are of the same order. On the other hand, heat transfer in low-viscosity liquids by convection and conduction are the same order since the Prandtl number is approximately 1. In highly viscous fluids where the Prandtl number is large, heat transfer by convection predominates over conduction, provided the Reynolds number is not small. The opposite is true for liquid metals, where the Prandtl number is very small, so conduction heat transfer is dominant.

Table 3.5.1
Prandtl Numbers for Some Common Fluids

Substance	Temperature		Prandtl Number
	K	°C	$Pr = \nu/\alpha$
Mercury	300	27	2.72×10^{-2}
Air	300	27	7.12×10^{-1}
Water	300	27	5.65
Ethyl alcohol	293	20	1.70×10
Glycerin	293	20	1.16×10^{4}

Table 3.5.2
Schmidt and Lewis Numbers for Dilute Gases and Dilute Solutions

Substance	Temperature		Schmidt Number	Lewis Number
	K	°C	$Sc = \nu/D$	$Le = \alpha/D$
$O_2 - N_2$	273	0	7.3×10^{-1}	1.0
Dilute gases	293	20	~ 1	~ 1
NaCl aqueous	293	20	7.0×10^{2}	1.0×10^{2}
Dilute solutions	293	20	$\sim 10^{3}$	$\sim 10^{2}$

In Table 3.5.2 values are given for a third dimensionless fluid parameter, which is the ratio of the Schmidt number to the Prandtl number. This parameter, *Lewis number*, is defined by

$$\text{Le} = \text{Lewis number} = \frac{\alpha}{D} \qquad (3.5.22)$$

The Lewis number appears in double diffusive problems of combined heat and mass transfer. From Table 3.5.2 it can be seen that in gases all the transport effects are of the same order, but in liquids conduction heat transfer is the controlling mechanism on a large scale.

References

BIRD, R.B., STEWART, W.E. & LIGHTFOOT, E.N. 1960. *Transport Phenomena*. New York: Wiley.

HIRSCHFELDER, J.O., CURTISS, C.F. & BIRD, R.B. 1954. *Molecular Theory of Gases and Liquids*. New York: Wiley.

HOWARTH, L. 1953. The equations of flow in gases. In *Modern Developments in Fluid Dynamics: High Speed Flow*, vol. 1 (ed. L. Howarth), pp. 34–70. London: Oxford.

KORYTA, J. & DVORAK, J. 1987. *Principles of Electrochemistry*. New York: Wiley.

LANDAU, L.D. & LIFSHITZ, E.M. 1987. *Fluid Mechanics*, 2nd edn. Oxford: Pergamon Press.

NEWMAN, J.S. 1991. *Electrochemical Systems*, 2nd edn. Englewood Cliffs, N.J.: Prentice-Hall.

VAN DE REE, J. 1967. On the definition of the diffusion coefficient in reacting gases. *Physica* 36, 118–126.

Problems

3.1 Equation (3.2.10) is an expression for the rate of change of entropy following a fluid element. By dividing this equation by the temperature T and integrating throughout the interior of a closed control surface moving with the fluid, show that if the fluid is completely enclosed inside a heat-insulating boundary that the entropy of the entire fluid can only increase.

3.2 In the conservation of species equation for a binary system (Eq. 3.3.17), it is sometimes assumed that the velocity **u** is given by the solution to the momentum equation for the solvent. Determine a criterion for this assumption to be valid that is dependent on the solute volume fraction ϕ, the ratio of the density of the solvent particle to the density of the solute particle, and the ratio of the velocity of the solute species to the velocity of the solvent species. Is the condition $\phi \ll 1$ generally a satisfactory basis for the validity of the assumption?

3.3 Two infinite plates are held parallel to each other a distance $h = 0.02$ m apart with a dilute singly ionized electrolyte solution contained between them that has a uniform species concentration $c_0 = 1$ mol m^{-3}. The solution temperature is 300 K. A potential of 2 V is then applied across the gap between the plates.

a.   Assuming the electric field between the plates is constant, and considering only the positive ions, write the governing differential equation for the steady-state concentration distribution in the gap subsequent to the application of the electric field.

b.   Obtain an expression for the steady-state concentration distribution as a function of distance across the gap, and evaluate this expression using the numerical values given.

3.4   Gas flows steadily and at low speed past a semi-infinite flat plate. The fluid properties are constant. Far from the plate the gas has a velocity U parallel to the plate, a temperature T_0, and a species concentration c_0. The temperature of the plate surface T_w and species concentration there c_w are constant, and the pressure is everywhere uniform. For what fluid property condition is there an analogy between the spatial distribution of temperature, concentration, and velocity component parallel to the plate? Is the fluid property condition realistic, and what is a statement of the analogy?

4 Solutions of Uncharged Molecules

4.1 Diffusion and Reaction Kinetics

In this chapter we consider the transport of mass by convective diffusion in an isothermal solution containing one or more uncharged molecular species. The system may involve a chemical, physical-chemical, or biological reaction, or it may be nonreacting. If it is reacting and the chemical, physical, or biological change takes place in the bulk of the fluid the reaction is termed *homogeneous*. If it takes place only in a restricted region, such as at bounding surfaces or phase interfaces, it is termed *heterogeneous*. In homogeneous reactions species are produced, the production rates of which enter into the conservation of mass equation for a multicomponent flow. On the other hand, for a heterogeneous reaction the species production enters only in the boundary conditions at the reaction surfaces.

Homogeneous reactions often lead to significant heat release accompanied by nonisothermal conditions that require an appropriate heat source term to be added to the energy equation. Significant density changes with attendant alteration of the flow pattern frequently occur, and the interaction is "strong" in the sense described in our introductory remarks to Chapter 1. Here and throughout the text, whenever homogeneous reactions are considered, isothermal conditions will be assumed to prevail. However, in problem application primary attention will be paid to heterogeneous reactions. In some instances, heterogeneous reactions may also be accompanied by significant heat release, but again we shall only examine cases where the heat release is sufficiently small that the flow remains isothermal.

A wide range of phenomena are incorporated under the umbrella of heterogeneous reactions, which are defined to include any chemical, physico-chemical, or biophysicochemical reaction or transformation that takes place at a surface or interface. Examples would be chemically catalyzed reactions at a solid surface, adsorption and desorption at solid and liquid surfaces, including membranes, dissolution and precipitation of materials from solutions and melts, enzyme-substrate reactions at surfaces, and so on. With electrolyte solutions, electrode and electrochemical reactions take place at surfaces, and this will be discussed in Chapter 6.

Heterogeneous reactions involve several steps. The first is the transfer of the reacting species to the surface on which the reaction occurs (reaction surface). The second step is the heterogeneous reaction itself. This step is often composed of a series of substeps that may include diffusion of the reactants through the material, adsorption on the surface, chemical reaction, desorption of products and diffusion of products out of the material. The third step is the transfer of the products away from the reaction surface into the bulk phase.

The overall rate is controlled by the rate of the slowest step. This is then called the *rate determining step* or *rate limiting step*. If the rate limiting step is either steps one or three, which involves the introduction or removal of reactants, then the reaction is said to be *diffusion controlled*, with the rate governed by the mass transport relations previously set out. On the other hand, if step two, involving the chemical, physical, or biological transformation, is the slow step, then the rate is determined by the kinetics of the given process. As noted, within this step there may in turn be a distinction between diffusion and chemical, physical, or biochemical rates. Those cases where the rates of the diffusion and reaction steps are comparable are sometimes termed *mixed heterogeneous reactions*.

For homogeneous reactions the molar rate of production of species i per unit volume R_i is defined by

$$R_i \equiv \frac{1}{V} \left(\frac{dn_i}{dt} \right)_{\text{by reaction}} \tag{4.1.1}$$

where V is the mixture volume and $n_i = c_i V$ is the number of moles of the species i having a concentration c_i. In general, the reaction rate is a function of temperature, pressure, and concentrations of the substances participating in the reaction and may also depend on the concentrations of species such as catalysts or inhibitors that may not appear in the overall reaction.

For heterogeneous reactions the molar rate of species production refers to a surface rather than a volume source, and we write

$$R_i' \equiv \frac{1}{A} \left(\frac{dn_i}{dt} \right)_{\text{by reaction}} \tag{4.1.2}$$

where A is the reaction surface area and the prime denotes heterogeneous reactions. In the discussion of reaction rate fundamentals that follows, we shall make reference to homogeneous reactions, although the functional form of the reaction rate expressions for heterogeneous reactions is exactly the same, and it is only the constants of proportionality and dimensions that change, since $R_i V = R_i' A$ (Levenspiel 1972).

A *simple reaction* is one in which the reaction rate at a given temperature depends only on the rate of collision of the reacting molecules or, according to the *law of mass action*, is proportional to the active masses (concentrations) of the reacting substances:

$$R_1 = k \prod (c_i^{\nu_i})_{\text{reactants}} \qquad \nu = \sum \nu_i \qquad (4.1.3)$$

where k is the rate constant and ν_i the stoichiometric coefficient of the species i. The coefficient ν_i is also called the *order of the reaction* with respect to the species i, and ν is called the *overall order of the reaction*. Most known reactions are of first or second order.

In a reversible reaction the net rate of the reaction R_1 is the difference between the forward and reverse reaction rates. At equilibrium there is no net rate of reaction, so the forward and reverse reaction rates are equal and

$$\frac{k_f}{k_r} = K \qquad (4.1.4)$$

where k_f and k_r are the rate constants for the forward and reverse reactions, respectively, and K is the thermodynamic equilibrium constant for the reaction.

A *complex reaction* is one in which the reaction rates depend on the concentrations of the reacting substances and on the concentrations of the final or intermediate products of the reactions. For complex reactions the overall stoichiometry is frequently not known, so the rate cannot be related to the stoichiometry. However, the stoichiometric-based rate expression of Eq. (4.1.3) is found to be generally applicable to all reactions, although the order of the reaction ν_i with respect to the species i is not necessarily its stoichiometric coefficient and need not be integer or positive.

When the mechanism of the reaction is not known, the species reaction rate is often expressed empirically by the power law relation

$$R_i = k c_i^{\nu} \qquad (4.1.5)$$

The limitations of this form are manifested by integrating the above expression, from which it can be seen that for $\nu < 1$ the reactant concentration becomes negative at a finite time. The relation is inapplicable beyond the time at which this occurs (Levenspiel 1972).

When the rate expression is written in the form of either Eqs. (4.1.3) or (4.1.5), the dimensions of the rate constant depend on the order of the reaction; for a νth-order homogeneous reaction the dimensions are $(\text{mol m}^{-3})^{1-\nu}\,\text{s}^{-1}$. In the special case of a first-order homogeneous reaction, the dimensions become inverse time. For a zero-order reaction the reaction rate is independent of concentration, and the dimensions become $\text{mol m}^{-3}\,\text{s}^{-1}$. For complex reactions, as for example catalytic reactions, there is often no well-defined reaction order with respect to the reacting species.

The rate constant k is not a true constant but is temperature dependent. For many reactions the rate constant has been found to be well represented by the Arrhenius law (Levenspiel 1972)

$$k = A \exp\left(-\frac{E}{RT}\right) \tag{4.1.6}$$

where E is the activation energy and A is the frequency factor, a constant for a given reaction with the same dimensions as k. The logarithm of k varies as T^{-1}; a plot of $\ln k$ versus T^{-1} is called an *Arrhenius plot*, and a T^{-1} scale is called an *Arrhenius scale*.

According to Eq. (4.1.6), the reaction rate is a function of the activation energy E. For homogeneous gas phase chemical reactions the activation energy is normally rather high, around 80 to 250 kJ mol^{-1}. On the other hand, the activation energy for diffusion is low, ranging from 4 kJ mol^{-1} at room temperature to about 17 kJ mol^{-1} at 1300 K. Transformations with low activation energies are only moderately accelerated by increases in temperature, but the increase for those with relatively high activation energies can be dramatic. For example, a doubling of reaction rate for a 10° increase in temperature at around ambient conditions is typical for many common homogeneous gas phase reactions. Activation energies for enzyme-catalyzed reactions are high, typically around 170 to 300 kJ mol^{-1}, but the temperature range of these reactions is limited to the usual biological range. In general, uncatalyzed heterogeneous reactions have low to moderate activation energies, with values generally from 25 to 40 kJ mol^{-1}.

For heterogeneous reactions all of the expressions given above are written in the same form except that the rate constant refers to a reaction surface (Eq. 4.1.2). With a prime denoting a heterogeneous condition, the empirical rate expression Eq. (4.1.5) is written

$$(R_i')_w = (k_i' c_i^\nu)_w \tag{4.1.7}$$

where the subscript w is used to indicate surface (wall) conditions. Note that the units of R_i' are mol m^{-2} s^{-1} and of k', mol$^{1-\nu}$ m$^{3\nu-2}$ s^{-1}, so for $\nu = 1$ the rate constant has the dimensions of a "reaction velocity."

In heterogeneous reactions at a solid surface, molecules adsorbed on the surface often play a fundamental role in the reaction mechanism. For a homogeneous surface where only a monolayer of adsorbed material is formed, the surface adsorption is often described by the *Langmuir adsorption isotherm*. For adsorption of a single substance from solution, the isotherm expresses a relation between the amount of substance adsorbed and the concentration c of the adsorbing species in solution next to the surface

$$\theta = \frac{c}{c + b} \tag{4.1.8}$$

In this expression θ is the fraction of surface adsorption sites occupied, and b is a constant for the adsorbent.

The reaction rate for adsorbed molecules is proportional to θ, and for molecules reacting by direct collision the rate is proportional to c. As an example of the application of the Langmuir isotherm to heterogeneous kinetics, consider a unimolecular reaction in the adsorbed state where the rate of reaction is proportional to the degree of coverage (Frank-Kamenetskii 1969):

$$A \rightarrow B \qquad (4.1.9)$$

If adsorption is rapid, as in the case of physical adsorption due to intermolecular forces, and the surface reaction is slow, then with the extent of coverage given by the Langmuir isotherm the reaction rate is

$$R' = k'_\alpha \theta = k'_\alpha \frac{c}{c+b} \qquad (4.1.10)$$

where k'_α is the rate constant of the surface reaction in appropriate units. Thus at low concentrations of the adsorbing species the reaction is of first order in c whereas at high concentrations it is of zero order. Therefore, depending on the concentration, the effective order of the reaction is generally not integer and will lie between 0 and 1.

The opposite situation of slow adsorption more frequently prevails in heterogeneous catalysis, where chemically adsorbed molecules undergo reaction on the surface. The rate of the process is determined by the chemical adsorption or *chemisorption*, where the adsorbed molecules react chemically with the surface. Since chemical bonds are broken, an activation energy is required and the adsorption is relatively slow. Because of the activation energy requirement, chemisorption is also often called *activated adsorption*. For our unimolecular example the reaction rate is equal to the chemical adsorption rate or

$$R' = k'_\beta c(1 - \theta) = k'_\beta b \frac{c}{c+b} \qquad (4.1.11)$$

where k'_β is the rate constant for chemisorption in appropriate units and $1 - \theta$ is the fraction of vacant sites on the surface. Simple enzyme-catalyzed chemical reactions have the same general type of kinetic behavior. In the context of enzyme catalysis the reactant is termed the *substrate*, the rate law is called the *Michaelis-Menten law*, and the constant b is the *Michaelis constant* (Castellan 1983).

The molar flux to the surface of the ith species if the species is uncharged is given by

$$(j_i^*)_w = (c_i \mathbf{u} - D_i \nabla c_i)_w \qquad (4.1.12)$$

If the surface is impermeable and any heterogeneous reaction is not accompanied by a local density change, then from the no-slip condition $\mathbf{u} = 0$. If a heterogeneous reaction is accompanied by a volume change, there will be a general convective flow of the reacting mixture in a direction normal to the surface at which the reaction is taking place. This convective flux can result in a "strong" fluid interaction and is termed *Stefan flow*. However, it is generally a small effect for most chemical and biochemical heterogeneous reactions and is normally important only in the presence of strong ablation or condensation. For our purposes here we shall neglect Stefan flow.

If the surface is permeable to the flow as, for example, with a membrane, then there is also a finite flux normal to the surface. The extent to which the species permeates will depend on the rejection characteristics of the membrane, and we shall discuss this boundary condition later in connection with membrane filtration. Here, we suppose the surface to be impermeable so that $\mathbf{u}_w = 0$, in which case at the reaction surface

$$(j_i^*)_w = D_i \left(\frac{\partial c_i}{\partial y} \right)_w \qquad (4.1.13)$$

where y is the outward normal direction (opposite to ∇c_i). At steady state the incoming flux at the wall must be balanced by the species reaction rate. Employing for simplicity the empirical rate law, Eq. (4.1.7), we obtain

$$D \left(\frac{\partial c}{\partial y} \right)_w = k' c_w^\nu \qquad (4.1.14)$$

where the subscript i is here understood.

Let us nondimensionalize the above boundary condition using the reduced variables

$$y^* = \frac{y}{L} \qquad c^* = \frac{c}{c_0} \qquad (4.1.15)$$

where c_0 is the free stream or bulk value of the concentration far from the surface and L is a characteristic length. In terms of these reduced variables and with D and k' constant, Eq. (4.1.14) takes the form

$$\frac{1}{\mathrm{Da}} \left(\frac{\partial c^*}{\partial y^*} \right)_w = c_w^{*\nu} \qquad (4.1.16)$$

The dimensionless similarity parameter Da is a measure of the reaction velocity and is termed the *Damköhler number*, where

$$\mathrm{Da} = \frac{k' c_0^{\nu-1}}{D/L} = \text{Damköhler number}$$

$$= \frac{\text{reaction velocity}}{\text{diffusion velocity}} \qquad (4.1.17)$$

If $Da \gg 1$, this implies the rate of species production by reaction is large compared to the rate of mass transfer by diffusion, and the boundary condition at the reaction surface becomes

$$c_w \approx 0 \qquad (4.1.18)$$

The meaning of this condition is that all particles approaching the surface react instantaneously. Since the solute species flux to the surface is proportional to the characteristic driving concentration difference $c_0 - c_w$, then, with $c_w = 0$, the driving force is a maximum and the flux so obtained is termed a *limiting flux*. We shall run into this concept of a limiting flux in many diffusion-controlled problems.

In the opposite limit where $Da \ll 1$, the rate of species production by reaction is small compared with the diffusional flux, and the boundary condition at the reaction surface reduces to

$$\left(\frac{\partial c}{\partial y}\right)_w \approx 0 \qquad (4.1.19)$$

In this limit the concentration is everywhere constant and equal to c_0, with the overall reaction rate specified by $k'c_0^\nu$.

When the removal of the reaction products from the surface is the slowest step, the conditions described above must be modified (Levich 1962). The concentration far from the surface is small compared with that at the surface, and the direction of the diffusion flux is reversed from that given by Eq. (4.1.14). Moreover, under some conditions a steady solution for slow product removal may not be achievable as, for example, with an autocatalytic reaction where the product catalyzes the reaction, causing the process to accelerate.

4.2   Convective Diffusion Layer Characteristics

We have already observed that for dilute solutions the Schmidt number is very large, as a consequence of which the diffusion Peclet number

$$Pe_D = Re\,Sc \qquad (4.2.1)$$

is generally large. This is true even at moderate Reynolds numbers, the consequence of which is that in the bulk of a convective flow past a solid surface convection dominates over diffusion.

In the limit $Pe_D \to \infty$,

$$\frac{Dc}{Dt} = 0 \qquad (4.2.2)$$

indicating $c = $ constant following a fluid particle. With $c = c_0$ in the free stream, this solution cannot satisfy the boundary conditions at a reaction surface discussed in the last section, for example, $c_w = 0$. Evidently, near the surface there must be a thin diffusion boundary layer of thickness δ_D within which the concentration changes rapidly. Within this layer the derivatives of the concentration in the direction normal to the surface (y direction) are much larger (of order δ_D^{-1}) than the derivatives in the streamwise direction (x direction). The reasoning parallels the Prandtl boundary layer argument for viscous flow past a solid boundary at high Reynolds number.

We recall that an estimate of the characteristic Prandtl viscous boundary layer thickness δ_U for steady unbounded flow is given by

$$\frac{\delta_U}{L} \sim (\text{Re})^{-1/2} \sim \left(\frac{\nu}{UL}\right)^{1/2} \tag{4.2.3}$$

Since D plays the same role as the kinematic viscosity ν, we may expect for large Schmidt numbers ($\nu \gg D$) that the viscous boundary layer thickness should be considerably larger than the diffusion boundary layer thickness. A consequence of this is that the velocity seen by the concentration layer at its "edge" is not the free stream velocity U but something much less, which is more characteristic of the velocity close to the wall (Fig. 4.2.1). We note also that since c is understood to be c_i, then in a multicomponent solution there may be as many distinct boundary layers as there are species, with the thickness of each defined by the appropriate diffusion coefficient. With this caveat in mind, we may write the convective diffusion equation for a two-dimensional diffusion boundary layer and estimate the diffusion layer thickness.

Since the derivatives normal to the surface are order δ_D^{-1} larger than those in the streamwise direction,

$$\frac{\partial^2 c}{\partial y^2} \gg \frac{\partial^2 c}{\partial x^2} \tag{4.2.4}$$

and the convective diffusion equation becomes

$$\frac{\partial c}{\partial t} + u \frac{\partial c}{\partial x} + v \frac{\partial c}{\partial y} = D \frac{\partial^2 c}{\partial y^2} \tag{4.2.5}$$

The equation as written is valid for two-dimensional, unsteady diffusion boundary layers. The basic features of this result and those that follow are unaltered if the diffusion coefficient is not constant or if the flow is generalized to axial symmetry or three dimensions, although some details will differ.

We now estimate the diffusion layer thickness, employing the fact that from viscous boundary layer analysis, $v \sim O(\delta)$ with $u \sim O(1)$, so

$$u \frac{\partial c}{\partial x} \sim v \frac{\partial c}{\partial y} \sim \frac{\partial c}{\partial t} \tag{4.2.6}$$

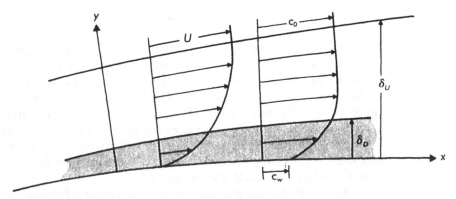

Figure 4.2.1 Relative thicknesses and profiles in diffusion and viscous boundary layers.

where if the flow is unsteady it is assumed that $t \sim x/u$. The flow may be steady, in which case $\partial c/\partial t = 0$ and we suppose this to be so here. The estimate of Eq. (4.2.6) is a statement of the fact that all the convective diffusion terms are of the same order.

For $\delta_D \ll \delta_U$ the velocity profile in the diffusion layer must be that corresponding to the viscous layer profile near the wall, which at small distances is linear in y:

$$\frac{u}{U} \sim \frac{y}{\delta_U} \qquad (4.2.7)$$

Therefore

$$u \frac{\partial c}{\partial x} \sim U \frac{\delta_D}{\delta_U} \frac{c_0 - c_w}{x} \sim \frac{\nu \delta_D (c_0 - c_w)}{\delta_U^3} \qquad (4.2.8)$$

where we have set $y \sim \delta_D$ and $\delta_U^2/x \sim \nu/U$. Similar viscous boundary layer arguments can be used, which show that $v(\partial c/\partial y)$ is of the same order. This follows using $\nu y^2/\delta_U^3$ for the value of v close to an impermeable surface.

In the above estimate it is supposed that a characteristic driving concentration difference $c_0 - c_w$ can be defined. In this sense there is an implicit restriction to no species production by reaction at the surface or no permeation through the surface, so the value of c at the surface can be specified independent of the solution. This would include, for example, the limiting flux condition of a large rate of species production by reaction compared with the diffusive transfer rate, so $c_w \approx 0$. Another example would be a soluble surface where the dissolution process is much more rapid than the removal of the dissolved particles so that

$$c_w = c_{sat} \qquad (4.2.9)$$

where c_{sat} is the equilibrium concentration of the saturated solution in the liquid layer at the surface. The other boundary conditions of mixed heterogeneous reactions and surface permeation will be discussed in the context of specific examples.

An estimate for δ_D is now obtained from the condition that for $y > \delta_D$ diffusion is negligible but becomes comparable with convection at $y \sim \delta_D$, or

$$D \frac{\partial^2 c}{\partial y^2} \sim u \frac{\partial c}{\partial x} \qquad (4.2.10)$$

whence

$$D \frac{c_0 - c_w}{\delta_D^2} \sim \frac{\nu \delta_D (c_0 - c_w)}{\delta_U^3} \qquad (4.2.11)$$

to give

$$\delta_D \sim \left(\frac{D}{\nu} \right)^{1/3} \delta_U \equiv \frac{\delta_U}{Sc^{1/3}} \qquad (4.2.12)$$

This result shows that for Schmidt numbers appropriate to liquid diffusion problems, which are of the order of 10^3, that the diffusion layer thickness is of the order of one tenth that of the viscous layer. However, the diffusion layer growth is parabolic in the streamwise distance x like that of the viscous boundary layer. Remember that the arguments used with respect to the velocity at the edge of the concentration layer are only appropriate for large Schmidt numbers.

A somewhat different behavior is found for the initial rate of growth of a diffusion layer in a flow where inertia is absent, as in the steady fully developed flow in a pipe or channel or in a fully developed thin film falling under gravity. To characterize this functional dependence, we must first define the velocity field within which the diffusion layer grows.

The length for a fully developed Poiseuille velocity profile to develop in a channel so that there is no longer an inviscid core may be estimated with reference to Fig. 4.2.2. The characteristic time for viscosity to diffuse to the center of the channel is of the order of h^2/ν, whereas the characteristic time for a particle to be convected the distance L_U to where the viscosity has diffused to the channel center is of the order of L_U/U. The profile will be fully developed and unchanging when these times are of the same order, or

$$\frac{L_U}{h} \sim \frac{Uh}{\nu} \equiv \mathrm{Re} \qquad (4.2.13)$$

The coefficient multiplying the Reynolds number for a straight channel is 0.16 (Schlichting 1979). Therefore, for a Reynolds number based on a channel width $(2h)$ of 1000, the Poiseuille profile would develop in about 40 channel widths. Again, since $\nu \gg D$ for diffusion in dilute solution, we may expect that the development length is very much longer for the concentration profile than for the velocity profile.

The concentration profile development in a channel is sketched in Fig. 4.2.3 for surface dissolution, where $c = 0$ at the entrance $x = 0$ and along the centerline $y = h$, and where at the wall $c = c_w$, corresponding to saturation concentration. The profile development for a reacting surface with $c_w \approx 0$ would be similar to that for the velocity profile shown in Fig. 4.2.2 except that the

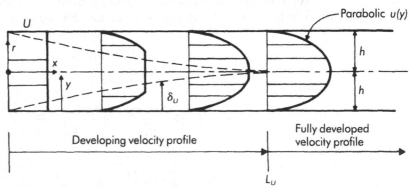

Figure 4.2.2 Development of Poiseuille velocity profile in a channel.

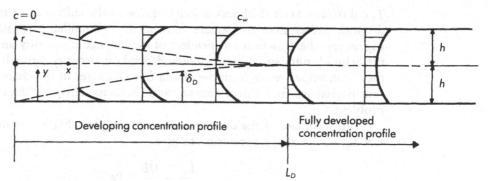

Figure 4.2.3 Development of solute concentration profile in a channel with soluble walls.

centerline concentration would remain constant at the entrance value, say c_0. Note that the developed concentration profile, unlike the velocity profile, will alter as long as solute is being redistributed.

For fully developed channel flow $v = 0$ and $u = u(y)$, where

$$u = u_{max}\left(1 - \frac{r^2}{h^2}\right) = u_{max}\left(1 - \frac{(y-h)^2}{h^2}\right) \qquad (4.2.14)$$

and

$$u_{max} = \frac{3}{2} U = -\frac{h^2}{2\mu}\frac{dp}{dx} \qquad (4.2.15)$$

For a pipe of circular cross section the velocity profile is the same as that given by Eq. (4.2.14). However, the maximum velocity at the centerline u_{max} equals twice the mean velocity U, so the value of the maximum velocity is four thirds that of Eq. (4.2.15). In either case, the velocity distribution close to the wall is

$$u \approx u_{max}\left(\frac{2y}{h}\right) \qquad (4.2.16)$$

Based on our estimates of the velocity and concentration profile development lengths, it is assumed that the velocity profile is fully developed and the developing diffusion boundary layer thickness δ_D is small in comparison with the channel half-width. We then estimate the diffusion layer thickness by inserting the linear velocity profile given above into Eq. (4.2.10):

$$u_{max}\frac{2\delta_D}{h}\frac{c_0 - c_w}{x} \sim D\frac{c_0 - c_w}{\delta_D^2} \qquad (4.2.17)$$

from which

$$\frac{\delta_D}{x} \sim \left(\frac{h}{x}\right)^{2/3}\left(\frac{D}{u_{max}h}\right)^{1/3} \qquad (4.2.18)$$

The diffusion layer thickness is seen to grow as the cube root of the streamwise distance rather than the square root, as in unbounded flow past a surface. Moreover, the growth is independent of the kinematic viscosity and therefore of the Schmidt number, although it does depend on viscosity through $u_{\max}$. Both of these conditions result from the inertia free character of the channel flow itself; that is, viscous forces predominate, and the mass density ρ does not enter the problem.

An estimate of the development length is provided by setting $\delta_D = h$ and $x = L_D$ in the above expression to give

$$\frac{L_D}{h} \sim \frac{Uh}{D} \equiv \text{Pe} \tag{4.2.19}$$

where we have used the fact that $u_{\max} \sim U$. As may have been expected, this result parallels that for the velocity profile development length with Re replaced by Pe.

For both of the flow types considered, we may also estimate the species flux from the relation $j^* = D(\partial c / \partial y)$ as

$$j^* \sim \frac{D(c_0 - c_w)}{\delta_D} \tag{4.2.20}$$

assuming a linear concentration profile in c with $c = c_w$ at $y = 0$ and $c = c_0$ at $y = \delta_D$. The use of a linear approximation for the concentration gradient is referred to as the *Nernst layer approximation*, although δ_D is not necessarily a constant as is sometimes assumed with the use of the *Nernst film theory*. Here, δ_D varies with both fluid properties and streamwise distance in accord with the behaviors derived.

Everything we have said concerning diffusion layer growth at high Schmidt number has a direct analogue with thermal layer growth at high Prandtl number. For the thermal problem the whole of the temperature field is confined within that portion of the velocity field where the velocity is linear in distance from the wall. The appropriate Peclet number is the thermal one Uh/α in place of Uh/D. In the following sections, keep this analogue in mind. Thus when we speak of a uniform concentration at the surface, the analogue is uniform surface temperature, whereas the analogue of constant surface mass transfer is constant heat flux. For the interested reader the *thermal boundary layer* and *thermal entry region* problems are examined in Schlichting (1979) and Kays & Crawford (1980).

4.3   Channel Flow with Soluble or Rapidly Reacting Walls

In the last section an order-of-magnitude estimate was given for the thickness of a high Schmidt number, developing diffusion layer in a channel with a fully developed velocity profile. Here, we will illustrate the similarity character of the developing diffusion layer close to the entrance and derive the mathematical solution for soluble channel walls. In this case the solute concentration at the wall is the saturated solution concentration ($c_w = c_{sat}$) with the concentration vanishing at the channel center ($c = 0$). The solution to this problem is complementary to the one for the reacting wall in the fast reaction limit where $c_w = 0$, and the concentration at the channel center is the bulk or free stream value c_0.

For specificity let us first consider the soluble wall problem sketched in Fig. 4.2.3. Now in a channel (or pipe) flow $v = 0$, and near the surface the streamwise velocity component is given by $u_{max}(2y/h)$. This behavior of the velocity profile is the same as for a "fully developed" thin liquid film on a vertical wall, falling under gravity with a free surface at atmospheric pressure. The velocity profile is parabolic with the fall velocity and has a maximum at the free surface equal to

$$u_{max} = \frac{g\delta^2}{2\nu} \tag{4.3.1}$$

The film thickness δ is given by $(3\nu\tilde{Q}/g)^{1/3}$, where $\tilde{Q}$ is the volume flow rate per unit width. As in plane channel flow the mean velocity is two thirds of the maximum value.

From the diffusion equation for steady flow and the velocity profile near the wall,

$$2\frac{y}{h} u_{max} \frac{\partial c}{\partial x} = D \frac{\partial^2 c}{\partial y^2} \tag{4.3.2}$$

with the lateral convection associated with the surface mass transfer taken to be negligible. The soluble wall boundary conditions are

$$c = c_{sat} \qquad \text{at } y = 0 \tag{4.3.3a}$$

$$c = 0 \qquad \text{as } y \to \infty \tag{4.3.3b}$$

Note that conditions close to the entrance have been assumed, with $y/h \ll 1$. This implies that not only is the velocity profile approximated by the profile near the wall, but the boundary condition on the concentration profile of $c = 0$ at $y = h$ is replaced by the asymptotic one of $c = 0$ as $y \to \infty$.

From physical reasoning, as well as by examination of the differential equation and boundary conditions, it is clear that if c_{sat} is doubled the only effect on the concentration distribution would be to double the local value of the concentration at each value of y. Therefore from Eq. (4.3.2),

$$\frac{c}{c_{sat}} \equiv c^* = \text{fcn}\left(x, y, \frac{u_{max}}{Dh}\right) \tag{4.3.4}$$

Since the left side of this equation is dimensionless, the right side must also be a function of a dimensionless quantity. The only nondimensional combination possible is the similarity variable

$$\eta = y\left(\frac{2u_{max}}{hDx}\right)^{1/3} \tag{4.3.5}$$

where the numerical factor of 2 has been introduced for convenience.

The solution for c^* must therefore have the similarity form $c^* = \text{fcn}(\eta)$, whence

$$\frac{\partial c^*}{\partial x} = \frac{dc^*}{d\eta}\frac{\partial\eta}{\partial x} = -\frac{\eta}{3x}\frac{dc^*}{d\eta} \tag{4.3.6a}$$

$$\frac{\partial c^*}{\partial y} = \frac{dc^*}{d\eta}\frac{\partial\eta}{\partial y} = \left(\frac{2u_{max}}{hDx}\right)^{1/3}\frac{dc^*}{d\eta} \tag{4.3.6b}$$

$$\frac{\partial^2 c^*}{\partial y^2} = \left(\frac{2u_{max}}{hDx}\right)^{2/3}\frac{d^2 c^*}{d\eta^2} \tag{4.3.6c}$$

Substituting these relations into the diffusion equation reduces this partial differential equation to the ordinary differential equation

$$\frac{d^2 c^*}{d\eta^2} + \frac{\eta^2}{3}\frac{dc^*}{d\eta} = 0 \tag{4.3.7}$$

with the boundary conditions

$$c^* = 1 \qquad \text{at } \eta = 0 \tag{4.3.8a}$$

$$c^* = 0 \qquad \text{as } \eta \to \infty \tag{4.3.8b}$$

The ordinary differential equation can also be written

$$\frac{d}{d\eta}\left(\ln\frac{dc^*}{d\eta}\right) = -\frac{\eta^2}{3} \tag{4.3.9}$$

Integrating twice gives

$$c^* = A\int_0^\eta e^{-\eta^{3/9}}\,d\eta + B \tag{4.3.10}$$

which, on evaluating the constants of integration, gives

$$c^* = 1 - \frac{\displaystyle\int_0^\eta e^{-\eta^{3/9}}\,d\eta}{\displaystyle\int_0^\infty e^{-\eta^{3/9}}\,d\eta} \tag{4.3.11}$$

The complete gamma function is defined by the definite integral

$$\Gamma(n) = \int_0^\infty z^{n-1}e^{-z}\,dz \qquad (n > 0) \tag{4.3.12}$$

with the function satisfying the recursion relation

$$\Gamma(n + 1) = n\Gamma(n) \tag{4.3.13}$$

where

$$\Gamma(n + 1) = \int_0^\infty z^n e^{-z} \, dz \tag{4.3.14}$$

The integration constant defined by the integral in the denominator of the solution is thus $(9)^{1/3}\Gamma(\frac{4}{3})$, the value of which is 1.858. It follows that

$$\frac{c}{c_{\text{sat}}} = 1 - 0.538 \int_0^\eta e^{-\eta^{3/9}} \, d\eta \tag{4.3.15}$$

The diffusion flux from the channel wall is

$$j^* = -D\left(\frac{\partial c}{\partial y}\right)_w = -c_{\text{sat}} D\left(\frac{2u_{\max}}{hDx}\right)^{1/3}\left(\frac{dc^*}{d\eta}\right)_w \tag{4.3.16}$$

or, from the above solution for c,

$$j^* = 0.678 c_{\text{sat}} D\left(\frac{u_{\max}}{hDx}\right)^{1/3} \tag{4.3.17}$$

Using this result and the Nernst relation $j^* = D(c_0 - c_w)/\delta_D$, we estimate the diffusion layer thickness to be

$$\frac{\delta_D}{x} = 1.475\left(\frac{h}{x}\right)^{2/3}\left(\frac{D}{u_{\max}h}\right)^{1/3} \tag{4.3.18}$$

This procedure for estimating the diffusion layer thickness was employed by Levich (1962), but recall that any definition of a "boundary layer thickness" is to some extent arbitrary. All of the results derived are seen to be consistent with the order-of-magnitude estimates given in the previous section.

In concluding this section, we point out that the solution for the concentration profile in the case of a rapidly reacting wall, where $c_w = 0$ and $c = c_0$ at the channel center, is simply the complement of the solution given by Eq. (4.3.15):

$$\frac{c}{c_0} = 0.538 \int_0^\eta e^{-\eta^{3/9}} \, d\eta \tag{4.3.19}$$

The quantity c/c_0 represents the concentration defect in the soluble wall problem. All other results are unchanged except that c_{sat} is replaced by c_0 and the diffusional flux is toward the wall.

4.4   Reverse Osmosis and Mixed Heterogeneous Reactions

Reverse osmosis is a pressure-driven membrane process used to separate relatively pure solvents, most often water, from solutions containing salts and dissolved organic molecules. Under the action of a hydrostatic pressure, typically in the range of 3 to 7 MPa, applied across the membrane separating the feed from the relatively pure solvent, the solvent passes through the membrane and the dissolved materials remain behind.

Most reverse osmosis systems in use today employ semipermeable *asymmetric membranes* made from cellulose acetate or polyamide, or *composite membranes* made with a dense, thin, polymer coating on a polysulfone support film. The asymmetric cellulose acetate membrane is a high-water-content gel structure consisting of a thin rejecting "skin" about 0.1 to 0.5 μm thick integral with a much thicker porous substrate 50 to 100 μm thick. The skin offers the main hydraulic resistance to the flow. The porous substructure gives the membrane strength but offers almost no hydraulic resistance. The dense rejecting skin of the composite membranes can be up to 10 times thinner than the skin of the cellulose acetate membranes.

The rejection of the dissolved materials is not complete, the incompleteness depending not only on the size of the rejected species but also on the chemistry of the membrane and the rejected species. With the use of the correct membrane the rejections, measured in percent, of inorganic salts will be in the high nineties, and there is almost complete rejection of most species with molecular weights greater than 150. The rejection of low molar mass nonelectrolytes, such as small organic molecules, is generally low with the asymmetric cellulose acetate and polyamide membranes. However, with the use of newer composite membranes and by proper pH adjustment, moderate to good rejections can be obtained with many intermediate and even low molar mass organics. Because the percent rejections tend to be constant over a wide range of concentration, the concentration in the solvent passing through the membrane will be proportional to the concentration retained. The higher the fraction of feed that passes through the membrane, that is, the higher the "recovery" of solvent, the higher will be the concentration in the product. The retained solvent, usually water, is often termed the *concentrate*, and the product is the *permeate*.

If an ideal semipermeable membrane separates an aqueous organic or inorganic solution from pure water, the tendency to equalize concentrations would result in the flow of the pure water through the membrane to the solution. The pressure needed to stop the flow is called the *osmotic pressure*. If the pressure on the solution is increased beyond the osmotic pressure, then the flow would be reversed and the fresh water would pass from the solution through the membrane, whence the name *reverse osmosis*. In actual reverse osmosis systems the applied pressure must be sufficient to overcome the osmotic pressure of the solution and to provide the driving force for adequate flow rates.

Osmotic pressure is a property of the solution and does not in any way depend on the properties of the membrane. For dilute solutions the osmotic pressure is independent of the solute species (a colligative property) and is given by the *van't Hoff equation*

$$\pi = \nu cRT \tag{4.4.1}$$

where π is the osmotic pressure and ν is the number of ions formed if the solute dissociates. Table 4.4.1 shows the osmotic pressures for some aqueous solutions at standard temperature as a function of mass concentration. The molar mass is expressed in molecular weight units. Evidently at the same mass concentration sugar will have a much lower osmotic pressure than salt because it has a higher molar mass.

The amount of solvent that will pass through a membrane is proportional to the excess of the hydrostatic pressure p over the osmotic pressure π, and for useful flow rates p should be large compared with π. The solvent (water) flux through the membrane may be written approximately as

$$j_A^* = A(\Delta p - \Delta \pi) \tag{4.4.2}$$

where Δp is the hydrostatic pressure difference across the membrane and $\Delta \pi$ is the osmotic pressure difference corresponding to the solute concentrations immediately adjacent to the membrane surface on both sides. The coefficient A is the membrane solvent permeability coefficient. It is inversely proportional to the thickness of the solute-rejecting portion of the membrane, a quantity that is generally not known precisely. The value of A is determined empirically by using Eq. (4.4.2).

Although the solvent flux is inversely proportional to the "active" thickness of the membrane, the solute rejection is independent of this thickness. The driving force for the solute flux is mainly the difference in solute concentration across the membrane between the feed and product and is given approximately by

$$j_s^* = B\Delta c_w = Bc_w R_s \tag{4.4.3}$$

Table 4.4.1
Osmotic Pressure Data for Some Aqueous Solutions at Standard Temperature

Dissolved Species	Concentration kg m^{-3}	Osmotic Pressure MPa
NaCl ($M = 58.5$)	50	4.609
	10	0.844
	5	0.421
Urea ($M = 60$)	50	2.127
	10	0.425
	5	0.213
Sucrose ($M = 342$)	50	0.380
	10	0.076
	5	0.038

Here, B is the solute permeability coefficient, Δc_w is the difference in solute concentrations immediately adjacent to the membrane (wall) on the feed and product sides, c_w is the concentration on the feed side, and R_s is the solute rejection coefficient, which is usually taken to be constant.

Combining the preceding two equations and using the fact that the osmotic pressure is proportional to the solute concentration, we can show that the permeation velocity of the solution through the membrane $v_w = j/\rho$ may be written (Gill et al. 1971)

$$v_w = A'\Delta p\left(1 - \frac{c_w}{c_0}R_s\frac{\pi_0}{\Delta p}\right) \tag{4.4.4}$$

where the membrane constant $A' = A/c_A$, where c_A is the molar concentration of the solvent. The molar concentration c_0 is the solute concentration in the bulk solution far from the membrane, $\pi_0/\Delta p$ representing the fraction of the osmotic pressure drop required to overcome the osmotic pressure of the bulk solution. Note that the rejection coefficient $R_s \to 1$ for $\Delta p \gg \pi_0$.

As a consequence of the passage of solvent, say water, through the membrane, solute is carried to the membrane surface, and the concentration at the membrane surface tends to be higher than in the bulk of the liquid. This phenomenon is called *concentration polarization*. Several deleterious effects arise from concentration polarization, one of which is the local increase in osmotic pressure due to the increased solute concentration at the membrane. The result is that the solvent flux is decreased because the effective driving pressure is reduced. Another effect is an increase in the solute concentration in the product, for with real leaky membranes the flux of solute across the membrane is proportional to the difference in solute concentration on both sides. The extent of concentration polarization depends on the hydrodynamics and geometry of the system, with increased flow speed of the solution past the membrane tending to reduce the effect.

To illustrate the concentration polarization problem, we again consider laminar flow in a parallel plate, reverse osmosis channel, where the channel walls are a porous support for the membrane and where the velocity profile is taken to be fully developed from the channel inlet. Let us first examine the qualitative behavior of the solute concentration distribution along the channel shown schematically in Fig. 4.4.1. The membranes are assumed to be perfectly rejecting ($R_s = 1$), that is, totally impermeable to solute.

When the feed enters the channel, solute is convected toward the membranes, as a result of the convective transport through them, while being diffused away. Because of the basic longitudinal character of the flow and the solute-rejecting property of the membranes, the net result is the streamwise development of diffusion boundary layers with enhanced solute concentration adjacent to the membranes. These polarization boundary layers at first grow rapidly in the immediate neighborhood of the inlet, with the excess solute concentration and thickness increasing as the one-third power of the streamwise distance, as was shown in the preceding two sections to be characteristic of entry length diffusion layers. Beyond the immediate entrance region the excess

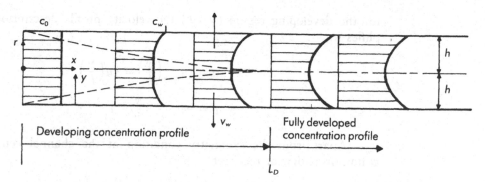

Figure 4.4.1 Solute concentration development in a reverse osmosis channel.

concentration growth becomes more nearly linear with downstream distance until the diffusion layers from both membrane walls eventually fill the channel. At this location the concentration profile adjusts to an asymptotic fully developed and unchanging profile characterized by the convective salt transport toward the wall, just balancing the diffusive transport away from it.

For the fully developed velocity profile the mathematical problem again reduces to the solution of the steady form of the convective diffusion equation for the solute concentration. In contrast to the diffusion equation treated in the channel flow problem with soluble or rapidly reacting walls, it is necessary to include here the lateral convection term to account for the product removal through the membrane walls, putting

$$u\frac{\partial c}{\partial x} + v\frac{\partial c}{\partial y} = D\frac{\partial^2 c}{\partial y^2} \qquad (4.4.5)$$

Because of the transverse velocity component, the velocity profile is a modification of the usual Poiseuille distribution. This problem has been solved by Berman (1953), including the effect of a constant permeation velocity v_w in altering the velocity profile in the x direction and in causing a streamwise variation in the bulk average velocity. However, when the Reynolds number based on the permeation velocity is small, as it generally is, the streamwise component has the same form as for an impermeable wall and the transverse component is proportional to the constant permeation velocity v_w; that is,

$$u = u_{max}\left(1 - \frac{r^2}{h^2}\right) = \frac{3}{2}\,\bar{u}\left(1 - \frac{r^2}{h^2}\right) \qquad (4.4.6a)$$

$$v = v_w\left(\frac{r}{2h}\right)\left(3 - \frac{r^2}{h^2}\right) \qquad \frac{hv_w}{\nu} \ll 1 \qquad (4.4.6b)$$

Note here that $\bar{u}$ is the average longitudinal velocity, which in this instance does decrease along the channel from the mean inlet value U because of the solvent withdrawal through the membrane. Setting $r = y - h$ (Fig. 4.4.1), we see that for the developing region ($y \ll h$) the velocity profile description further simplifies to

$$u = 2u_{max}\left(\frac{y}{h}\right) = 3\bar{u}\left(\frac{y}{h}\right) \tag{4.4.7a}$$

$$v = -v_w \qquad \frac{y}{h} \ll 1 \tag{4.4.7b}$$

Apart from the symmetry condition at the channel center, the initial condition is that at the inlet

$$c = c_0 \qquad \text{at } x = 0 \tag{4.4.8}$$

The boundary condition at the membrane wall derives from conservation of solute flux applied across the membrane. Stated in words: At steady state the bulk flow of solute toward the membrane minus the diffusional flux of the solute away from the membrane toward the bulk of the fluid must equal the solute permeation through the membrane. Mathematically, taking into account that the y coordinate is in the direction of the concentration gradient and opposite to the permeation velocity,

$$-v_w c_w - D\left(\frac{\partial c}{\partial y}\right)_w = -(1 - R_s)v_w c_w \tag{4.4.9a}$$

or

$$R_s v_w c_w = -D\left(\frac{\partial c}{\partial y}\right)_w \tag{4.4.9b}$$

In writing the above boundary condition, we have allowed for the possibility of a rejection coefficient less than unity, although R_s is supposed constant. With a constant permeation velocity v_w the problem is a linear one. However, were v_w to depend on the concentration at the membrane, as characterized by, say, Eq. (4.4.4), a nonlinearity would be introduced through the boundary condition. Let us assume that R_s is not only constant but equal to 1, in which case the boundary condition becomes

$$v_w c_w = -D\left(\frac{\partial c}{\partial y}\right)_w \tag{4.4.10}$$

Equation (4.4.9b) may be compared with the boundary condition for a mixed heterogeneous reaction $k'c_w^v = D(\partial c/\partial y)_w$. It is evident that $R_s = 1$ corresponds to a first-order reaction, whereas if $R_s \sim c_w^m$, there is a correspondence as well for reactions other than first order. For simplicity, let us consider the perfectly rejecting case of $R_s = 1$. Here, the appropriate dimensionless parameter characterizing the boundary condition is the ratio of the permeation velocity to the diffusion velocity

$$\text{Da} = \frac{v_w}{D/h} = \frac{\text{permeation velocity}}{\text{diffusion velocity}} \tag{4.4.11}$$

We could refer to this as a Damköhler number, by analogy with Eq. (4.1.17), or recognize that it is interpretable as a Peclet number for mass transfer:

$$\text{Pe} = \frac{v_w h}{D} = \frac{\text{mass transported by permeation}}{\text{mass transported by diffusion}} \qquad (4.4.12)$$

It has been common in the literature to interpret the parameter as a Peclet number, and we shall follow that practice.

What is important to recognize from the discussion is that the boundary condition for mass transfer through a semipermeable membrane is directly analogous to that for a mixed heterogeneous reaction. A consequence of this is that what is said about the one problem can be translated to the other, despite the somewhat different physics and chemistry. The example of reverse osmosis is therefore used as an illustration of a mixed heterogeneous reaction. The major part of the discussion will, however, be confined to the developing layer, where

$$\text{Pe} = \frac{v_w \delta_D}{D} \ll 1 \qquad (4.4.13)$$

In the analogous chemical reaction problem this corresponds to a small Damköhler number, wherein the rate of species production is small compared with the diffusional flux.

Before presenting a solution for the developing layer problem, let us give an order-of-magnitude estimate for the behavior of the solute concentration at the wall. From Eq. (4.4.10) we estimate

$$\frac{c_w - c_0}{c_0} \sim \frac{v_w \delta_D}{D} \ll 1 \qquad (4.4.14)$$

Thus, for the relatively small concentration changes characteristic of conditions at the inlet, the normal convection term in the diffusion equation can be neglected and the estimate of δ_D from Eq. (4.2.18) may be applied to give

$$\frac{c_w - c_0}{c_0} \sim \xi^{1/3} \qquad (4.4.15)$$

where

$$\xi = \left(\frac{v_w h}{D}\right)^3 \left(\frac{D}{3\bar{u}h}\right)\left(\frac{x}{h}\right) \qquad (4.4.16)$$

the factor of 3 having been inserted for later convenience.

The parameter $(c_w - c_0)/c_0$ is a measure of the extent of concentration polarization. A straightforward dimensional analysis of the full diffusion equation and boundary conditions would indicate a dependence on the three parameters in parentheses in Eq. (4.4.16). However, it can be seen that for the developing region this dependence is reduced to only one parameter. Equation (4.4.15) suggests the introduction of a new dependent variable that is a measure of the concentration defect and defined by

$$\frac{c - c_0}{c_0} \equiv c^* = \xi^{1/3} f(\eta) \tag{4.4.17}$$

where, as before (Eq. 4.3.5),

$$\eta = y \left(\frac{3\bar{u}}{hDx} \right)^{1/3} = \frac{y}{\xi^{1/3}} \left(\frac{v_w}{D} \right) \tag{4.4.18}$$

From the chain rule

$$\frac{\partial c^*}{\partial x} = \frac{\partial c^*}{\partial \xi} \frac{d\xi}{dx} + \frac{\partial c^*}{\partial \eta} \frac{d\eta}{dx} \tag{4.4.19}$$

and the assumed form of c^* we may write

$$\frac{\partial c^*}{\partial x} = \frac{\xi^{1/3}}{3x} \left(f - \eta \frac{df}{d\eta} \right) \tag{4.4.20a}$$

and from Eqs. (4.3.6b) and (4.3.6c)

$$\frac{\partial c^*}{\partial y} = \left(\frac{3\bar{u}}{hDx} \right)^{1/3} \xi^{1/3} \frac{df}{d\eta} \tag{4.4.20b}$$

$$\frac{\partial^2 c^*}{\partial y^2} = \left(\frac{3\bar{u}}{hDx} \right)^{2/3} \xi^{1/3} \frac{d^2 f}{d\eta^2} \tag{4.4.20c}$$

Transforming the derivatives in the convective diffusion equation, with the velocity components replaced by the expressions of Eq. (4.4.7), reduces the partial differential equation to the ordinary linear differential form

$$\frac{d^2 f}{d\eta^2} + \frac{\eta^2}{3} \frac{df}{d\eta} - \frac{\eta}{3} f = 0 \tag{4.4.21}$$

under the assumption that $\xi^{1/3} \to 0$. With this assumption, which is appropriate to the developing layer near the inlet, the normal convection term can be neglected.

The boundary conditions become

$$f \to 0 \qquad \text{as } \eta \to \infty \tag{4.4.22a}$$

$$\frac{df}{d\eta} = -1 \qquad \text{at } \eta = 0 \tag{4.4.22b}$$

The second condition as written also requires that $\xi^{1/3} \to 0$. The requirement of small ξ has as its analogue the slow reaction rate condition, the limit of which is the zero mass transfer statement $\partial c / \partial y \to 0$.

Except for the last term, the differential equation is the same as that for the soluble channel wall problem where the concentration is uniform at the surface. Observing that $f = \eta$ is a solution of Eq. (4.4.21) suggests the form

$$f = \eta g \tag{4.4.23}$$

which on substituting in Eq. (4.4.21) gives

$$\eta \frac{d^2 g}{d\eta^2} + \left(\frac{\eta^3}{3} + 2 \right) \frac{dg}{d\eta} = 0 \tag{4.4.24}$$

This may be integrated twice to yield

$$g = -A \left(\frac{e^{-\eta^{3/9}}}{\eta} + \frac{1}{3} \int_{\infty}^{\eta} \eta e^{-\eta^{3/9}} \, d\eta \right) \tag{4.4.25}$$

where the boundary condition, Eq. (4.4.22a), has been applied. From the boundary condition at $\eta = 0$ it can be shown that

$$A = -\left(\frac{1}{3} \int_0^{\infty} \eta e^{-\eta^{3/9}} \, d\eta \right)^{-1} = \frac{9^{1/3}}{\Gamma(\frac{2}{3})} = 1.536 \tag{4.4.26}$$

The solution for $f(\eta)$ is therefore

$$f(\eta) = 1.536 \left(e^{-\eta^{3/9}} - \frac{\eta}{3} \int_{\eta}^{\infty} \eta e^{-\eta^{3/9}} \, d\eta \right) \tag{4.4.27}$$

where $f(0) = 1.536$. This result is expressible in terms of an incomplete Gamma function, and a graph of $f(\eta)$ is shown in Fig. 4.4.2, a curve obtained from the

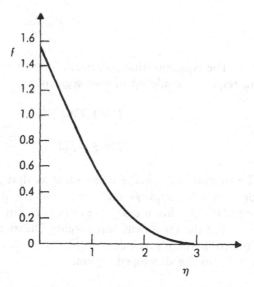

Figure 4.4.2 Similarity solution for concentration defect in developing layer near inlet (after Solan & Winograd 1969).

solution of a heterogeneous reaction problem dealing with electrodialysis (Solan & Winograd 1969). The solute concentration at the wall corresponding to the solution for $f(\eta)$ is

$$\frac{c_w}{c_0} = 1 + 1.536 \xi^{1/3} \tag{4.4.28}$$

This result is valid for $\xi \leq 0.02$, as discussed below.

Dresner (see Sherwood et al. 1965) obtained the approximate analytic solution given by Eq. (4.4.28) and one valid further downstream in the developing layer. He expressed his results in terms of a concentration polarization parameter Γ, which measures the excess solute concentration at the membrane relative to a so-called *mixing-cup concentration* c_m. This concentration varies along the channel and is defined as the concentration that would be measured at a streamwise location "if the channel were chopped off at that point and the fluid issuing forth were collected in a container and thoroughly mixed." For parallel membranes at a downsteam station x, with a constant withdrawal velocity the fraction of solvent removed is $(v_w/\bar{u})(x/h)$. The fraction of solvent remaining is therefore 1 minus this quantity. Now the solute concentration at the inlet is c_0, and if the membranes are completely rejecting then, at a given downstream position, this concentration must be increased in proportion to the amount of water removed. Therefore, we may express c_m at a location x along the channel by

$$c_m = c_0 \left(1 - \frac{v_w}{\bar{u}} \frac{x}{h} \right)^{-1} \tag{4.4.29}$$

The dimensionless polarization parameter denoted by Γ is then defined as

$$\Gamma = \frac{c_w}{c_m} - 1 \tag{4.4.30}$$

The concentration polarization result that Dresner derived for the developing region is made up of two segments:

$$\Gamma = 1.536 \xi^{1/3} \qquad \xi \leq 0.02 \tag{4.4.31}$$

$$\Gamma = \xi + 5(1 - e^{-\sqrt{\xi/3}}) \quad \xi \geq 0.02 \tag{4.4.32}$$

The formula for small ξ is the same as that given by Eq. (4.4.28). The segment for $\xi \geq 0.02$ employs the same velocity profile (Eq. 4.4.7) but applies the boundary condition $v_w c_w = -D(\partial c/\partial y)_w$ in a form valid for large ξ.

For the far downstream region Dresner used the more accurate approximtion to Berman's velocity profile given by Eq. (4.4.6) and showed that, for $x > L_D$ in the developed region,

$$\Gamma \approx \tfrac{1}{3} \, \text{Pe}^2 \qquad \text{for Pe} \equiv \frac{v_w h}{D} \gg 1 \tag{4.4.33}$$

Closed-form expressions that are applicable over the entire range of Pe have been derived by Gill et al. (1971).

Finite difference solutions of the full diffusion equation using the more accurate velocity profile of Berman and the boundary condition $v_w c_w = -D(\partial c/\partial y)_w$ were obtained by Sherwood et al. (1965). The results are shown in Fig. 4.4.3, where the comparison is seen to be excellent. The curve labeled "Analytic (developing)" is made up from overlapping the large and small ξ formulas given above.

The downstream extent of the developing region is characterized approximately by the intersection of the solid and dashed curves in Fig. 4.4.3. From equating Eqs. (4.4.32) and (4.4.33) for large ξ, we obtain

$$\frac{L_D}{h} \sim \frac{\bar{u}}{v_w} \qquad \xi \gg 1 \qquad (4.4.34)$$

The quantity $(v_w/\bar{u})(L_D/h)$ is just the fraction of solvent removed over the length L_D. On the other hand, for $\xi \sim O(1)$, from the definition of ξ (Eq. 4.4.16), we have

$$\frac{L_D}{h} \sim \frac{1}{\mathrm{Pe}^2} \frac{\bar{u}}{v_w} \qquad \xi \sim O(1) \qquad (4.4.35)$$

This result is obtained by neglecting the effect of streamwise convection so that the thickness δ_D follows from a balance of the transverse convective solute flux

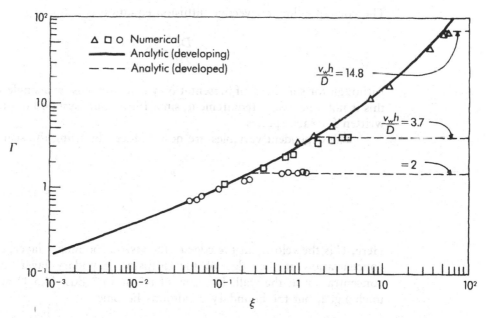

Figure 4.4.3 Concentration polarization in a parallel membrane reverse osmosis channel with fully developed laminar flow and complete solute rejection (after Sherwood et al. 1965).

toward the membrane against the diffusive flux away from it, or $D/\delta_D \sim v_w$. The development length is then given by the product of the characteristic time for the solute to diffuse from the wall to the edge of the diffusion layer δ_D^2/D and the average streamwise velocity in the layer, $\bar{u}\delta_D/h$. Substitution of D/v_w for δ_D gives Eq. (4.4.35).

4.5   Flow Past a Reacting Flat Plate

High Reynolds number boundary layer flow past a reacting flat plate has been examined at some length by Levich (1962). We looked at it in Section 4.2, using order-of-magnitude arguments. Because much of the physics and many of the general features do not differ from those for channel flows, we shall not discuss this problem in great detail. Instead, we will concentrate on general similarity criteria and behavior that can be deduced from the well-known Blasius solution for flow past a nonreacting flat plate. Moreover, mixed heterogeneous reactions will not be considered, and the discussion will be limited to those cases where the concentration at the wall is uniform. See Levich (1962) for more detail.

For high Reynolds number boundary layer flow past a flat plate with gravitational forces absent, the equation governing the streamwise velocity component u for constant viscosity is

$$\frac{Du}{Dt} = \nu \frac{\partial^2 u}{\partial y^2} \tag{4.5.1}$$

The corresponding convective diffusion equation is

$$\frac{Dc}{Dt} = D \frac{\partial^2 c}{\partial y^2} \tag{4.5.2}$$

Although for simplicity of presentation we again focus on a single solute species, this is not a necessary requirement, since for a dilute system Eq. (4.5.2) could be written for each species.

The dependent variables are now reduced by using the scalings

$$u^* = \frac{u}{U} \tag{4.5.3a}$$

$$c^* = \frac{c - c_w}{c_0 - c_w} \tag{4.5.3b}$$

Here, U is the velocity at the edge of the viscous boundary layer, c_0 is the solute concentration at the edge of the diffusion boundary layer, and c_w is the concentration at the wall. In terms of u^* and c^* Eqs. (4.5.1) and (4.5.2) are unchanged, but the boundary conditions become

$$u^* = 1 \quad c^* = 1 \quad \text{as } y \to \infty \tag{4.5.4a}$$

$$u^* = 0 \quad c^* = 0 \quad \text{at } y = 0 \tag{4.5.4b}$$

In the special case when $\nu = D$ (Sc = 1), a condition appropriate to dilute gases, the equations and boundary conditions are identical and the system admits the particular integral

$$c^* = u^* \tag{4.5.5}$$

or

$$c = c_w + (c_0 - c_w)\frac{u}{U} \tag{4.5.6}$$

This general result is usually presented in a more restricted sense as an analogy between skin friction and mass transfer. In particular, a skin friction coefficient may be defined as

$$C_f = \frac{(\tau_{yx})_w}{\tfrac{1}{2}\rho U^2} = \frac{\mu}{\tfrac{1}{2}\rho U^2}\left(\frac{\partial u}{\partial y}\right)_w = \frac{2\nu}{U}\left(\frac{\partial u^*}{\partial y}\right)_w \tag{4.5.7}$$

and a mass transfer coefficient as

$$C_{\text{diff}} = \frac{j_w^*}{(c_0 - c_w)U} = \frac{D}{U}\left(\frac{\partial c^*}{\partial y}\right)_w \tag{4.5.8}$$

With $u^* = c^*$ and $\nu = D$ it follows that

$$C_{\text{diff}} = \frac{C_f}{2} \tag{4.5.9}$$

This relation between mass and momentum transfer is termed *Reynolds analogy*, although the terminology more frequently is used to denote a correspondence between heat and momentum transfer. It is evident that the same argument used above could have been applied to heat transfer for a constant wall temperature, and with $\nu = \alpha$ a relation identical to Eq. (4.5.9) could be obtained with a dimensionless heat transfer coefficient in place of the mass transfer coefficient.

Of much greater interest is the case where Sc $= \nu/D \gg 1$, since our principal concern is with dilute solutions. For this situation the diffusion boundary layer is imbedded in the viscous boundary layer, and the velocity it sees is that close to the wall. Solution to the steady, Blasius, flat plate, viscous boundary layer equation shows the velocity components close to the wall ($y \ll \delta_U$) to be expressible in the series form

$$u = U\left[0.332\eta - \frac{0.028}{12}\eta^4 + O(\eta^7)\right] \tag{4.5.10a}$$

$$v = \left(\frac{\nu U}{x}\right)^{1/2}\left[0.083\eta^2 + O(\eta^5)\right] \tag{4.5.10b}$$

where the similarity coordinate η is defined by

$$\eta = y\left(\frac{U}{\nu x}\right)^{1/2} \tag{4.5.11}$$

Provided the mass transfer rate is small so that the velocity profile is unaltered by diffusion, the interaction is weak and the series expressions for the velocity components can be applied in the convective diffusion equation. It is just this situation with which we have concerned ourselves to this point. A criterion for the weak interaction is that $\nu\delta_D/D \gg 1$, which can be shown to translate to $Sc^{2/3} > 1$, a criterion satisfied for the case examined.

Replacing the velocity components in the boundary layer momentum and diffusion equations by the series expansions valid near the wall reduces these equations to the linear forms

$$a(x)\, y\, \frac{\partial u^*}{\partial x} + b(x)\, y^2\, \frac{\partial u^*}{\partial y} = \nu\, \frac{\partial^2 u^*}{\partial y^2} \tag{4.5.12}$$

$$a(x)\, y\, \frac{\partial c^*}{\partial x} + b(x)\, y^2\, \frac{\partial c^*}{\partial y} = D\, \frac{\partial^2 c^*}{\partial y^2} \tag{4.5.13}$$

To consistent order, Eq. (4.5.12) is, by definition, satisfied identically by the series solution of Eq. (4.5.10). This may be verified by substitution. If the y coordinate in Eq. (4.5.13) is transformed following

$$\zeta = y\left(\frac{\nu}{D}\right)^{1/3} \tag{4.5.14}$$

the diffusion equation reduces identically in form to the momentum equation and the system is seen to admit the particular integral

$$c^*(x, \zeta) = u^*(x, y) \tag{4.5.15}$$

The boundary conditions are satisfied by the integral in both the original and the transformed coordinates.

Using the definitions of the skin friction and mass transfer coefficients, Eqs. (4.5.7) and (4.5.8), and the fact that $\partial c^*/\partial \zeta = \partial u^*/\partial y$, we find

$$C_{\text{diff}} = \frac{C_f}{2}\, \frac{1}{Sc^{2/3}} \tag{4.5.16}$$

This generalization of the Reynolds analogy is termed the *Chilton-Colburn analogy*. Although derived for large Schmidt numbers, it is found to be quite accurate down to Schmidt numbers of 0.5, far below where it might be expected to apply (Bird et al. 1960). Again observe that the argument as outlined could also have been applied to heat and momentum transfer, with c and D replaced by T and α, respectively. The resulting formula would be the same as that above, but the Schmidt number is replaced by the Prandtl number, and the dimensionless diffusion coefficient is replaced by a dimensionless heat transfer coefficient.

The diffusion layer thickness can be estimated from the Nernst relation

$$\delta_D = \frac{D(c_0 - c_w)}{j_w^*} \tag{4.5.17}$$

where, from the Chilton-Colburn analogy,

$$\frac{j_w^*}{U(c_0 - c_w)} = \frac{C_f}{2} \frac{1}{Sc^{2/3}} \tag{4.5.18}$$

The Blasius flat plate solution gives the well-known result

$$C_f = \frac{0.664}{Re_x^{1/2}} \tag{4.5.19}$$

for the skin friction coefficient, where $Re_x = Ux/\nu$.

Combining the last three relations, we get

$$\frac{\delta_D}{x} = \frac{3}{Sc^{1/3}} \frac{1}{Re_x^{1/2}} \tag{4.5.20}$$

We also recall that the "boundary layer thickness" for the flat plate viscous layer is

$$\frac{\delta_U}{x} = \frac{5}{Re_x^{1/2}} \tag{4.5.21}$$

Thus,

$$\delta_D = \frac{0.6}{Sc^{1/3}} \delta_U \tag{4.5.22}$$

confirming the earlier order-of-magnitude estimate.

Finally, we note that the concentration profile will be self-similar like the Blasius, flat plate velocity profile, that is, a function only of the stretched normal coordinate

$$z = \zeta \left(\frac{U}{\nu x} \right)^{1/2} \tag{4.5.23}$$

We can obtain the concentration profiles simply by rescaling the velocity profiles, using Eqs. (4.5.14) and (4.5.15); therefore we do not explicitly discuss the similarity solution of the boundary layer diffusion equation. However, in terms of the similarity variable given above and from the series expressions for the velocity profile near the wall, the equation is reducible to the linear form

$$\frac{d^2 c^*}{dz^2} + 0.083 z^2 \frac{dc^*}{dz} = 0 \tag{4.5.24}$$

which is the same type as that in channel flow with soluble or rapidly reacting walls.

4.6   Taylor Dispersion in a Capillary Tube

In this section we consider what happens when two miscible liquids are contacted and mixed in a flow. The equilibrium for such a case is simply that of one phase uniformly distributed throughout the second, so any phenomena of interest are unsteady ones. The process whereby one phase is distributed in the second is termed *miscible dispersion*. In a laminar flow both convection and molecular diffusion will contribute to this dispersion. Other factors can also enter, including the geometry and any forced unsteadiness.

G.I. Taylor (1953, 1954) first analyzed the dispersion of one fluid injected into a circular capillary tube in which a second fluid was flowing. He showed that the dispersion could be characterized by an unsteady diffusion process with an "effective" diffusion coefficient, termed a *dispersion coefficient*, which is not a physical constant but depends on the flow and its properties. The value of the dispersion coefficient is proportional to the ratio of the axial convection to the radial molecular diffusion; that is, it is a measure of the rate at which material will spread out axially in the system. Because of Taylor's contribution to the understanding of the process of miscible dispersion, we shall, as is often done, refer to it as *Taylor dispersion*.

We illustrate the mechanisms of Taylor dispersion by considering what happens to a slug of solute A injected at time zero into a fully developed laminar flow of a solvent B as shown in Fig. 4.6.1 (Nunge & Gill 1969). At very short times after injection, except at high flow rates, the dispersion of A into B takes place by pure axial molecular diffusion because of the high axial concentration gradients. The dispersion thus takes place as if the two phases were stagnant. The time over which this type of dispersion occurs is usually very short and not of practical interest.

The characteristic diffusion length $(Dt)^{1/2}$ is proportional to the square root of time, and the characteristic convection length Ut is linear in time. It follows that for times slightly greater than considered above, axial convection

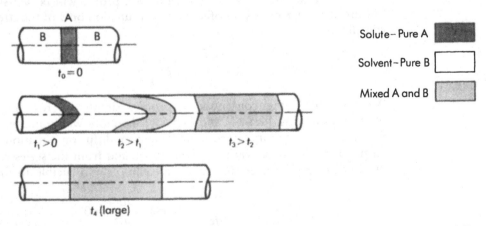

Figure 4.6.1   Schematic as a function of time of dispersion of a solute slug A in a fully developed laminar flow of a solvent B (after Nunge & Gill 1969).

enters in a significant way into the dispersion process. Indeed, at sufficiently high flow rates axial convection will be the controlling mechanism, and the solute slug will be distorted into a parabolic shape by the parabolic velocity field, as shown in Fig. 4.6.1 at time t_1. Thus convection, or rather the velocity variation across the cross section, enlarges the axial region where A is present and hence contributes to enhancing the axial dispersion or spreading of the region occupied by A. But it is clear that axial convection establishes large *radial concentration gradients*. As a result, at slightly higher times t_2 radial molecular diffusion contributes to the dispersion process, as illustrated in Fig. 4.6.1. The radial molecular diffusion acting at the front end of the slug moves A from the high-velocity central regions of the tube, where it is present, to the low-velocity wall region, thereby slowing the front end down. At the rear end of the slug, molecular diffusion moves A from the low-velocity wall region to the higher-velocity central regions, thereby speeding up the rear. The net effect of the radial diffusion is therefore to compress the mixing zone, which the mechanism of axial convection tends to elongate. Contrary to the usual perception of the effect of diffusion, in this case the radial molecular diffusion inhibits the axial dispersion of A. The result is that there is a mixed zone of varying concentration.

As time goes on, the action of radial diffusion continues to inhibit axial dispersion by diffusion and convection and makes the mixed zone more uniform, as shown at time t_3. Finally, at still larger times a quasi equilibrium is established. Here, convection, radial diffusion, and axial diffusion all contribute to the dispersion, with the net effect appearing as if the fluid were in plug flow, whereas in fact the velocity is radially distributed. With a further increase in time, the effect is only to increase the length of the mixed zone.

Following Taylor (1953), let us first consider the case of dispersion by convection alone, illustrated in Fig. 4.6.1 at time t_1. Taylor asked the question, what would be the distribution of mean concentration of the solute averaged over the cross section of the tube as a function of axial position? To define the answer, he first looked at the somewhat simpler case of a semi-infinite slug of solute with its leading edge located at $x = 0$ at time zero, as shown in Fig. 4.6.2. This figure shows the solute distributions at $t = 0$ and at a later time large enough that convection is dominant but not so large that radial diffusion has entered the picture. The leading edge of the solute will be distorted into the paraboloid

$$x = 2Ut\left(1 - \frac{r^2}{a^2}\right) \tag{4.6.1}$$

This relation derives from the velocity distribution for fully developed flow in a circular tube, Eq. (4.2.14) with u_{max} equal to twice the mean velocity U and h equal to the tube radius a. Now the mean concentration $\bar{c}$ at any cross section x is simply

$$\bar{c} = c_0\left(\frac{\text{area of section of paraboloid}}{\text{area of cross section of tube}}\right) = c_0\left(\frac{\pi r^2}{\pi a^2}\right) \tag{4.6.2}$$

From the last two relations

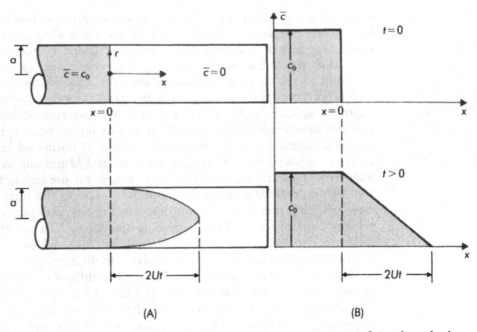

Figure 4.6.2 Dispersion by convection alone of an initially semi-infinite slug of solute: (A) solute location along tube; (B) mean concentration distribution along tube (after Taylor 1953).

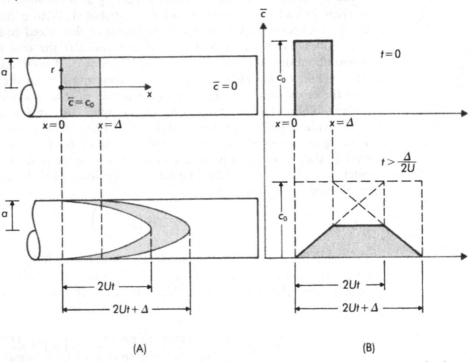

Figure 4.6.3 Dispersion by convection alone of an initially finite-width slug of solute: (A) solute location along tube; (B) mean concentration distribution along tube (after Taylor 1953).

$$\bar{c} = c_0 \qquad\qquad\qquad x < 0$$

$$\bar{c} = c_0\left(1 - \frac{x}{2Ut}\right) \qquad 0 < x < 2Ut \qquad\qquad (4.6.3)$$

$$\bar{c} = 0 \qquad\qquad\qquad x > 2Ut$$

This distribution is shown in Fig. 4.6.2B.

The slug of solute of finite width Δ located initially at time zero at $x = 0$ and the resulting location at time $t > \Delta/2U$ are shown in Fig. 4.6.3A. The corresponding mean concentration distributions are shown in Fig. 4.6.3B. The solution for the mean concentration distribution can be obtained simply by superposing two examples of the preceding case, namely

$$\bar{c} = c_0 \qquad x < \Delta \quad \text{and} \quad \bar{c} = 0 \quad x > \Delta \qquad (4.6.4a)$$

$$\bar{c} = -c_0 \qquad x < 0 \quad \text{and} \quad \bar{c} = 0 \quad x > 0 \qquad (4.6.4b)$$

For $t > \Delta/2U$ the solution is

$$\bar{c} = 0 \qquad\qquad\qquad\qquad x < 0$$

$$\bar{c} = c_0\left(\frac{x}{2Ut}\right) \qquad\qquad 0 < x < \Delta$$

$$\bar{c} = c_0\left(\frac{\Delta}{2Ut}\right) \qquad\qquad \Delta < x < 2Ut \qquad\qquad (4.6.5)$$

$$\bar{c} = c_0\left(\frac{\Delta + 2Ut - x}{2Ut}\right) \qquad 2Ut < x < 2Ut + \Delta$$

$$\bar{c} = 0 \qquad\qquad\qquad\qquad x > \Delta + 2Ut$$

Having looked at the effect of the change in concentration resulting from convection alone, we next consider somewhat larger times where the radial concentration gradients have become large enough that the effect of molecular diffusion in the radial direction becomes important, as discussed in connection with Fig. 4.6.1 for time t_2. It was pointed out by Taylor (1953) that as far back as 1911 A. Griffiths had observed that a tracer injected into a stream of water in a tube spreads out in a symmetrical manner about a plane in the cross section that moves with the mean speed of the flow. That is, at some downstream point in the flow the tracer concentration would increase from zero to a maximum and then decrease to zero again. This is clearly not the result shown in Fig. 4.6.3B, where if we let $\Delta \to 0$ to correspond to a thin input tracer, there would simply be a rectangular concentration spread out over a distance $2Ut$.

As Taylor remarked, Griffiths result is indeed surprising for two reasons. First, since the water near the center of the tube moves with twice the mean speed of the flow but the tracer moves at the mean speed, this means the water near the center must approach the column of tracer, absorb the tracer as it passes through the column, and then reject the tracer as it leaves on the other side of the column. Second, the velocity with respect to a plane moving at the speed U is

$$2U\left(1 - \frac{r^2}{a^2}\right) - U = U\left(1 - \frac{2r^2}{a^2}\right) \tag{4.6.6}$$

that is, it is unsymmetrical about the plane moving at the mean speed, but yet the column of tracer spreads out symmetrically. The explanation, as we have discussed, lies in the effects of the radial molecular diffusion.

Because of the symmetry of the problem, we employ the binary, unsteady, axially symmetric convective diffusion equation with a constant diffusion coefficient:

$$\frac{\partial c}{\partial t} + u(r)\,\frac{\partial c}{\partial x} = D\left(\frac{\partial^2 c}{\partial x^2} + \frac{1}{r}\,\frac{\partial}{\partial r}\left(r\,\frac{\partial c}{\partial r}\right)\right) \tag{4.6.7}$$

The criterion for the "pure convection" case discussed earlier is the smallness of both diffusional terms on the right side of Eq. (4.6.7) compared with the convection term on the left side. Alternatively, this may be expressed as the largeness of the radial and axial diffusion times, a^2/D and L^2/D, respectively, compared with the convection time L/U. Here, a and L are the characteristic lengths over which there is an appreciable concentration change in the radial and axial directions, respectively.

With the diffusion Peclet number

$$\mathrm{Pe} = \frac{Ua}{D} \tag{4.6.8}$$

the criterion to neglect radial diffusion may be written

$$\mathrm{Pe} \gg \frac{L}{a} \quad \text{(pure convection)} \tag{4.6.9a}$$

and to neglect axial diffusion,

$$\mathrm{Pe} \gg \frac{a}{L} \quad \text{(pure convection)} \tag{4.6.9b}$$

If $L > a$ and Eq. (4.6.9a) is satisfied, then the above inequality is automatically satisfied. The inequalities may also be interpreted as indicating that the diffusion length $(Dt)^{1/2}$ is small with respect to lengths a and L, respectively. Mathematically, this limit is associated with a singular perturbation problem, since the highest derivatives of the system are neglected.

Another limit is that of simple axial diffusion in a stationary medium (Fick's second law). In that situation the radial diffusion time a^2/D must be short compared with the convection time L/U or

$$\mathrm{Pe} \ll \frac{L}{a} \quad \text{(pure axial diffusion)} \tag{4.6.10a}$$

This implies that the relative changes in radial concentration are small because of the rapid "washout" of any radial gradients that appear. On the other hand, the axial diffusion term must dominate the convection term. Alternatively, the axial diffusion time L^2/D must be short compared with the convection time L/U. Both statements lead to the condition

$$\text{Pe} \ll \frac{a}{L} \qquad \text{(pure axial diffusion)} \qquad (4.6.10b)$$

If $L > a$ and Eq. (4.6.10b) is satisfied, then Eq. (4.6.10a) is automatically satisfied.

A.A. Sonin has observed that the domains of these two limiting cases may be conveniently illustrated in a log-log plot of L/a versus Peclet number, as sketched in Fig. 4.6.4. In the diagram, regions below the curves $\text{Pe} = a/L$ and $\text{Pe} = L/a$ have been left blank to indicate the transition to other behaviors, which we examine below.

The dispersion model discussed next is that treated by Taylor, where the times are such that axial convection is important, but where radial diffusion can be assumed large in comparison with axial diffusion. For the unsteady concentration term $\partial c/\partial t$ in the axially symmetric convective diffusion equation to be of the same order as $D(\partial^2 c/\partial r^2)$, we have with $r \sim a$ that

$$t \sim \frac{a^2}{D} \qquad (4.6.11)$$

The dispersion calculation is considerably simplified when the radial molecular diffusion $D(\partial^2 c/\partial r^2)$ is supposed large in comparison with the axial molecular diffusion $D(\partial^2 c/\partial x^2)$. With the characteristic lengths as before, this implies $(L/a)^2 \gg 1$. But with $L = Ut$ and $t \sim a^2/D$, the criterion for the neglect of the axial gradients is

$$\text{Pe} \gg 1 \qquad (4.6.12)$$

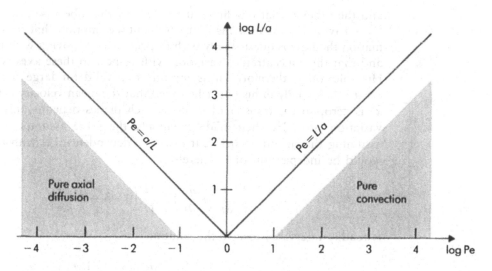

Figure 4.6.4 Axial diffusion and convection limits.

At still larger times, $t \gg a^2/D$, the radial diffusion term continues to play a role, but the smaller radial concentration change is now controlled by convection rather than radial diffusion. The requirement to encompass this larger time scale obtains from replacing t by L/U to give

$$\frac{L}{a} \gg \text{Pe} \qquad (4.6.13)$$

We shall show a posteriori that the criteria derived for the applicability of Taylor's model, as defined by Eqs. (4.6.12) and (4.6.13), are correct to within constants of $O(1)$.

With axial diffusion neglected, the axial transfer of c can be due only to convection, in which case it is convenient to consider the convection with respect to axes that move with the mean speed of the flow U. The concentration and velocity are then defined with respect to these axes. The velocity so defined is given by $U(1 - 2r^2/a^2)$, whence the diffusion equation transforms to

$$\frac{\partial c}{\partial t} + U\left(1 - 2\frac{r^2}{a^2}\right)\frac{\partial c}{\partial x'} = \frac{D}{r}\frac{\partial}{\partial r}\left(r\frac{\partial c}{\partial r}\right) \qquad (4.6.14)$$

The coordinate x' denotes the moving axis defined by

$$x' = x - Ut \qquad (4.6.15)$$

and $\partial/\partial t$ denotes differentiation with respect to time at points fixed relative to the axes moving with velocity U. One boundary condition is that for an impermeable wall

$$\frac{\partial c}{\partial r} = 0 \qquad \text{at } r = a \qquad (4.6.16)$$

and the other is that c is finite at the axis of the tube $r = 0$.

Taylor (1953) now made the brilliant assumption that to a first approximation the flow is quasi-steady with respect to axes moving with mean speed U, and that the concentration variation with respect to these axes is purely radial. His solution is therefore an asymptotic one valid for large time, that is, for $t \gg a^2/D$. Actually in his paper he noted that the mean velocity across any plane x' is zero, so the transfer of c across such planes depends only on the radial variation of c. He then made essentially the same assumption as above by assuming that in Eq. (4.6.14), if c were independent of x with t large, $\partial c/\partial x'$ would be independent of r. Therefore

$$\frac{1}{r}\frac{\partial}{\partial r}\left(r\frac{\partial c}{\partial r}\right) = \frac{U}{D}\left(1 - \frac{2r^2}{a^2}\right)\frac{\partial c}{\partial x'} \qquad (4.6.17)$$

is readily integrated to give the solution

$$c = c_0 + \frac{Ua^2}{4D}\frac{\partial c}{\partial x'}\left(\frac{r^2}{a^2} - \frac{1}{2}\frac{r^4}{a^4}\right) \qquad (4.6.18)$$

which satisfies the boundary condition at $r = a$ that $\partial c / \partial r = 0$. Here, c_0 is the finite concentration at the tube axis.

The average concentration over the tube cross section is defined by

$$\bar{c} = \frac{1}{\pi a^2} \int_0^a c \cdot 2\pi r \, dr = \frac{2}{a^2} \int_0^a cr \, dr \qquad (4.6.19)$$

On integrating the solution for c over r, we obtain

$$\bar{c} = c_0 + \frac{1}{3} \frac{Ua^2}{4D} \frac{\partial c}{\partial x'} \qquad (4.6.20)$$

enabling c to be expressed in terms of the mean concentration as

$$c = \bar{c} + \frac{Ua^2}{4D} \frac{\partial c}{\partial x'} \left(-\frac{1}{3} + \frac{r^2}{a^2} - \frac{1}{2} \frac{r^4}{a^4} \right) \qquad (4.6.21)$$

which says c approaches the mean value $\bar{c}$, the radial variation being small. This form is more appropriate than Eq. (4.6.18), since in problems of transport along a tube the mean concentration $\bar{c}$ is more significant than c_0.

Differentiating Eq. (4.6.21) gives

$$\frac{\partial c}{\partial x'} \approx \frac{\partial \bar{c}}{\partial x'} \qquad (4.6.22)$$

whence, following Taylor, we have

$$\frac{a^2 U}{4D} \left(\frac{1}{L} \right) \ll 1 \qquad (4.6.23)$$

where again L is the characteristic distance over which the greatest change in c occurs. This inequality can be written

$$\frac{4L}{a} \gg \mathrm{Pe} \qquad (4.6.24)$$

which may be compared with the earlier approximate result $L/a \gg \mathrm{Pe}$. Note that the inequality of Eq. (4.6.23) is satisfied for times large compared with $a^2/4D$, which was the basis upon which the earlier result was derived. It follows that the gradient of $\partial c / \partial x'$ is approximately equal to the gradient of the mean $\partial \bar{c} / \partial x'$, when the radial dispersion model applies.

Now the average mass flux of solute across any section x' is

$$\bar{J} = \frac{1}{\pi a^2} \int_0^a \rho u' \cdot 2\pi r \, dr = 2M \int_0^1 cu' \frac{r}{a} d\left(\frac{r}{a} \right) \qquad (4.6.25)$$

where u' is the velocity with respect to the moving axis x' (Eq. 4.6.6) and $\rho = Mc$. Both the mass concentration ρ and molar concentration c refer to the solute, although a subscript 1 has been appended. Substituting Eq. (4.6.21) for c in Eq. (4.6.25), with $\partial c/\partial x'$ replaced by $\partial \bar{c}/\partial x'$, and integrating, we find that the average mass flux relative to the moving axes is given by

$$\bar{J} = -\left(\frac{a^2 U^2}{48 D}\right)\frac{\partial \bar{\rho}}{\partial x'} \qquad (4.6.26)$$

where $\bar{\rho} = M\bar{c}$ (not to be confused with a mean density averaged over a number of species). This rather remarkable result says that the average solute concentration is dispersed relative to a plane that moves with the mean velocity U exactly as though it were being diffused by a molecular diffusion process with an effective diffusion coefficient

$$D_{\text{eff}} = \frac{a^2 U^2}{48 D} \qquad (4.6.27)$$

This is frequently termed the *Taylor dispersion coefficient*.

Now from conservation of mass applied in the moving reference frame,

$$\frac{\partial \bar{\rho}}{\partial t} = -\frac{\partial \bar{J}}{\partial x'} \qquad (4.6.28)$$

where $\partial/\partial t$ is differentiation with respect to time at a point where x' is constant. Substituting the result of Eq. (4.6.26) into Eq. (4.6.28), we obtain the equation for longitudinal dispersion

$$\frac{\partial \bar{c}}{\partial t} = D_{\text{eff}} \frac{\partial^2 \bar{c}}{\partial x'^2} \qquad (4.6.29)$$

or the same form with respect to $\bar{\rho}$. In the original fixed coordinate system x the equation, here termed the *Taylor dispersion equation*, is written

$$\frac{\partial \bar{c}}{\partial t} + U\frac{\partial \bar{c}}{\partial x} = D_{\text{eff}} \frac{\partial^2 \bar{c}}{\partial x^2} \qquad (4.6.30)$$

The range of validity of the last two equations is defined not only by the inequality $4L/a \gg \text{Pe}$ but also by the condition that the axial molecular diffusion must be negligible compared with the dispersion effect, or

$$D \ll \frac{a^2 U^2}{48 D} \qquad (4.6.31a)$$

This criterion can be written

$$\text{Pe} \gg 7 \qquad (4.6.31b)$$

which may be compared with the more approximate criterion derived earlier of $\text{Pe} \gg 1$. Combining the above inequality with the requirement that $4L/a \gg \text{Pe}$ defines a region in the L/a-Pe plane where the Taylor model is valid, namely

$$4 \frac{L}{a} \gg \text{Pe} \gg 7 \qquad (4.6.32)$$

Just as "pure axial diffusion" is in a sense an opposite limit to "pure convection," so also is "convective axial diffusion" an opposite limit to "convective radial diffusion" (Taylor dispersion). The convective axial diffusion limit, as with the Taylor dispersion limit, characterizes the convection at the mean flow speed U, whereas true convection is at the actual local speed. The criterion for the axial convection term to be of the same order as the axial diffusion term is

$$t \sim \frac{D}{U^2} \qquad (4.6.33)$$

where t is characteristic time. For larger times, $t \gg D/U^2$, the axial diffusion still cannot be neglected in comparison with axial convection, although the smaller axial concentration change is now controlled by convection rather than axial diffusion. In all cases, radial diffusion occurs so quickly by comparison, in a time of the order of a^2/D, that any changes in radial concentration are always small compared with the mean concentration.

Using the above criterion and arguments similar to those to define the Taylor solution range, but with the constants chosen to agree with the Taylor solution, we find that for convective axial diffusion to be the dominant mode the inequality

$$4 \frac{L}{a} \gg \text{Pe} \ll 7 \qquad (4.6.34a)$$

must be satisfied. An additional criterion comes from the fact that the convective term $U(\partial \bar{c}/\partial x)$ enters whenever it is greater than, say, 0.1 of the diffusion term $D(\partial^2 \bar{c}/\partial x^2)$, from which we also require

$$\text{Pe} > \frac{1}{10} \frac{a}{L} \qquad (4.6.34b)$$

Clearly there is a region between $\text{Pe} \ll 7$ and $\text{Pe} \gg 7$, with $4L/a \gg \text{Pe}$, where both radial and axial diffusion are important. Aris (1956), in a mathematically elegant paper, showed that the governing equation for the mean concentration distribution averaged over the tube cross section can be written in the form of the Taylor dispersion equation, with

$$D_{\text{eff}} = D \left(1 + \frac{\text{Pe}^2}{48} \right) \qquad (4.6.35)$$

This is termed the *Taylor-Aris dispersion coefficient*, and is simply the sum of the axial molecular diffusion coefficient and the Taylor radial dispersion coefficient. As can be seen, at large Peclet numbers D_{eff}/D increases as the square of the Peclet number (the Taylor dispersion limit), and at small Peclet numbers D_{eff}/D approaches 1 (the convective axial diffusion limit).

Aris obtained his result by using a moment method in which he calculated the total moments of the concentration $c(x, r, t)$ and the average concentration $\bar{c}(x, t)$ in the respective forms

$$M_n(t) = \frac{2}{a^2} \int_{-\infty}^{+\infty} x^n \, dx \int_0^a c(x, r, t) r \, dr \qquad n = 0, 1, 2, \ldots \qquad (4.6.36)$$

and

$$\bar{M}_n(t) = \int_{-\infty}^{+\infty} x^n \bar{c}(x, t) \, dx \qquad n = 0, 1, 2, \ldots \qquad (4.6.37)$$

The functions $M_n(t)$ were determined from the complete unsteady axially symmetric convective diffusion equation (Eq. 4.6.7), and $\bar{M}_n(t)$ were obtained from the Taylor dispersion equation, which was used as the "model" equation. The "phenomenological coefficients" U and D_{eff} in the equation were determined by matching the first three moments of the infinite sequence $M_n(t)$ to $\bar{M}_n(t)$ for asymptotically large times ($t \gg a^2/D$). Applying his scheme to the circular capillary problem, Aris showed that D_{eff}, where axial molecular diffusion is not neglected, is given by Eq. (4.6.35). Fried & Combarnous (1971) later showed that the satisfaction of the first three moments for $t \to \infty$ implies that $\bar{c}(x, t)$, obtained as a solution of the Taylor dispersion equation with $D_{\text{eff}} = D(1 + \text{Pe}^2/48)$, is asymptotically the solution of the complete, unsteady, axially symmetric convective diffusion equation averaged over the cross section.

The Taylor-Aris result can be shown in a somewhat simpler mathematical way by starting with the complete convective diffusion equation (Eq. 4.6.7), including the axial diffusion term. The procedure is essentially the same as Taylor's. Equation (4.6.7) is integrated over the tube cross section, since what is of interest is the average concentration, and the radial concentration distribution is given by Eq. (4.6.21). The replacement of $\partial c/\partial x'$ by $\partial \bar{c}/\partial x'$ is still made. The analysis follows through as before. In addition to requiring $t \gg a^2/D$, we must also require $t \gg D/U^2$. Aris's more formal procedure shows that only the requirement $t \gg a^2/D$ is actually needed.

Following A.A. Sonin, Fig. 4.6.5 indicates the regions of the various solution forms in a log-log plot of L/a versus Pe. The regions are distinguished by using a ratio of 10:1 to define "large compared with" and 1:10 to define "small compared with". As an exception, we have taken the liberty of defining the convective axial diffusion region in the figure by the criterion $\text{Pe} > 0.4a/L$ rather than $0.1a/L$ as in Eq. (4.6.34b) in order to match the corresponding Taylor-Aris limits. This small difference is not important because of the order-of-magnitude arguments used. There is no sharp delineation between regions. In fact, with D_{eff} equal to the Taylor-Aris value, the convective diffusion equation is appropriate everywhere in the regions covered by what we termed the convective axial diffusion, Taylor-Aris, and Taylor solution regions. Note also, that the pure axial diffusion region is simply a defined limit and that the convective and pure axial diffusion regions form a continuum.

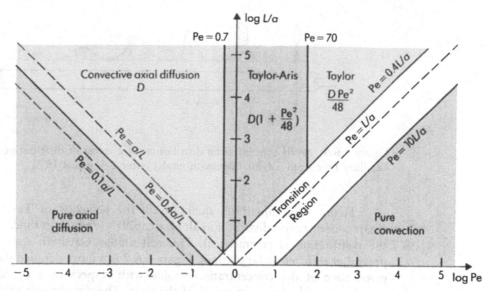

Figure 4.6.5 Regions of applicability of various dispersion solutions for a tracer in a capillary tube (after A.A. Sonin, M.I.T., personal communication).

In connection with the present and following discussions, it is of interest to show some experimental results that Taylor (1953) obtained on the dispersion of a tracer introduced into water flowing in a capillary tube. Taylor used an approximately 1.5-m-long circular glass tube with a 0.5-mm bore and water flow speeds as low as about 0.05 mm s^{-1}. As a tracer he employed potassium permanganate ($KMnO_4$), which has a very strong, dark purple color in solution, so it is easily seen. Its concentration in the glass capillary could be measured by comparison of the color intensity with a solution of known concentration. The test conditions were within the range where Taylor's dispersion model would be expected to hold.

One of the remarkable predictions of the theory that Taylor sought to check by his experiments was that an initially concentrated mass would be dispersed symmetrically about the point $x = Ut$. To this end the introduction of the small volume of permanganate solution at the capillary entrance could be modeled as a delta function input at time zero, namely

$$\bar{c} = \frac{n_0}{\pi a^2}\,\delta(x) \qquad \text{at } t = 0 \tag{4.6.38}$$

where n_0 is the number of moles of substance introduced. For this initial condition the solution to the dispersion equation is easily shown to be

$$\bar{c} = \frac{n_0/\pi a^2}{2(\pi D_{\text{eff}} t)^{1/2}}\exp\left(-\frac{(x - Ut)^2}{4 D_{\text{eff}} t}\right) \tag{4.6.39}$$

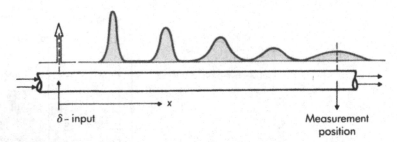

Figure 4.6.6 Mean concentration distribution of a tracer at different positions along a capillary tube from Taylor dispersion model (after Levenspiel 1972).

Figure 4.6.6 illustrates qualitatively the behavior of Taylor's predicted mean concentration distribution along a capillary, that is, in time. At any instant the distribution is symmetrical. This self-similar Gaussian solution will only prevail at sufficiently large times. Figure 4.6.7 is a more detailed illustration at a given time of the concentration profile with respect to a coordinate system translating with the mean speed of the flow. The distribution sketched follows from Eqs. (4.6.39) and (4.6.21) with $\partial c/\partial x' = \partial \bar{c}/\partial x'$. Also sketched in are the

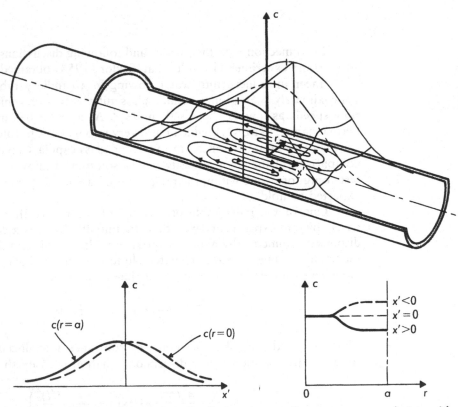

Figure 4.6.7 Solute concentration distribution in coordinate system translating with mean flow speed from Taylor dispersion model. Solute paths denoted by arrows.

solute paths showing the motion relative to the higher-velocity regions at the tube center and the lower-velocity regions at the tube wall.

In Fig. 4.6.8 are shown the results from the set of Taylor's experiments that give the distributions of mean concentration at three stages of dispersion. Taylor fitted data like this at a given position (time) to the Gaussian function solution, Eq. (4.6.39), and in this way determined a best-fit value of D_{eff}. The coefficient of the exponential, which is equal to the maximum value of the Gaussian, is the best-fit value of the maximum concentration.

What is of particular interest in the results of Fig. 4.6.8 is a comparison of the measured distributions with those that would have obtained if there were no diffusion and the material had been dispersed convectively and, therefore, uniformly. Taylor calculated the total amount of dispersed material by measuring the area of the curve labeled III, the peak of which is at 1.22 m. If this material were uniformly distributed, according to Eq. (4.6.5) for an initial slug of solvent of finite width Δ, it would have the constant mean value $\bar{c}_0 \Delta / 2Ut$, where $\bar{c}_0$ is the initial mean concentration. Taylor pointed out that for the material distributed through 2×1.22 m, this constant mean value would only have been $0.018\bar{c}_0$, which may be compared with the observed maximum concentration of $0.4\bar{c}_0$ shown by curve III, indicating the marked effect of diffusion in preventing a dissolved material from being dispersed.

In concluding this discussion of Taylor dispersion, we mention Brenner's generalization to nonrectilinear spatially and time-periodic flows (Brenner &

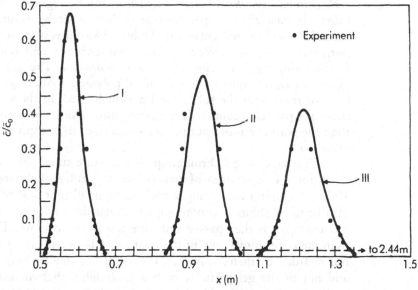

Figure 4.6.8 Measured mean concentration distributions at three positions along a capillary tube. Dashed line is distribution that would be due to convection alone for comparison with curve III. [After Taylor, G.I. 1953. Dispersion of soluble matter in solvent flowing slowly through a tube. *Proc. Roy. Soc.* A219, 186–203. With permission.]

Edwards 1993). Brenner's work generalizes the moment analysis techniques introduced by Aris. The idea behind the approach is that the media considered can generally be characterized at two different length scales, a local and global one (cf. Section 1.1). For the Taylor dispersion problem in a circular tube, the local or *microscale* is the tube radius a and the global or *macroscale* is L, where $L \gg a$. Generally our interest lies in describing the average transport process at the length L, which is the characteristic distance over which the mean solute concentration field has a sensible change. Generalized Taylor dispersion is valid only for times large enough that the particles have had an opportunity to sample the spatially inhomogeneous flow at the microscale. For dispersion in a circular tube, this flow is the parabolic velocity profile across the tube.

In general in complex media, the convection and diffusion of a solute is a difficult problem to analyze at the microscale and the moment method enables the analysis to be carried out at the macroscale leading to the replacement of the convective-diffusion problem by an effective global velocity and an effective dispersion tensor as in Eq. (4.6.30). Brenner's procedure analyzes the time evolution of the spatial moments (cf. Eq. 4.6.36) of the conditional probability density that a Brownian particle is located at a given position at a specific time knowing the position from which it was initially released into the fluid.

4.7   Gel Chromatography and Capillary Models of Porous Media

Gel chromatography or *size exclusion chromatography*, as it is frequently called, is basically a separation procedure whereby solutes are fractionated according to their molecular size (Dubin 1988). It is often used in the large-scale purification of macromolecules, including enzymes and other proteins, and in fractionating nucleic acids and small molecules. The gels or resins used as molecular sieves consist of cross-linked polymers that are generally inert, do not bind or react with the solute, and are uncharged. In the simplest steric picture these gel particles contain small pores into which molecules of a size smaller than the pore can move but particles of a size larger cannot (hence the term *size exclusion*).

The basis of gel chromatography is quite simple and is illustrated in Fig. 4.7.1 for the separation of two solute molecules of different sizes (Freifelder 1982). A column containing a packing of small spherical resin beads is used. A thin band of solution containing the molecules of various dimensions (two in our example) is then passed onto the top of the column. The molecules larger than the pores move only in the void space between the particles and hence are not retarded by them. However, the molecules smaller than the pores diffuse in and out of the gel particles with a probability that increases with decreasing molecular size and in this way are slowed down in their movement down the column. Solvent is passed through the column at a slow enough rate to allow the molecules to equilibrate with the gel particles at each level. Molecules are thus eluted from the column in order of decreasing size or, if their shape is relatively constant, decreasing molar mass.

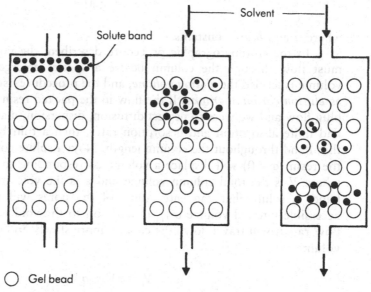

Figure 4.7.1 Separation of two molecules by passage through a column containing gel particles that are accessible to the small molecules but inaccessible to the large ones. [After Freifelder, D.M. 1982. *Physical Biochemistry*, 2nd edn. San Francisco: W.H. Freeman. Copyright © 1976, 1982 W.H. Freeman and Company. With permission.]

Introduced into the analysis of the procedure is the concept of a partition coefficient σ that characterizes the probability of gel accessibility to a molecule. It is defined as the ratio of the solute mass concentration inside the gel to that outside:

$$\sigma = \frac{\rho_{in}}{\rho} = \frac{m_{in}}{\rho V_{in}} \tag{4.7.1}$$

where ρ = solute mass concentration in the feed
ρ_{in} = solute concentration in the gel
m_{in} = solute mass in the gel
V_{in} = internal volume of pores accessible to the solvent

Note that accessibility to the solvent does not imply accessibility to the solute. As before, the species subscript i is to be understood, there being a different partition coefficient for each species. If the pores of the gel were all of uniform size and the accessibility were only a steric one, then σ would be 1 for molecules smaller than the pore size and 0 for molecules bigger. For a distribution of pore sizes, σ will vary gradually with molecular size. For similar shapes, size is related to molar mass, and for a chromatographic separation of various proteins on any particular resin a general correlation has been shown to be of the form (Cantor & Schimmel 1980)

$$\sigma = -a \log M + b \qquad (4.7.2)$$

where a and b are constants.

In the chromatographic procedure described, the volume of solvent that must flow through the column before a species emerges is measured. This volume is termed the *elution volume*, and the procedure itself is often referred to as *elution chromatography*. If the flow in the packed resin bed were completely plug flow and we were to neglect diffusion, dispersion, mass transfer resistance, and finite absorption and desorption rates, then a thin band would remain a thin band throughout the column length. Now a solute completely excluded by the resin ($\sigma = 0$) will displace a volume corresponding to the void volume εV, where V is the total column volume and ε is the void fraction. On the other hand, a solute that can enter some of the pores will have to displace the accessible internal volume σV_{in} in addition to the void volume. At a constant flow rate it will travel down the column more slowly and emerge after eluting a volume

$$V_e = \varepsilon V + \sigma V_{in} \qquad (4.7.3)$$

The procedure used to find σ is to determine $\varepsilon V + V_{in}$ by measuring the mass of water taken up by dry resin, and to determine εV by measuring the elution volume of a particle much larger than the gel pores.

In terms of the fraction of column cross-sectional area available to the solute, α, the elution volume may be written as

$$V_e = \alpha A L \qquad (4.7.4)$$

where A is the volume cross-sectional area and L is the column length, and, from Eq. (4.7.3),

$$\alpha = \frac{1}{A} \left(\frac{\varepsilon V}{L} + \frac{\sigma V_{in}}{L} \right) \qquad (4.7.5)$$

Here, the first term in parentheses is the cross-sectional void area, and the second term is the cross-sectional internal area of the gel available to the solute. The mean *interstitial velocity* or *effective velocity* through the column is

$$U_e = \frac{Q}{\alpha A} = \frac{L}{t} \qquad (4.7.6)$$

where Q is the constant volume flow rate at which the column is eluted and t is the time it takes for the band to move down the column.

What we have described is an idealized picture, and the bands of solute will not remain thin but will spread out as they move down the column for the reasons described above. The error, for example, in using the molar mass determination described in conjunction with Eq. (4.7.2) is primarily a result of the band breadth during elution. By way of illustration, we ask the question, what would be the actual shape of an eluted band of solute if dispersion were to dominate, recognizing that with a number of solutes the more the bands are spread out the greater will be the loss of resolution in the separation of the different solutes?

The flow situation in the porous medium comprising the column of packed resin beads is a complex one. One approach long used to model flow through porous media has been to consider the medium as made up of bundles of straight capillaries or assemblages of randomly oriented straight pores or capillaries in which the flow is of Poiseuille type.

It has been well established by experiment that for low Reynolds number flow through porous media the pressure drop follows *Darcy's law*. In one dimension Darcy's law may be written

$$\frac{dp}{dx} = -\frac{\mu}{k} U \tag{4.7.7}$$

where the *superficial velocity* U is the volume flow rate through the medium divided by its total cross-sectional area; that is, it is the uniform velocity upstream of the medium. Note that this velocity is lower than the interstitial velocity U_e because of the volume taken up by the solids. The constant k is called the *permeability* and has the dimensions of length squared. In terms of the capillary model, comparison of Darcy's law with the low Reynolds number, inertia free Poiseuille solution (Eq. 4.2.14) shows that dimensionally $k \sim a^2$, with a the characteristic radius of the channels.

In what follows we derive an empirical relation for the permeability, known as the *Kozeny-Carman equation*, which supposes the porous medium to be equivalent to a series of channels. The permeability is identified with the square of the characteristic diameter of the channels, which is taken to be a *hydraulic diameter* or *equivalent diameter*, d_e. This diameter is conventionally defined as four times the flow cross-sectional area divided by the wetted perimeter, and measures the ratio of volume to surface of the pore space. In terms of the porous medium characteristics,

$$d_e = \frac{4V_{\text{void}}}{A} \tag{4.7.8}$$

where V_{void} is the volume of voids and A is the total surface area. For a straight circular capillary, d_e is the capillary diameter. Since the porosity ε is defined by

$$\varepsilon = \frac{V_{\text{void}}}{V} \tag{4.7.9}$$

where V is the total volume of the medium, we may write

$$d_e = \frac{4\varepsilon V}{A} \tag{4.7.10}$$

It is common to express the total surface area in terms of an inverse length, termed the *specific area S*, which is the ratio of the surface area to the volume of the solid's fraction of the porous medium:

$$S = \frac{A}{(1 - \varepsilon)V} \tag{4.7.11}$$

Substituting the above definition of the specific area into Eq. (4.7.10), we obtain the following expression for the equivalent diameter:

$$d_e = \frac{4\varepsilon}{S(1 - \varepsilon)} \tag{4.7.12}$$

As noted, the average interstitial or effective pore velocity in the channels, U_e, is greater than the superficial velocity U, due to the volume occupied by the solids. It is supposed that

$$U_e = \frac{U}{\varepsilon} \tag{4.7.13}$$

Replacing U by U_e in Eq. (4.7.7) and assuming that $k \sim d_e^2$, where d_e is given by Eq. (4.7.12), we can write Darcy's equation in the form

$$\frac{dp}{dx} = -\mu U \frac{KS^2(1 - \varepsilon)^2}{\varepsilon^3} \tag{4.7.14}$$

where K is the *Kozeny constant*. The permeability is thus

$$k = \frac{\varepsilon^3}{KS^2(1 - \varepsilon)^2} \tag{4.7.15}$$

which is termed the *Kozeny-Carman equation*. P.C. Carman proposed a value of 5.0 for K. Using this value of K and assuming the porous medium to be composed of uniform spheres of diameter d, we obtain that $S = 6/d$ and Eq. (4.7.15) reduces to a commonly written form of the Kozeny-Carman equation:

$$k = \frac{d^2 \varepsilon^3}{180(1 - \varepsilon)^2} \tag{4.7.16}$$

It is now understood that the Kozeny "constant" is by no means a universal constant but varies with porosity, particle shape, and orientation. Moreover, measurements have indicated that the effective interstitial velocity may be larger than U/ε because the void volume available for the bulk flow can be smaller as a result of stagnant voidage. In addition, it is common to account for the sinuous nature of the capillaries—that is, a larger actual flow length than the straight-through thickness of the porous medium—by introducing a tortuosity factor as proposed by Carman. However, these correction factors are largely empirical and difficult to quantify from purely theoretical considerations.

Despite our use of a capillary model to characterize a porous medium, most porous beds employed for chromatographic purposes are random and generally the medium is isotropic. In such media, the effective solute dispersivity still arises from the nonuniform pore velocity coupled with molecular diffusion resulting in a dispersive mixing process at the macroscale characterized by an effective dispersion tensor and global velocity as for Taylor dispersion in a capillary (Section 4.6).

In a circular bed, the effective dispersion tensor is anisotropic and is composed of the longitudinal and lateral dispersion coefficients $D_{eff}^{\parallel}$ and $D_{eff}^{\perp}$, respectively. The longitudinal dispersion coincides with the direction of the mean fluid flow with the lateral dispersion normal to this direction. At high Peclet numbers, the longitudinal dispersion is large in comparison with the lateral dispersion, since the component of the fluid velocity parallel to the mean flow direction has the largest gradients. The lateral dispersion $D_{eff}^{\perp}$ is associated with the weaker lateral fluid motion, whence $D_{eff}^{\parallel} \gg D_{eff}^{\perp}$.

The Taylor-Aris result for the dispersion coefficient (Eq. 4.6.35) has been applied to the empirical correlation of measured and calculated longitudinal dispersion coefficients in flow through packed beds and porous media (see Eidsath et al. 1983). Typically, the velocity in the Peclet number of the Taylor-Aris formula is identified with the superficial velocity, and the capillary diameter with the hydraulic diameter for spherical particles. An alternative velocity suggested by the capillary model is the interstitial velocity, and an alternative length is the square root of the permeability. In an isotropic packing of particles $(k)^{1/2}$ is about one-tenth the particle diameter (Probstein & Hicks 1982).

At large Peclet numbers where diffusive mixing is dominant, as for dispersion in a capillary tube, from Eq. (4.6.35)

$$D_{eff}^{\parallel} \sim Pe^2 \qquad (4.7.17)$$

If, however, the dispersive effects are due principally to velocity variations within the medium, then D_{eff} should be proportional to U_e (Adler 1992) and

$$D_{eff}^{\parallel} \sim Pe \qquad (4.7.18)$$

There is lack of agreement in the literature on the behavior of the longitudinal dispersion coefficient with Peclet number, other than it tends to increase monotonically with Pe. For large Peclet numbers the increase is generally found to have the power law behavior Pe^n, with $1 < n < 2$ (Plumb & Whitaker 1990, Adler 1992, Brenner & Edwards 1993).

To the extent that dispersion in an inertia free porous medium flow arises from a nonuniform velocity distribution, its physical basis is the same as that of Taylor dispersion within a capillary. Data on solute dispersions in such flows show the long-time behavior to be Gaussian, as in capillaries. The Taylor dispersion equation for circular capillaries (Eq. 4.6.30) has therefore been applied empirically as a model equation to characterize the dispersion process in chromatographic separations in packed beds and porous media, with the mean velocity identified with the interstitial velocity. In so doing it is implicitly assumed that the mean interstitial velocity and flow pattern is independent of the flow rate, a condition that would, for example, not prevail when inertial effects become important.

In chromatographic separations, as with Taylor's capillary tube measurements, the dispersion coefficients are determined empirically. One procedure is matching a solution of the dispersion equation, such as that for a step change in inlet concentration at the top of the column, with the corresponding measured change in average concentration of the displacing fluid that is observed with time, say, at the bottom of the column. Direct scanning procedures of the solute distribution along gel columns have also been developed for determining the dispersion coefficient by appropriate matching with a theoretical solution (Cantor & Schimmel 1980).

For a sharp band of solute run onto the top of a packed gel column the solution is the Gaussian of Eq. (4.6.39). To compare with the experiments of the eluting column, we replace the variables as follows:

$$\pi a^2 \to \alpha A , \qquad x - Ut \to x - \frac{Qt}{\alpha A}$$

$$t \to \frac{V}{Q} \qquad\qquad x \to L = \frac{V_e}{\alpha A} \qquad\qquad (4.7.19)$$

Here, V_e is the elution volume, and V is the total volume that has flowed through the column. Since the measurements are usually made at the bottom of the column, x has been replaced by the column length L. Therefore, for the concentration of solute eluting from the column as a function of the volume flow rate and volume of flow, dropping the longitudinal dispersion notation we can write

$$\bar{c} = \frac{1}{2} \frac{n_0}{\alpha A} \left(\frac{Q}{\pi D_{\text{eff}} V} \right)^{1/2} \exp\left(-\frac{Q(V_e - V)^2}{\alpha^2 A^2 V 4 D_{\text{eff}}} \right) \qquad (4.7.20)$$

As discussed, the determination of D_{eff} is normally empirical, although it might be deduced from semiempirical or theoretical considerations.

From the above formula, bearing in mind the caveats mentioned in its derivation, we can see that decreasing D_{eff} will decrease the spread of the solute concentration distribution and, in this way, increase the resolution. Since D_{eff} increases with Pe, then one way to decrease D_{eff} is to decrease the permeability k by going to smaller resin beads. Decreasing the resin size will also reduce any mass transfer resistance within the beads and, hence, tend to reduce band spreading. A lower permeability, however, means that to maintain the same flow rate we must increase the pressure drop along the column and not just employ gravity drainage. It is clear that operating at the highest flow rate possible for equilibration reduces the time for spreading by dispersion and hence increases the resolution. The use of very fine particles, about 10 μm in diameter, and high pressure to maintain an adequate flow rate is called *high-performance or high-pressure liquid chromatography* (HPLC). Figure 4.7.2 shows a typical difference in the separation of four components by HPLC and ordinary chromatography with larger particles (Freifelder 1982). Finally, any alteration of the velocity distribution, such as might result from inertial effects or non-Newtonian behavior, will alter the dispersion characteristics.

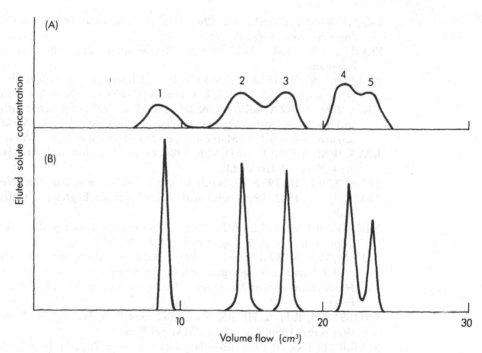

Fig. 4.7.2 Comparison of normal chromatography (A) and high-performance liquid chromatography (B) in the separation of several components of the citric acid cycle: (1) α-ketoglutaric acid; (2) citric acid; (3) malic acid; (4) fumaric acid; and (5) succinic acid. The times to obtain the separations are 180 and 20 minutes for normal chromatography and HPLC, respectively. [After Freifelder, D.M. 1982. *Physical Biochemistry*, 2nd edn. San Francisco: W.H. Freeman. Copyright © 1976, 1982 W.H. Freeman and Company. With permission.]

References

ADLER, P.M. 1992. *Porous Media: Geometry and Transports*. Boston: Butterworth-Heinemann.

ARIS, R. 1956. On the dispersion of a solute in a fluid flowing through a tube. *Proc. Roy. Soc.* **A235**, 67–77.

BERMAN, A.S. 1953. Laminar flow in channels with porous walls. *J. Appl. Phys.* **24**, 1232–1235.

BIRD, R.B., STEWART, W.E. & LIGHTFOOT, E.N. 1960. *Transport Phenomena*. New York: Wiley.

BRENNER, H. & EDWARDS, D.A. 1993. *Macrotransport Processes*. Boston: Butterworth-Heinemann.

CANTOR, C.R. & SCHIMMEL, P.R. 1980. *Biophysical Chemistry. Part II: Techniques for the Study of Biological Structure and Function*. San Francisco: W.H. Freeman.

CASTELLAN, G.W. 1983. *Physical Chemistry*, 3rd edn. Reading, Mass.: Addison-Wesley.

DUBIN, P.L. (ed.) 1988. *Aqueous Size-Exclusion Chromatography*. Amsterdam: Elsevier.

EIDSATH, A., CARBONELL, R.G., WHITAKER, S. & HERRMANN, L.R. 1983. Dispersion in pulsed systems–III. Comparison between theory and experiments for packed beds. *Chem. Eng. Sci.* **38**, 1803–1816.

FRANK-KAMENETSKII, D.A. 1969. *Diffusion and Heat Transfer in Chemical Kinetics*, 2nd edn. New York: Plenum.

FREIFELDER, D.M. 1982. *Physical Biochemistry*, 2nd edn. San Francisco: W.H. Freeman.

FRIED, J.J. & COMBARNOUS, M.A. 1971. Dispersion in porous media. In *Advances in Hydroscience*, vol. 7 (ed. V.T. Chow), pp. 169–282. New York: Academic.

GILL, W.N., DERZANSKY, L.J. & DOSHI, M.R. 1971. Convective diffusion in laminar and turbulent hyperfiltration (reverse osmosis) systems. In *Surface and Colloid Science*, vol. 4 (ed. E. Matijevic), pp. 261–360. New York: Wiley.

KAYS, W.M. & CRAWFORD, M.E. 1980. *Convective Heat and Mass Transfer*, 2nd edn. New York: McGraw-Hill.

LEVENSPIEL, O. 1972. *Chemical Reaction Engineering*, 2nd edn. New York: Wiley.

LEVICH, V.G. 1962. *Physicochemical Hydrodynamics*. Englewood Cliffs, N.J.: Prentice-Hall.

NUNGE, R.J. & GILL, W.N. 1969. Mechanisms affecting dispersion and miscible displacement. *Ind. & Eng. Chem.* **61**(9), 33–49.

PLUMB, O.A. & WHITAKER, S. 1990. Diffusion, adsorption and dispersion in porous media: Small-scale averaging and local volume averaging. In *Dynamics of Fluids in Hierarchical Porous Media* (ed. J.H. Cushman), pp. 97–149. San Diego: Academic Press.

PROBSTEIN, R.F. & HICKS, R.E. 1982. *Synthetic Fuels*. New York: McGraw-Hill. (Reprinted 1990, Cambridge, MA: pH Press.)

SCHLICHTING, H. 1979. *Boundary-Layer Theory*, 7th edn. New York: McGraw-Hill.

SHERWOOD, T.K., BRIAN, P.L.T., FISHER, R.E. & DRESNER, L. 1965. Salt concentration at phase boundaries in desalination by reverse osmosis. *Ind. & Eng. Chem. Fundamentals* **4**, 113–118.

SOLAN, A. & WINOGRAD, Y. 1969. Boundary-layer analysis of polarization in electrodialysis in a two-dimensional laminar flow. *Phys. Fluids* **12**, 1372–1377.

TAYLOR, G.I. 1953. Dispersion of soluble matter in solvent flowing slowly through a tube. *Proc. Roy. Soc.* **A219**, 186–203.

TAYLOR, G.I. 1954. Conditions under which dispersion of a solute in a stream of solvent can be used to measure molecular diffusion. *Proc. Roy. Soc.* **A225**, 473–477.

Problems

4.1 Derive the Langmuir adsorption isotherm (Eq. 4.1.8) under the assumptions that there are a finite number of free adsorption sites, the adsorption rate is second order in c and the available adsorption sites, and the desorption rate is first order in the material adsorbed. Show what the constant b represents.

4.2 The estimate given in Section 4.2 and the solution presented in Section 4.3 describe the behavior of the mass transfer from a channel wall at constant concentration into a flowing solvent with a fully developed velocity profile that is initially free of solute. In deriving this solution, we assume the solute concentration zero outside the concentration boundary layer near the wall. Sufficiently far downstream of the channel entrance this will no longer be the case. By considering a control volume of axial length L along a long circular tube of radius a into which a solute-free liquid flows and out of which flows liquid plus solute, derive a dimensionless parameter that must be *large* to ensure the validity of the solution presented. This parameter is known as the *Graetz number.*

4.3. Liquid with a zero initial solute concentration enters a long cylindrical tube of radius a along which a single species solute is supplied at a *constant flux* j_w^* (mol m^{-2} s^{-1}). It is assumed that the velocity profile is fully developed at the entrance to the tube, and we are interested in determining the behavior of the concentration profile from the axial position at which the concentration boundary layer becomes fully developed.

a. Taking the origin of the axial coordinate x to be at the position where the concentration profile becomes fully developed, find an expression for the concentration c in terms of x, r, j_w^*, a, $\bar{u}$, and D, where r is the radial coordinate measured from the tube center, D is the diffusion coefficient, and $\bar{u}$ is the mean fluid velocity. The velocity profile is given by

$$u = 2\bar{u}\left(1 - \frac{r^2}{a^2}\right)$$

b. Determine the *Nusselt number for mass transfer*, often called the *Sherwood number*, defined by

$$\mathrm{Nu}_D = \frac{2aj_w^*}{D(c_w - c_m)}$$

where c_w is the concentration at the wall and c_m is the mixing-cup or bulk average concentration at any cross section defined by

$$c_m = \frac{1}{\pi a^2 \bar{u}} \int_0^a uc \cdot 2\pi r \, dr$$

4.4 In a so-called unstirred, batch-operated, reverse osmosis system a long cylinder holding a salt solution is closed by a semipermeable membrane at one end and a piston at the other. The pressure applied by the piston initially is that corresponding to osmotic equilibrium; then at time $t = 0$, the pressure is suddenly increased to a predetermined value at which it is maintained. The result is that there will be a flow through the membrane, which in general will be time dependent, and neglecting any wall effects there will be a concentration variation in the solution that will depend on time and distance into the solution measured from the membrane.

a. With $v_w(t)$ the permeation velocity through the membrane, c_0 the initial salt concentration in the solution, c_w the salt concentration at the membrane, and c_p the salt concentration on the product side of the membrane, write down the equation governing the salt diffusion, together with the boundary and initial conditions. Assume that the rejection coefficient $R_s = 1 - c_p/c_w$ is constant but that it may be different from unity.

b. Suppose that, following Eq. (4.4.4),

$$v_w = A'(\Delta p - B R_s c_w)$$

where B is a constant. With the characteristic velocity taken to be $A'\Delta p$, determine what the other appropriate characteristic variables would be and write the governing differential equation, together with the boundary and initial conditions for the concentration in dimensionless form using the characteristic variables. Upon what independent dimensionless parameters will the dimensionless concentration, wall concentration, and permeation velocity depend?

4.5 Water is contained between two infinite parallel plates separated by a small distance $h = 10^{-3}$ m. The bottom plate is held stationary, and the top plate is moved at a constant velocity $U = 10^{-3}$ m s^{-1} so that a simple shear flow is generated between the plates. A thin band of a dye of thickness $\Delta = 10^{-4}$ m is injected between and perpendicular to the plates extending fully across the gap. The band depth is very deep and may be supposed to be infinite. The dye concentration is $c_0 = 10$ mol m^{-3}, and its molecular diffusion coefficient in water is $D = 10^{-9}$ m^2 s^{-1}.

a. From the convective diffusion equation for the dye concentration, make an order-of-magnitude estimate of the time for which molecular diffusion in the direction of motion will no longer be important and the times for which lateral molecular diffusion is negligible.

b. Let x be the Cartesian coordinate in the direction of motion with origin at the intersection of the stationary plate and center of the band Δ when it is injected into the water, and let $\bar{c}$ be the average concentration per unit depth defined by

$$\bar{c} = \frac{1}{h} \int_0^h c \, dy$$

where y, which is measured positive in the direction toward the moving plate, is the coordinate normal to x. Within the time interval for which the dispersion of c takes place essentially by convection, what is the solution for $\bar{c}$?

4.6 a. For the simple shear flow of Problem 4.5, determine the dispersion coefficient by following Taylor's general approach. Neglect molecular diffusion in the direction of motion. Estimate the times, analytically and numerically, for which any approximations used in your analysis are valid, and the times for which diffusion in the direction of motion can be neglected.

b. How is the result for the dispersion coefficient changed if molecular diffusion in the direction of motion is included? Derive your answer, justifying any approximation not made in part a.

4.7 A porous medium is modeled as made up of uniformly distributed straight circular capillaries of the same diameter. The flow through each capillary is an inertia free Poiseuille flow. By comparing the Poiseuille pressure drop and the Darcy pressure drop formulas, deduce an expression for the permeability. Discuss the difference between the result obtained and the Kozeny-Carman permeability.

5 Solutions of Uncharged Macromolecules and Particles

5.1 Microhydrodynamics of Macromolecules and Particles

To this point the book has dealt principally with solutions of "simple" low molar mass molecules. The aim of this chapter is to examine the hydrodynamics of solutions of small uncharged macromolecules, that is, molecules of molar mass greater than about 10^5. Apart from an interest in determining the general flow characteristics of such solutions, it is also possible to obtain information on the properties of macromolecules and colloidal particles from suitable experiments on their behavior in solution. Moreover, the separation of mixtures of suspended matter of differing physical and chemical properties, so important in technical and biological applications, can be accomplished by hydrodynamic means.

The particle sizes of concern are in the range typically between 0.1 and 10 μm, although the band may be extended by a factor of 10 on either side in some circumstances. The distinction between the lower size limit at which one distinguishes a colloidal particle and a dissolved molecule is somewhat vague. Of importance, however, is that the characteristic size of the dispersed phase be large compared with simple molecules. In this regard note that the diameter of a water molecule is about 3×10^{-4} μm. Sizes of some particles of interest were shown in Fig. 1.3.1 of the Introduction.

Batchelor (1976) has termed the study of flow systems of the small scale considered as *microhydrodynamics* because of the numerous distinctive features, many of which are physical-chemical in nature. Among the more important characteristics are

1. Inertia forces are normally small compared with viscous forces, and the equation of motion for the fluid reduces to the linear Stokes form.

2. The displacements of free particles due to thermal (Brownian) motion in time intervals characteristic of the imposed flow are often significant.

3. Free colloidal particles about a micrometer or less in size settle out so slowly under gravity that they may often be considered as suspended in the flow and moving with it. As a result, the principal concern is to determine the rheological properties of the colloidal suspension.

4. Interfacial surface effects are important. This is a consequence of the fact that interfacial forces between a dispersed particle and the surrounding phase are proportional to the square of the particle's characteristic dimension and body forces are proportional to the cube, so the ratio is inversely proportional to the particle size.

5. Electrokinetic effects are important because solid and liquid particles normally acquire a charge in aqueous solutions. However, our discussion of these effects will be reserved for Chapters 7 and 8, with only uncharged particles examined in this chapter.

To understand the forces acting on small particles in solution, let us first review some low Reynolds number flow results in which inertial forces are negligible. The basic text on this topic is that of Happel & Brenner (1983) with more recent advances to be found in Kim & Karrila (1991), who also include computational and variational considerations, and in Leal (1992).

With body forces in addition to inertial forces neglected, the incompressible Navier-Stokes equation reduces to the *Stokes equation*

$$\nabla p = \mu \nabla^2 \mathbf{u} \tag{5.1.1}$$

which, together with the continuity equation

$$\nabla \cdot \mathbf{u} = 0 \tag{5.1.2}$$

defines the flow. This system of equations governing low Reynolds number flow is linear, so solutions are superposable. Moreover, since time does not explicitly enter the solutions, they are kinematically reversible.

An important result from low Reynolds number theory is that for a body of arbitrary shape in translational motion with a velocity U the resultant force F exerted by the body depends on its orientation and may be written in Cartesian tensor notation as (Happel & Brenner 1983, Batchelor 1967)

$$F_i = 6\pi\mu R_{ij} U_j = f_{ij} U_j \tag{5.1.3a}$$

or, in inverse form,

$$U_i = v_{ij} F_j \tag{5.1.3b}$$

The tensor R_{ij}, termed the *translation tensor*, for a rigid body depends solely on the size and shape of the body. The translation tensor has the dimensions of length and may be interpreted as an *equivalent radius*. In the polymer literature the force is usually expressed in terms of a *translational friction tensor* f_{ij}. The components are called *translational friction coefficients* or, usually, just *friction coefficients*. (Often the term "frictional" is used instead.) The inverse of the friction tensor is essentially the mobility introduced earlier in Section 2.5. Accordingly, in Eq. (5.1.3b) we have also introduced a *mobility tensor* v_{ij}, which has units $N^{-1}\, m\, s^{-1}$ or $s\, kg^{-1}$.

For immiscible *fluid* drops held spherical by surface tension,

$$R_{ij} = \frac{2}{3}\, \delta_{ij} \left(\frac{\mu + \frac{3}{2}\, \mu_{in}}{\mu + \mu_{in}} \right) a \qquad (5.1.4)$$

where μ_{in} is the internal viscosity of the fluid drop, which is different from μ, and a is its radius. The limit $\mu_{in}/\mu \to 0$, for which $R_{ij} \to \frac{2}{3} a \delta_{ij}$, approximately characterizes a spherical gas bubble moving through a liquid. The limit $\mu_{in}/\mu \to \infty$, for which $R_{ij} \to a\delta_{ij}$, corresponds to a rigid sphere moving through a viscous fluid. Thus for a rigid sphere the drag force parallel to the direction of translation is

$$F = -6\pi\mu a U \qquad (5.1.5)$$

which is the classical *Stokes drag law*.

Many particles and macromolecules that cannot be modeled as spherical are often amenable to being modeled as ellipsoidal. The problem of the steady low-speed motion of a rigid ellipsoid through a viscous liquid was first treated by A. Oberbeck in 1876. It was analyzed subsequently by a number of authors, with several independent presentations (Happel & Brenner 1983). For our purposes we give the results as presented by Perrin (1934) for prolate and oblate spheroids. The prolate spheroid is a rodlike shape generated by rotating an ellipse around its long semiaxis a, with the two shorter semiaxes b identical. An oblate spheroid is a disk shape generated by rotating an ellipse about its short semiaxis a, the two long semiaxes b being identical (Fig. 5.1.1). Perrin's work, although widely quoted in polymer studies, has largely bypassed the fluid mechanics community.

Let R_1, R_2, R_3 be the translation coefficients for translation of an ellipsoid parallel to its semiaxes a_1, a_2, a_3. For rigid spheroids generated by rotating an ellipse about the semiaxis a, Perrin's solution with $a_1 = a$ and $a_2 = a_3 = b$ may be written

$$R_1 = \frac{8}{3} \frac{a^2 - b^2}{(2a^2 - b^2)S - 2a} \qquad (5.1.6)$$

$$R_2 = R_3 = \frac{16}{3} \frac{a^2 - b^2}{(2a^2 - 3b^2)S + 2a} \qquad (5.1.7)$$

where, if $a > b$ (prolate spheroid),

$$S = 2(a^2 - b^2)^{-1/2} \ln\left[\frac{a + (a^2 - b^2)^{1/2}}{b} \right] \qquad (5.1.8)$$

while if $a < b$ (oblate spheroid)

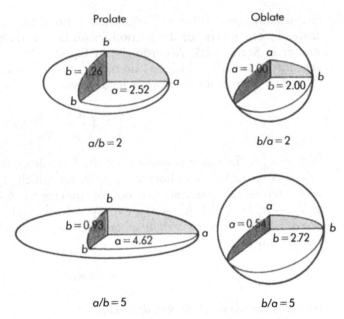

Figure 5.1.1 Four spheroids with equal volumes (after Cantor & Schimmel 1980).

$$S = 2(b^2 - a^2)^{-1/2} \tan^{-1}\left[\frac{(b^2 - a^2)^{1/2}}{a}\right] \tag{5.1.9}$$

When $a = b$, the Stokes result for a sphere is obtained.

   When Brownian motion and its attendant randomizing effect is important, an ellipsoid, or colloidal particles of arbitrary shape, will have various orientations. As we have seen, there is a translation coefficient for each orientation. It can be shown that the *mean translation coefficient, mean friction coefficient*, and *mean mobility* are given, respectively, by the formulas (Perrin 1936, Happel & Brenner 1983)

$$\frac{1}{\bar{R}} = \frac{1}{3}\left(\frac{1}{R_1} + \frac{1}{R_2} + \frac{1}{R_3}\right) \tag{5.1.10a}$$

$$\frac{1}{\bar{f}} = \frac{1}{3}\left(\frac{1}{f_1} + \frac{1}{f_2} + \frac{1}{f_3}\right) \tag{5.1.10b}$$

$$\bar{v} = \frac{1}{3}(v_1 + v_2 + v_3) \tag{5.1.10c}$$

The subscripts 1, 2, 3 refer to the principal axes of translation, which are three mutually perpendicular axes fixed to the body defined such that if the body translates without rotation parallel to one of them it will experience a force only in that direction.

   From Eqs. (5.1.6) and (5.1.7) the mean translation coefficient for a prolate or oblate spheroid is simply $2/S$, whereas for a sphere it is just the sphere radius. A quantity of interest, sometimes termed the *Perrin factor*, is the ratio of the mean translation coefficient (or mean friction coefficient) of a prolate or oblate spheroid to that of a sphere of equal volume. With the volume of a prolate spheroid given by $\frac{4}{3}\pi ab^2$ and an oblate spheroid by $\frac{4}{3}\pi a^2 b$ the Perrin factor $\mathscr{F}$ is readily found to be

$$\mathscr{F} = \frac{\bar{R}}{R_{sph}} = \frac{\bar{f}}{f_{sph}} = \frac{(p^2 - 1)^{1/2}}{p^{1/3} \ln[p + (p^2 - 1)^{1/2}]} \qquad p > 1 \qquad (5.1.11)$$

for a prolate spheroid and

$$\mathscr{F} = \frac{\bar{R}}{R_{sph}} = \frac{\bar{f}}{f_{sph}} = \frac{(1 - p^2)^{1/2}}{p^{1/3} \tan^{-1}[(1 - p^2)^{1/2} p^{-1}]} \qquad p < 1 \qquad (5.1.12)$$

for an oblate spheroid, where $p = a/b$.

In Fig. 5.1.2 the Perrin factor is plotted as a function of the *axial ratio*, defined as the ratio of the long semiaxis to short semiaxis, equal to p for the prolate spheroid and p^{-1} for the oblate spheroid. As seen, the Perrin factor is always greater than unity, which may have been anticipated, since for equal volumes the surface area of the spheroid will be greater than that of a sphere of the same volume, so the friction coefficients will be greater. Because the volume of a molecule is proportional to its molar mass, then, for a constant mass, the more a molecule deviates from a spherical shape, the larger will be its mean friction coefficient.

For the prolate spheroid an interesting approximate form of the translation coefficients is obtained for the case when the long semiaxis a is large compared with the short semiaxis b. From Eqs. (5.1.6) to (5.1.8), by expanding for $a/b \gg 1$ we find that for translation parallel to the a axis

$$R_a = \frac{\frac{2}{3}a}{\ln(2a/b) - 0.5} \qquad (5.1.13)$$

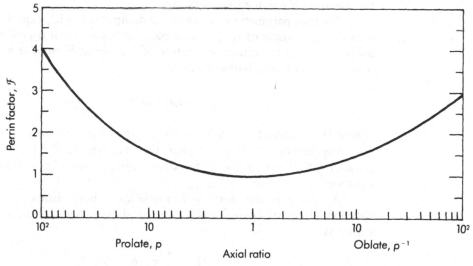

Figure 5.1.2 Perrin factor for prolate and oblate spheroids.

and for translation parallel to the b axis

$$R_b = \frac{\frac{4}{3}a}{\ln(2a/b) + 0.5} \qquad (5.1.14)$$

A uniformly valid asymptotic result for the translation coefficient of a long *but finite* straight cylinder moving parallel to its axis of symmetry, with b the cylinder radius and $2a$ its length, neglecting terms $O(a/(\ln a/b)^3)$ is similar to Eq. (5.1.13) with 0.5 replaced by 0.81 (Brenner 1974). The analogous result for the translation coefficient when the cylinder moves perpendicular to its axis is the same form as Eq. (5.1.14) with 0.5 replaced by 0.19. (For other finite axisymmetric geometries see Brenner.) In passing, we recall there is no finite drag solution of the inertia free Stokes equation for an infinite cylinder moving through an otherwise unbounded flow.

The consequence of the above results is

$$\frac{F_b}{F_a} \to 2 \qquad \frac{a}{b} \gg 1 \qquad (5.1.15)$$

That is, the drag approaches a limit of being twice as great when the direction of motion is perpendicular to the axis of symmetry as it is when moving along the axis. As Taylor (1969) has shown, this is a property of all long axisymmetric bodies in low Reynolds number, inertia free flow no matter what the cross-sectional distribution, provided that the center of gravity of the body is in such a position that the viscous forces do not exert a couple about it whatever the orientation of the body to the force. This leads to the condition, proved by Taylor, that a long axisymmetric body, for example when falling in a fluid at constant speed, can never fall along a path that is inclined to the vertical at more than 19.5°, and this occurs when the axis of the body lies at $\cos^{-1}(1/\sqrt{3})$ or $54°44'$ (cf. Happel & Brenner 1983). This low Reynolds number behavior might be utilized as an analytical tool for particle characterization in those cases where Brownian motions may be neglected.

Another parameter of interest for comparison with experimental measurements is the couple or torque on a particle arising from rotational motion. The torque $\mathbf{T}$ about the center of rotation of a body with angular velocity $\boldsymbol{\omega}$ can be written in Cartesian tensor form as

$$T_i = -6\mu\Omega_{ij}\omega_j = -(f_{\text{rot}})_{ij}\omega_j \qquad (5.1.16)$$

where Ω_{ij} is termed the *rotation tensor*. It has the dimensions of length cubed and may be interpreted as an *equivalent volume*. In the polymer literature, in parallel with the translational force, the torque is usually expressed in terms of a *rotational friction tensor* $(f_{\text{rot}})_{ij}$.

A sphere is the simplest example of a body isotropic with respect to rotation about its center, and it can be shown that the rotation tensor at its center is

$$\Omega_{ij} = \frac{4}{3}\pi a^3 \delta_{ij} = V_{\text{sph}}\delta_{ij} \qquad (5.1.17)$$

with V_{sph} the sphere volume. Perrin (1934) also derived the results for the rotation tensor for an ellipsoid of revolution, and they are

$$\Omega_1 = \frac{16}{9}\,\pi\,\frac{(a^2 - b^2)b^2}{2a - b^2 S} \tag{5.1.18}$$

$$\Omega_2 = \Omega_3 = \frac{16}{9}\,\pi\,\frac{a^4 - b^4}{(2a^2 - b^2)S - 2a} \tag{5.1.19}$$

with S given by Eqs. (5.1.8) or (5.1.9). In the case of a long slender prolate spheroid, where $a/b \gg 1$, the rotation coefficient around a transverse axis, Ω_b, simplifies to

$$\Omega_b = \frac{\frac{4}{9}\pi a^3}{\ln(2a/b) - 0.5} \tag{5.1.20}$$

The uniformly valid asymptotic result for a long cylindrical rod, with b the cylinder radius, is of the same form with 0.5 replaced by 1.14 (cf. Brenner 1974). Note that $\Omega_a \to 0$.

Polymer molecules will generally have more complex shapes than the simple ones considered so far. A method for computing the friction coefficients of a structure composed of identical subunits, and spheres in particular, was developed by J.G. Kirkwood and J. Riseman (Bird et al. 1977). For example, a linear or coiled polymer can be approximated by a string of spheres, the spheres representing groups of monomer units of which each is impenetrable. An oligomeric protein might, for example, be represented by a cluster of spheres. In either case, the problem is to deal with the hydrodynamic interactions between the spheres, since each sphere, as it moves through the fluid, perturbs the velocity distribution of the fluid nearby, and this perturbation is felt by the other spheres. These models are discussed further in Section 9.2.

Since the Stokes equations are linear and homogeneous, the velocities produced by different forces and boundaries at different points in the liquid are additive. Nevertheless, the solution to a multiparticle arrangement is still somewhat formidable. The important physical feature is that in a low Reynolds number flow the hydrodynamic influence of an applied force or boundary falls off relatively slowly with increasing distance. In particular, the velocity $\mathbf{u}$ produced at a point $\mathbf{r}$ by a force $\mathbf{F}$ acting at the origin, corresponding to a point particle, can be shown to be

$$\mathbf{u} = \frac{1}{8\pi\mu r}\left[\mathbf{F} + \frac{(\mathbf{F}\cdot\mathbf{r})\mathbf{r}}{r^2}\right] \tag{5.1.21}$$

The Kirkwood-Riseman calculation is sufficiently detailed that we shall not discuss it further here. The interested reader is referred to Bird et al. (1987). However, the issue raised in considering complex shapes is the same one that arises when one questions whether the particle is not an isolated particle in suspension, which is the only situation for which we have outlined results. It is evident in a multiparticle suspension that the hydrodynamic interactions between the particles become important. For example, from what we have shown it is clear that a particle in translation would drag along its neighbors. To handle this general problem, we could assume a "suspension" or "colloidal" phase with an effective viscosity and then assume that the friction coefficients would be proportional to this effective viscosity. The value of the friction coefficients could also be obtained from sedimentation experiments where the force is known and the resultant suspension velocity can be measured. These are among the approaches that we shall discuss in the following sections and in Chapter 9, dealing with multiparticle suspensions.

5.2   Brownian Motion

The random thermal motion of suspended particles that are sufficiently large to be observed is termed *Brownian motion*, after the Scottish botanist Robert Brown, who in 1828 described this phenomenon from microscopic observations he had made of pollen grains suspended in water. We have mentioned this phenomenon a number of times in previous sections, and our purpose here is to quantify it.

In the absence of external forces, all suspended particles regardless of their size have the same translational kinetic energy. The average translational kinetic energy for any particle is equal to $\frac{3}{2}kT$ (by equipartition $\frac{1}{2}kT$ per degree of freedom); therefore

$$\tfrac{1}{2}m\langle \mathbf{U}^2 \rangle = \tfrac{3}{2}kT \tag{5.2.1}$$

or

$$\langle \mathbf{U}^2 \rangle = \frac{3kT}{m} \tag{5.2.2}$$

Here, m is the mass of the particle, and $\langle \mathbf{U}^2 \rangle$ is the mean square velocity in a fluid medium at temperature T (the symbol $\langle \ \rangle$ denotes a time-average value). The velocities of Eq. (5.2.2) are very much greater than the mean square velocities of randomly moving particles observed under the microscope. For example, the root mean square value $\langle \mathbf{U}^2 \rangle^{1/2}$ is about $1.7 \, \text{mm s}^{-1}$ at room temperature for a particle of $1 \, \mu\text{m}$ radius and the same density as water. Einstein (1956) pointed out that the thermally induced fluctuations would cause a particle to vary in direction many millions of times per second, so the path traversed per second could not be resolved microscopically. He suggested instead that a quantity that could be more readily observed on a macroscopic scale and compared with the theoretical value is the rate of increase of the mean square displacement of a particle or, equivalently, the diffusivity. That is, as shown below, Brownian motion gives rise to a macroscopic diffusive flux.

Unknown to the physical and biological scientific communities, the ideas on the diffusion of probabilities had been published by the French mathematician Louis Bachelier (1900) 5 years before Einstein's 1905 *Ann. Phys.* paper on Brownian motion. The work constituted his thesis for the degree of Doctor of

Mathematical Sciences at the Sorbonne, carried out under the supervision of the renowned mathematician Poincaré and was published in *Ann. Sci. École Norm. Sup., 3.* The title of this relatively short thesis was *Theory of Speculation* and it dealt with the pricing of options (a much debated topic among US business executives today). Until the 1950s his work was unknown even to the economics community (Merton 1992). The importance of his thesis was emphasized by Paul Samuelson, the Nobel Laureate in economics, who is quoted by Merton as saying that "Bachelier's methods dominated Einstein's in every element of the vector." Bachelier's paper is available in English translation (Bachelier 1900) and our readers may draw their own conclusions. Although we cannot presume in this text to change the credited authorship on Brownian motion, we can recognize the debt owed to Bachelier.

Consider a suspension of rigid particles in thermal equilibrium, where the apparent mass of each particle (mass corrected for buoyancy) is sufficiently small to be neglected, so no external force is acting on the particles. The fluctuations of the particles are then a consequence of the random fluctuating force due to the response of the fluid to the thermal agitation of the particles. That is, the collisions of the solvent molecules with the particles cause the particles to execute a random walk. From random walk considerations the center of mass, say G, of a particle describes an irregular trajectory with successive positions G_0, G_1, $G_2,\ldots$ of G at times $t, t + \tau,\ t + 2\tau,\ldots$ independent of the direction of the preceding displacement, provided the time τ between displacements is not so small that the velocity changes become indefinitely large. This random walk behavior, shown in Fig. 5.2.1A, is referred to as *translational Brownian motion.*

Similarly, for an axis bound to the particle its orientation will change randomly with time. If we define the orientation of the particle by the intersection H of the particle axis with a sphere of unit radius described around the center (Fig. 5.2.1B), the motion of the point H on the spherical surface is random. Its successive positions H_0, H_1, H_2 at times $t, t + \tau, t + 2\tau, \ldots$ along the

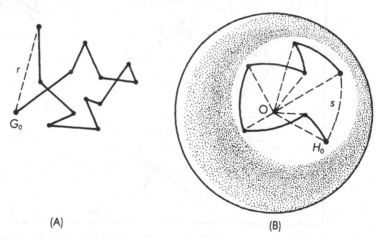

(A)   (B)

Figure 5.2.1 Brownian motion: (A) translational, (B) rotational (after Sadron 1953).

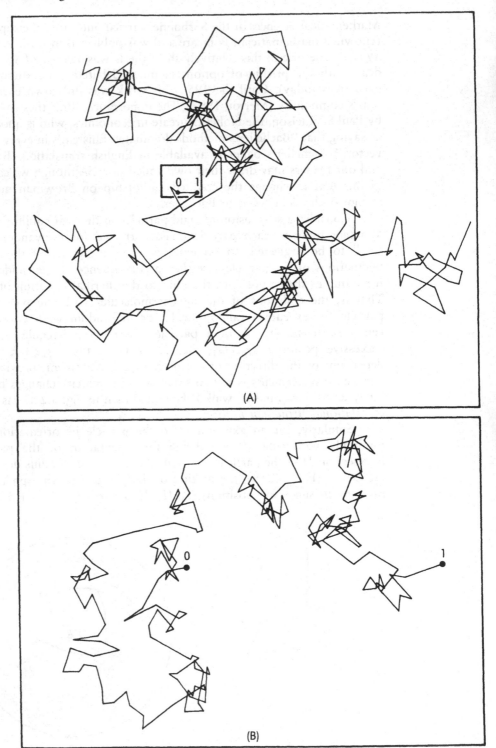

Figure 5.2.2   (see facing page for legend).

spherical surface are independent of the preceding displacement. This random walk behavior, shown in Fig. 5.2.1B, is referred to as *rotational Brownian motion*.

We digress here for a moment to observe that the jagged and irregular path of a particle in Brownian motion is scale-invariant under a change of length scale; that is, it is self-similar under dilation. In the definition of Mandelbrot (1982), as discussed in Section 1.3, the trajectory is geometrically a fractal. This is illustrated in Fig. 5.2.2. Figure 5.2.2A shows a redrawing of a tracing made in 1912 by the Nobel prize winner Jean B. Perrin (see Perrin 1923) of the path of a colloidal particle in water as seen under a microscope, with the successive positions marked every 30 s and then joined by straight lines. Jean Perrin, not to be confused with his son Francis, who was referred to earlier, fully recognized that the trajectory is practically plane filling, in the sense that if a portion of the trajectory is enlarged and the particle position marked more frequently, the irregular path would be reproduced qualitatively. This is seen in Fig. 5.2.2B (Lavenda 1985), which is a numerical simulation for the path between two points of Perrin's observations with a frequency of measurement 100 times greater than Perrin's.

Let us return to the analysis of Brownian motion. For simplicity we begin by considering the continuous one-dimensional translational Brownian motion as represented by a one-dimensional random walk problem. The probability of a displacement between x and $x + dx$ after n random steps of length l is given by the Gaussian distribution

$$P(n, x)\, dx = (2\pi nl^2)^{-1/2} e^{-x^2/2nl^2}\, dx \qquad (5.2.3)$$

Now consider the diffusion of a solute of concentration c_0 in a thin layer at the origin $x = 0$ at $t = 0$. The material diffuses out, and the number of steps is taken to be proportional to the time; that is,

$$n = Kt \qquad (5.2.4)$$

This statement assumes that the "mean" motion is uniform, but this cannot be exact because when the direction and velocity of particle movement change there must be an acceleration. It is shown below that the approximation is indeed very accurate, from which it follows that the concentration at any x and t is given by

$$c = c_0 P(x, t) \qquad (5.2.5)$$

Figure 5.2.2   Translational Brownian motion of a colloidal particle in water. (A) Particle motion as observed every 30 s under a microscope by Perrin (1923). (B) Numerical simulation of magnified portion of particle path observed 100 times more frequently. [After Lavenda, B.H. 1985. Brownian motion. *Sci. Amer.* 252(2), 70–85. Copyright © 1985 by Scientific American, Inc. All rights reserved. With permission.]

If Eqs. (5.2.3) to (5.2.5) are indeed a solution to the one-dimensional diffusion equation

$$\frac{\partial c}{\partial t} = D \frac{\partial^2 c}{\partial x^2} \qquad (5.2.6)$$

then we need only substitute these equations into Eq. (5.2.6) and see if it is satisfied. Doing this, we find it is satisfied with $K = 2D/l^2$, whence

$$c = \frac{c_0}{2\sqrt{\pi D t}} e^{-x^2/4Dt} \qquad (5.2.7)$$

But this is just the solution given earlier by Eq. (4.6.39) for the diffusion of a solute initially concentrated at the origin. We are therefore led to conclude that for dilute solutions of noninteracting particles the probability of finding a particle in a particular element of space is given by a solution of the continuum diffusion equation.

The corresponding three-dimensional solution for the particle concentration appropriate to translational Brownian motion is

$$c = \frac{c_0}{8(\pi D t)^{3/2}} e^{-r^2/4Dt} \qquad (5.2.8)$$

The probability of finding a particle at a distance between r and $r + dr$ from the origin at time t is obtained by multiplying the above solution for c/c_0 by the volume of the spherical shell $4\pi r^2\,dr$ or

$$P(r, t)\,dr = \frac{1}{2(\pi D^3 t^3)^{1/2}} e^{-r^2/4Dt} r^2\,dr \qquad (5.2.9)$$

The mean displacement is zero because the positive and negative displacements are equally probable, and it is therefore not a measure of the particle displacement. Such a measure is given by the root mean square displacement $\langle r^2 \rangle^{1/2}$ since the sign differences are eliminated. The mean square of the displacement is obtained by integrating the square of the displacement multiplied by the probability of displacement over all possible displacements; that is,

$$\langle r^2 \rangle = \int_0^\infty r^2 P(r, t)\,dr \qquad (5.2.10)$$

Substituting Eq. (5.2.9) in (5.2.10) and integrating gives

$$\langle r^2 \rangle = 6Dt \qquad (5.2.11)$$

where D is the *translational diffusion coefficient*.

A similar result is obtained for the particle motion associated with the random change of orientation with time. With s the length of the arc of the circle between successive positions, it can be shown that

$$\langle s^2 \rangle = 4D_{rot}t \tag{5.2.12}$$

where D_{rot} is the *rotational diffusion coefficient*. Note that s is taken on a sphere of unit radius (Fig. 5.2.1B) so that D_{rot} has the dimensions of inverse time.

To evaluate the diffusion coefficients following Einstein's original thermodynamic argument (Landau & Lifshitz 1987), we assume that for the suspension of particles equilibrium and steadiness of the distribution of the particle number density are established. For translation this equilibrium is brought about by the translational diffusion flux balancing the convective flux resulting from the application of a steady hydrodynamic force to each particle. Assuming the particle to be in inertia free flow in an infinite fluid, we can express the resultant drag force acting on a particle of arbitrary shape through the low Reynolds number relation

$$\mathbf{F} = -\bar{f}\mathbf{U} \tag{5.2.13}$$

The low Reynolds number, Stokes-type drag law has been written here in terms of the mean translational friction coefficient $\bar{f}$, since it is assumed that, because of the Brownian motion, all orientations are equally probable. We recall that $\bar{f}^{-1}$ is the mean mobility $\bar{v}$. The use of a steady drag law for a particle that is changing its velocity rapidly can be shown to be justified on the basis that the velocities are of such small magnitude that the fluid acceleration can be neglected.

From the equilibrium condition of the balance of diffusive and convective fluxes, we have

$$-D\nabla n = n\mathbf{U} \tag{5.2.14}$$

with $\mathbf{U}$ the velocity acquired by the particle from the hydrodynamic force and n the particle number density (not number of moles). Hence from Eq. (5.2.13),

$$\bar{f}D\nabla \ln n = \mathbf{F} \tag{5.2.15}$$

To ensure equilibrium, Einstein assumed that the applied hydrodynamic force must be balanced by a steady "thermodynamic" force acting on each particle. This force may be identified with the change in Gibbs free energy G of the suspension due to the addition of the particle. It follows from the expression for chemical potential (Eq. 3.3.8), equal to the Gibbs free energy per mole, that

$$\mathbf{F}_{therm} = -\nabla G = kT\nabla \ln n \tag{5.2.16}$$

where we have made the dilute suspension assumption and replaced the activity by the particle concentration. This statement is equivalent to the particle distribution satisfying the equilibrium Boltzmann relation.

Substituting the thermodynamic force for the hydrodynamic force in Eq. (5.2.15), we obtain the translational diffusion coefficient

$$D = \frac{kT}{\bar{f}} \qquad (5.2.17)$$

Similarly it can be shown, with $\bar{f}_{\text{rot}}$ the mean rotational friction coefficient, that the rotational diffusion coefficient is

$$D_{\text{rot}} = \frac{kT}{\bar{f}_{\text{rot}}} \qquad (5.2.18)$$

These fundamental relations were originally obtained by Einstein.

A standard alternative approach to evaluating the Brownian diffusion coefficients is to employ the Langevin equation (Chandrasekhar 1943, Batchelor 1976, Russel et al. 1989). Here, it is supposed that the "force" acting on a single isolated particle, say in translational motion, is a combination of a force $G(t)$ characterizing the very rapid motions associated with the molecular motion time scale ($\sim 10^{-13}$ s for water) plus a frictional drag force F associated with the much slower fluid response to the particle motion. The Langevin equation of motion for the single particle is then

$$m \frac{d^2 \mathbf{r}}{dt^2} = \mathbf{G}(t) - \mathbf{F} \qquad (5.2.19)$$

with F given by Eq. (5.2.13),

$$m \frac{d^2 \mathbf{r}}{dt^2} = \mathbf{G}(t) - \bar{f}\left(\frac{d\mathbf{r}}{dt}\right) \qquad (5.2.20)$$

Equation (5.2.20) cannot be used directly for evaluating the mean square displacement since the mean values of the velocity and acceleration are zero. We therefore take the scalar product of $\mathbf{r}$ with this equation and transform it to the squares of these quantities:

$$\frac{m}{2} \frac{d^2 (\mathbf{r}^2)}{dt^2} - m\left(\frac{d\mathbf{r}}{dt}\right)^2 = \mathbf{r} \cdot \mathbf{G} - \frac{\bar{f}}{2} \frac{d(\mathbf{r}^2)}{dt} \qquad (5.2.21)$$

For a large number of particles, averaging all of the terms in this expression, we obtain

$$\frac{m}{2} \frac{d^2 \langle \mathbf{r}^2 \rangle}{dt^2} + \frac{\bar{f}}{2} \frac{d \langle \mathbf{r}^2 \rangle}{dt} = 3kT \qquad (5.2.22)$$

Here, the term $\langle \mathbf{r} \cdot \mathbf{G} \rangle$ has been dropped as negligibly small because of the small time of fluctuations in the force $\mathbf{G}(t)$, and $m \langle (d\mathbf{r}/dt)^2 \rangle$ has been set equal to $3kT$.

Integrating Eq. (5.2.22) once gives

$$\frac{d \langle \mathbf{r}^2 \rangle}{dt} = \frac{6kT}{\bar{f}} \{ 1 - e^{-\bar{f}(t_0 - t)/m} \} \qquad (5.2.23)$$

with t_0 the time from which the particle displacement is measured. The characteristic time to attain the constant asymptotic value $6kT/\bar{f}$, representative of the "mean" motion, is approximately the viscous relaxation time $m/\bar{f}$. This time is about 2×10^{-7} s for a 1-μm radius particle ($\rho = 10^3$ kg m^{-3}). The diffusivity $\langle r^2 \rangle /6t = kT/\bar{f}$, derived by Einstein from random walk considerations and equilibrium thermodynamics, is therefore reached very rapidly. Thus for time intervals large compared with $m/\bar{f}$, the motion of a Brownian particle can be considered one of random walk governed by the diffusion equation.

From $D = kT/\bar{f}$, the translational diffusion coefficient for a sphere is

$$D = \frac{kT}{6\pi\mu a} \tag{5.2.24}$$

This is commonly referred to as the *Stokes-Einstein equation*. The corresponding rotational diffusion coefficient is

$$D_{rot} = \frac{kT}{8\pi\mu a^3} \tag{5.2.25}$$

where, as noted above, D_{rot} has the dimensions of inverse time.

In 1908 and subsequent years, J.B. Perrin (1923) reported consistent values of Avogadro's number N_A based on the Stokes-Einstein equation and experiment. Perrin determined experimentally values of $\langle r^2 \rangle$ for different colloidal particle sizes, temperatures, and liquid solutions, and substituted the measured values into the formula

$$N_A = \frac{RTt}{\langle r^2 \rangle \pi\mu a} \tag{5.2.26}$$

This formula follows from the Stokes-Einstein equation, the relation $\langle r^2 \rangle = 6Dt$, and the definition $N_A = R/k$.

Perrin's experiments consisted of microscopic measurements of the observed displacements in a fixed time of spherical particles whose radii were determined microscopically. Actually, Perrin projected his two-dimensional planar observations onto a straight line giving the Brownian motion in one direction, and the projections were then squared and averaged. For this one-dimensional case, Eq. (5.2.26) would be divided by 3 (because one is looking at a single degree of freedom rather than three degrees of freedom).

In concluding this section, we remind the reader of the brief discussion of flexible macromolecules given in the Introduction. In a fluid at rest such a macromolecule will continuously change its configuration due to forces associated with Brownian motion. As a result, geometric properties such as the end-to-end distance and radius of gyration will fluctuate rapidly with time. As with free-particle Brownian motion, however, what is of interest are not the instantaneous values but the time-average values of the squares of these quantities. In a state of flow hydrodynamic forces will also affect the configuration. The analysis of the behavior of randomly coiled macromolecules modeled, for example, as a necklace of beads connected by frictionless springs is discussed in Section 9.2.

5.3   Viscosity of Dilute Suspensions

In this section we examine the flow of a suspension of particles, particularly the apparent viscosity coefficient of the suspension. Our interest is in calculating the convective mass flux of a suspension as distinct from the diffusive flux of Brownian motion. As previously, we shall assume a very dilute suspension in which each particle behaves as if it were in a liquid of infinite extent. To simplify the calculation, we neglect Brownian motion, although, as we discuss later, in the very dilute limit considered and for spherical particles it has no effect on the suspension viscosity.

As previously, the particles are supposed sufficiently small that the effects of gravity are negligible, and they are in inertia free flow in the surrounding liquid so that they move locally with the ambient flow. Again, however, the particle size is large compared with the molecular dimensions of the liquid, so it may be regarded as a continuum. The presence of the particles in the flowing liquid will disturb the particle-free flow. The nature of this hydrodynamic interaction problem has been briefly indicated in discussing model polymer molecules made up of monomer units that interact hydrodynamically. The difficulty in taking this interaction into account is a consequence of the particles' long-range influence, for an isolated particle will generate a velocity that decays very slowly (r^{-1}). This introduces some mathematical difficulties, since the superposition of these disturbances will lead to a divergent sum at large distances (see Batchelor 1972 for handling such divergences).

It is evident that by adding particles in a flow the amount of energy dissipated will be increased, since the work done by the shearing stresses is increased because of the addition of the solid boundaries associated with the particles. The particles plus liquid, which we term the *suspension phase*, might therefore be looked upon as a Newtonian fluid but with a coefficient of viscosity larger than that of the pure liquid. To understand the relation between the particle and fluid characteristics, Einstein (1956) set himself the problem of a dilute suspension in a simple Couette flow viscometer and asked what would be the measured viscosity.

For a Couette flow (Fig. 2.2.1) the velocity distribution is linear and, in rectangular Cartesian coordinates, may be written

$$u = \dot{\gamma}y \qquad v = 0 \qquad w = 0 \qquad (5.3.1)$$

where $\dot{\gamma}$ is the shear rate:

$$\dot{\gamma} = 2\varepsilon_{yx} = \frac{du}{dy} = \frac{U}{h} \qquad (5.3.2)$$

Here h is the plate spacing, and U is the velocity of the upper plate. The shear is

$$\tau = \tau_{yx} = \mu\,\frac{du}{dy} = \mu\,\frac{U}{h} \qquad (5.3.3)$$

and the corresponding energy dissipation per unit volume is

$$\Phi = \tau\left(\frac{du}{dy}\right) = \mu\left(\frac{U}{h}\right)^2 \tag{5.3.4}$$

Unfortunately, with particles present in the liquid the mathematical problem is considerably more difficult than the usual simple Couette flow results would lead one to believe, principally because the liquid flow must now be calculated with the complicated set of boundary conditions determined by the surfaces of the particles. As defined, the problem is to calculate the additional flow brought about by the presence of all the particles contained in the suspension undergoing a shearing action. This is made simpler by the assumption of a very dilute suspension. In this case the perturbation in the shear flow brought about by a single particle can be calculated and the total perturbation for a uniform distribution of particles determined by integrating the component perturbations from all of them.

To further simplify the problem, we assume the particles are spherical and rigid. The new fluid velocity is obtained by integrating the Stokes equations with boundary conditions of no-slip at the sphere and undisturbed flow far from it. To satisfy the boundary conditions, we must know the motion assumed by the sphere in the shear flow. The particle is regarded as having a velocity of translation, say u_0, equal to the velocity of the undisturbed fluid at the point occupied by the center of the sphere. With respect to coordinate axes attached to the particle and moving with velocity u_0, the velocity components will again be given by Eq. (5.3.1), and by symmetry there is no tendency for the sphere to translate (Fig. 5.3.1). The sphere will, however, rotate at the local angular velocity of the fluid. One of the better-known results from fluid mechanics is that the vorticity is twice the mean angular velocity of a fluid particle. It follows that a small spherical particle must rotate around the z axis with an angular velocity $\dot{\gamma}/2$. Using the fact that the mean angular velocity is defined by

$$-\frac{\dot{\gamma}}{2} = \frac{1}{2}\left(\frac{\partial v}{\partial x} - \frac{\partial u}{\partial y}\right) \tag{5.3.5}$$

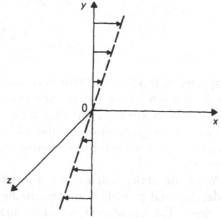

Figure 5.3.1 Translational motion with respect to coordinate axes translating with velocity of undisturbed fluid at center of sphere.

we have that the velocity of a point on the sphere surface is

$$u = \tfrac{1}{2}\dot{\gamma}y \qquad v = -\tfrac{1}{2}\dot{\gamma}x \qquad w = 0 \tag{5.3.6}$$

Now the introduction of the sphere has disturbed the flow, and the new components of the velocity will be

$$u + u', v', w' \tag{5.3.7}$$

with the primes denoting the perturbation components. From the above definition and Eq. (5.3.6) the values of the perturbation components at the sphere surface are defined by the condition $\mathbf{u} \cdot \mathbf{n} = 0$. Detailed analysis shows that the components of the perturbation flow satisfying the equations and boundary conditions are (Einstein 1956, Sadron 1953)

$$u' = -\frac{5}{2}\frac{a^3\dot{\gamma}x^2y}{r^5} + \frac{1}{6}\dot{\gamma}a^5\left(\frac{3y}{r^5} - \frac{15x^2y}{r^7}\right) \tag{5.3.8a}$$

$$v' = -\frac{5}{2}\frac{a^3\dot{\gamma}xy^2}{r^5} + \frac{1}{6}\dot{\gamma}a^5\left(\frac{3x}{r^5} - \frac{15xy^2}{r^7}\right) \tag{5.3.8b}$$

$$\omega' = -\frac{5}{2}\frac{a^3\dot{\gamma}xyz}{r^5} - \frac{1}{6}\dot{\gamma}a^5\left(\frac{15xyz}{r^7}\right) \tag{5.3.8c}$$

where a is the sphere radius and r is the radial spherical coordinate.

The first terms in Eqs. (5.3.8) are of the order a^3r^{-2}, the second ones of the order a^5r^{-4}. With $(a/r)^2 \ll 1$, which will generally be true, the second terms in Eq. (5.3.8) can be neglected and

$$u' \approx -\frac{5}{2}\frac{a^3\dot{\gamma}x^2y}{r^5} \tag{5.3.9a}$$

$$v' \approx -\frac{5}{2}\frac{a^3\dot{\gamma}xy^2}{r^5} \tag{5.3.9b}$$

$$w' \approx -\frac{5}{2}\frac{a^3\dot{\gamma}xyz}{r^5} \tag{5.3.9c}$$

This approximate additional flow is radial (partly inward, partly outward) with a dipole character near the surface (Fig. 5.3.2). The absolute value of the velocity has a maximum along directions making angles of 45° and 135° with the x axis. In a given direction the additional velocity decreases as $1/r^2$. Its magnitude is proportional to the volume of the particle and to the rate of shear (Sadron 1953).

With the above solution for the additional flow produced by a single particle, the total perturbation can now be calculated for a dilute suspension in shear flow. The calculation was first made by Einstein, and we follow an approach given by J.M. Burgers as outlined by Sadron (1953). In this approach the apparent, or effective, viscosity is found from the shearing stress on the

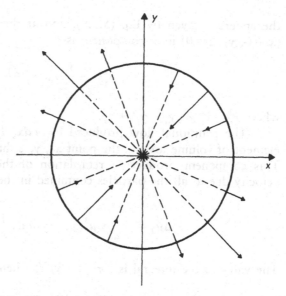

Figure 5.3.2   Perturbation flow according to Eq. (5.3.9) (Sadron 1953).

walls. The undisturbed flow of the pure liquid before introducing the spheres is given by Eq. (5.3.1) with the shear rate denoted by $\dot{\gamma}_0$ to indicate the undisturbed flow.

Consider two planes in the shear flow with coordinates y_1 and $-y_2$ (Fig. 5.3.3). The undisturbed velocities in these planes are $u_1 = \dot{\gamma}_0 y_1$ and $u_2 = -\dot{\gamma}_0 y_2$, respectively. It is now assumed that n particles per unit volume are added to the liquid to form a dilute suspension. In the suspension consider a layer parallel to xz, with ordinate y $(y_2 < y < y_1)$ and thickness dy. Let x_s, y, z_s be the coordinates of the center of a sphere in this layer. The perturbation flow due to

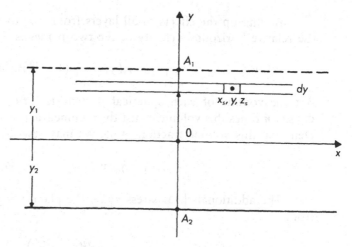

Figure 5.3.3   Geometry for shear flow of a dilute suspension.

the sphere is given by Eq. (5.3.9), and at the point A_1 shown in Fig. 5.3.3 $(x = 0, y_1, z = 0)$ its x component is

$$u'_s = -\frac{5}{2}\,\dot{\gamma}_0 a^3 x_s^2 \,\frac{y_1 - y}{r_s^5} \qquad (5.3.10)$$

where $r_s^2 = x_s^2 + (y_1 - y)^2 + z_s^2$.

The additional flow produced by $n\,dx_s\,dy\,dz_s$ particles contained in an element of volume around the point x_s, y, z_s has x component $u'_s n\,dx_s\,dy\,dz_s$. This component produces a retardation of the flow, and the retardation in velocity due to all the particles contained in the layer of thickness dy is

$$\Delta u_1 = -\frac{5}{2}\,\dot{\gamma}_0 a^3 (y_1 - y)n\,dy \int\int\limits_{-\infty}^{\infty} \frac{x_s^2}{r_s^5}\,dx_s\,dz_s \qquad (5.3.11)$$

The value of the integral is $\frac{2}{3}\pi(y_1 - y)^{-1}$, whence

$$\Delta u_1 = -\tfrac{5}{3}\pi\dot{\gamma}_0 a^3 n\,dy \qquad (5.3.12)$$

This retardation in velocity is independent of y and of the position of the point A_1 in the plane $y = y_1$.

Similarly, the retardation at the plane y_2 is

$$\Delta u_2 = +\tfrac{5}{3}\pi\dot{\gamma}_0 a^3 n\,dy \qquad (5.3.13)$$

so the relative horizontal velocity of the flow in the planes $y = y_1$ and $y = -y_2$, which in the absence of particles was equal to $\dot{\gamma}_0(y_1 + y_2)$, is now decreased by the amount

$$\Delta u_{1-2} = -\,\Delta u_1 + \Delta u_2 = \tfrac{10}{3}\pi\dot{\gamma}_0 a^3 n\,dy \qquad (5.3.14)$$

Adding up the effects of all layers from $-y_2$ to $+y_1$, we readily find that the relative horizontal velocity of the two planes is

$$u_{1-2} = \dot{\gamma}_0(y_1 + y_2)(1 - \tfrac{10}{3}\pi a^3 n) \qquad (5.3.15)$$

But the volume of each spherical particle is $\frac{4}{3}\pi a^3$, and the particle number density n times this volume is just the volume fraction of the spheres in the fluid. Denoting this volume fraction by ϕ, we may rewrite Eq. (5.3.15) in the form

$$u_{1-2} = \dot{\gamma}_0(y_1 + y_2)(1 - 2.5\phi) \qquad (5.3.16)$$

The additional shear stress τ'_1 on the plane $y = y_1$ can now be calculated from

$$\tau'_1 = \mu\left(\frac{\partial u'}{\partial y} + \frac{\partial v'}{\partial x}\right) \qquad (5.3.17)$$

where u' and v' are the components of additional flow produced by the spheres of all layers together. But u' and v' are independent of x, so $\partial v'/\partial x'$ is identically zero. The additional shear at $y = y_1$ produced by a single sphere at (x_s, y, z_s) obtained by differentiating Eq. (5.3.10) is

$$\tau_1' = \frac{5}{2} \, \mu \dot{\gamma}_0 a^3 \left(\frac{x_s^2}{r_s^5} - \frac{5x_s^2(y_1 - y)^2}{r_s^7} \right) \tag{5.3.18}$$

Integrating over all the spheres of the same layer and then over all the layers, we see that the resultant value vanishes. Therefore, the shear stress at the planes $y = y_1$ and $y = -y_2$ is the same as before the introduction of the particles, and the only effect of the perturbation flow from all the particles is to produce a decrease of the relative horizontal velocity in these planes.

Let us now take the planes y_1 and $-y_2$ to be the walls of the Couette viscometer so that $y_1 + y_2 = h$ and U is the relative velocity of the walls. When the viscometer is filled with pure liquid, the shear stress τ_0 is given by $\mu U_0/h$, the subscript 0 referring to the particle-free state. However, with particles in the fluid the planes move with the new relative velocity

$$U = h\dot{\gamma}_0(1 - 2.5\phi) = U_0(1 - 2.5\phi) \tag{5.3.19}$$

But by definition the shear stress at the wall is

$$\tau = \eta \, \frac{U}{h} = \eta \, \frac{U_0}{h} \, (1 - 2.5\phi) \tag{5.3.20}$$

where η is the *apparent viscosity* or *effective viscosity*. However, $\tau = \tau_0$ so

$$\mu \, \frac{U_0}{h} = \eta \, \frac{U_0}{h} \, (1 - 2.5\phi) \tag{5.3.21}$$

or

$$\mu = \eta(1 - 2.5\phi) \tag{5.3.22}$$

The volume fraction of particles is small because the solution is very dilute, so we may expand $(1 - 2.5\phi)^{-1}$ and write

$$\eta_r = \frac{\eta}{\mu} = 1 + 2.5\phi \tag{5.3.23}$$

where η_r is the *relative viscosity*. This is the much heralded formula obtained by Einstein for the viscosity of a dilute suspension. It has been widely verified experimentally (Hiemenz 1986). The Einstein formula is usually held to be valid for volume fractions $\phi < 0.02$ of suspensions of particles that can be approximated by hard spheres, although it is frequently applied up to $\phi \sim 0.1$.

An analogous result was subsequently obtained by Taylor (1932) for fluid drops having an internal viscosity μ_{in} different from μ and held spherical by surface tension. In this case the formula for the relative viscosity is

$$\eta_r = \frac{\eta}{\mu} = 1 + \left(\frac{2.5\mu_{in} + \mu}{\mu_{in} + \mu} \right) \phi \qquad (5.3.24)$$

so for a gas the factor 2.5 in Einstein's equation is replaced by 1.

Einstein's result is remarkable, since it says that for uniform shear the relative viscosity does not depend on the size or size distribution of the spheres but only on the volume fraction, provided the solution is very dilute. A physical explanation for this follows from the diluteness criterion, which may be restated as the interparticle distance being large enough that the motion of any particle is unaffected by that of any neighboring particles. As a result, the increased energy dissipation arising from the presence of the particles must be proportional to the particle number density. Therefore the relative viscosity is simply linear in the particle volume fraction.

One effect neglected in the calculation is the interaction of the particles with the wall; however, this can be shown to be negligible, provided $a/h \ll 1$. A second neglected effect is Brownian motion, which introduces a diffusive flux in addition to the convective viscous flux. So long as the solution is very dilute and the dispersed particles are rigid spheres, the Brownian motion will not alter the mean angular velocity $\dot{\gamma}/2$, and the Einstein result is unchanged. Although the translational Brownian motion does act on the particle microstructure in trying to uniformize the relative positions of the particles, the relative viscosity is unaffected, since any particle is still unaware of any other particle. The rotational Brownian motion plays no role because of the isotropic behavior of the spherical particles.

If the particles are not spherical, even in the very dilute limit where the translational Brownian motion would still be unimportant, rotational Brownian motion would come into play. This is a consequence of the fact that the rotational motion imparts to the particles a random orientation distribution, whereas in shear-dominated flows nonspherical particles tend toward preferred orientations. Since the excess energy dissipation by an *individual* anisotropic particle depends on its orientation with respect to the flow field, the suspension viscosity must be affected by the relative importance of rotational Brownian forces to viscous forces, although it should still vary linearly with particle volume fraction.

A measure of the importance of Brownian motion is given by the ratio of the Brownian diffusion time to the convection time. The diffusion time may be interpreted as the time taken for a particle to diffuse a distance equal to its radius, which is the characteristic time given by the reciprocal of D_{rot}. This time characterizes the time taken for the restoration of the equilibrium microstructure from a disturbance caused, for example, by viscous convection. The characteristic convection time is simply given by the reciprocal of the shear rate. We denote the ratio of these two times by the Peclet number symbol, since they measure viscous convection to Brownian diffusion, and we write

$$\text{Pe} = \frac{\mu \dot{\gamma} a^3}{kT} \qquad (5.3.25)$$

5.4   Sedimentation under Gravity

In this section we examine small particles, either solid or fluid, falling (or rising) freely under gravity in a liquid. When the particles are falling, the process is termed *sedimentation,* and when particles are rising, *flotation.* In the former case the particle density is greater than that of the liquid, and in the latter case it is less. We shall generally be concerned with sedimentation, which is used for the separation of dispersed particles from the carrier liquid, for the separation of polydisperse particles in solution according to their size, and for the determination of particle mass. The particle mass is assumed large enough that mass diffusion may be neglected.

For a particle falling freely under gravity, the net force acting on the particle is the difference between the gravitational force and buoyancy force. In Cartesian tensor notation the net force can be written

$$(F_i)_{\text{net}} = (\rho - \rho_{\text{fl}}) V g_i \tag{5.4.1}$$

where ρ = particle density
ρ_{fl} = fluid density
V = particle volume

The details of the particle shape are irrelevant, and it does not matter whether the particle continuously turns over and changes in orientation relative to the direction of gravity or whether it moves on a path that is not vertical. For $\rho > \rho_{\text{fl}}$ the force will pull the particle down (sedimentation), and for $\rho < \rho_{\text{fl}}$ the particle will move up (flotation).

As the particle velocity in a free-fall (or rise) increases, the viscous drag opposing the motion will also increase. For small particles, a steady terminal fall speed is reached very rapidly, in a time of the order of the viscous relaxation time. For a sphere of radius a this time is about a^2/ν for a density difference of $O(1)$. If the suspending liquid is water, $\nu \approx 10^{-6} \, \text{m}^2 \, \text{s}^{-1}$, so even for 100-$\mu$m particles this time is exceedingly short. With small particles the Reynolds numbers are generally sufficiently small that we may neglect inertia and, from Eq. (5.1.3), write for the steady drag force acting on the particle

$$(F_i)_{\text{visc}} = -f_{ij} U_j \tag{5.4.2}$$

Here, U_j is the terminal velocity and f_{ij} is the translational friction tensor; that is, the force depends on the orientation for a particle of arbitrary shape.

With $(F_i)_{\text{net}} = -(F_i)_{\text{visc}}$,

$$f_{ij} U_j = (\rho - \rho_{\text{fl}}) V g_i \tag{5.4.3}$$

or

$$f_{ij} U_j = m\left(1 - \frac{\rho_{\text{fl}}}{\rho}\right) g_i \tag{5.4.4}$$

where m is the particle mass. When the translational friction tensor can be replaced by a mean value $\bar{f}$, it follows from Eq. (5.4.4) that the viscous relaxation time $m/\bar{f}$ for the particle can be determined from a measurement of its fall speed. Note that m alone cannot be determined. This situation is comparable to the classical experiment of J.J. Thomson in which the charge-to-mass ratio of an electron could be determined, but neither by itself.

For a sphere the translational friction coefficient is independent of orientation, and the viscous drag force for a rigid particle of radius a is given by Stokes' drag law, Eq. (5.1.5), with the result

$$6\pi\mu a\mathbf{U} = \tfrac{4}{3}\pi a^3(\rho - \rho_{fl})\mathbf{g} \tag{5.4.5}$$

or

$$\mathbf{U} = \frac{2}{9}\frac{a^2}{\nu}\left(\frac{\rho}{\rho_{fl}} - 1\right)\mathbf{g} \tag{5.4.6}$$

This shows that the terminal velocity decreases as the square of the decrease in particle size and linearly with a decrease in the density difference. The corresponding Reynolds number based on the particle radius is

$$\mathrm{Re} = \frac{Ua}{\nu} = \frac{2}{9}\frac{a^3}{\nu^2}\left(\frac{\rho}{\rho_{fl}} - 1\right)g \tag{5.4.7}$$

For a density ratio of 2 the Reynolds number will be less than 1 for spherical particles with radii less than about 75 μm. We are therefore generally justified in neglecting inertia effects for the particle range of interest.

For a liquid drop held spherical by surface tension, the terminal speed from Eq. (5.1.4) is given by

$$\mathbf{U} = \frac{1}{3}\frac{a^2}{\nu}\frac{\mu + \mu_{in}}{\mu + \tfrac{3}{2}\mu_{in}}\left(\frac{\rho}{\rho_{fl}} - 1\right)\mathbf{g} \tag{5.4.8}$$

where μ_{in} is the viscosity of the fluid sphere. The result of Eq. (5.4.6) is recovered with $\mu_{in}/\mu \to \infty$. For a rising spherical gas bubble, where $\mu_{in}/\mu \to 0$ and $\rho/\rho_{fl} \to 0$, the flotation terminal speed is simply $\tfrac{1}{3}a^2g/\nu$.

Let us consider now a container with an initially homogeneous suspension of particles denser than the liquid with no specification at this stage on the degree of diluteness (Fig. 5.4.1A). If allowed to stand, the particles will settle to the bottom of the container, and at some later time a discrete boundary will be seen separating clarified liquid at the top from the suspension (Fig. 5.4.1B). This boundary will be moving downward. A second discrete boundary will be seen separating the sedimented particles at the bottom from the suspension, and this boundary will be moving upward. After a long enough time all the particles will have sedimented, and an equilibrium state will be reached, as shown in Fig. 5.4.1C.

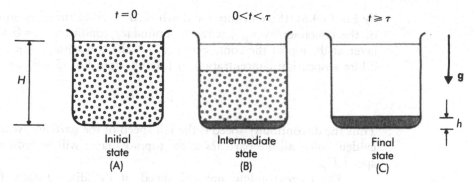

Figure 5.4.1 Batch sedimentation.

The discontinuities diagrammed in Fig. 5.4.1 are termed *kinematic shocks* in that they represent discontinuities in density. Let us calculate the speed at which the top discontinuity moves down and the bottom one up. For specificity consider a downward-moving shock. With respect to a coordinate system moving down with the speed of the discontinuity u (Fig. 5.4.2A), the flow is steady and conservation of mass for the one-dimensional picture considered gives

$$\rho_1(U_1 - u) = \rho_2(U_2 - u) \qquad (5.4.9)$$

Here, ρ is the particle concentration, with the subscript 1 denoting conditions above the discontinuity, and 2 denoting those below. The speed of the shock is therefore

$$u = \frac{j_2 - j_1}{\rho_2 - \rho_1} \qquad (5.4.10)$$

where j is the particle flux passing downward by gravity alone.

In the case of a dilute suspension $U_1 = U_2 = U_0$, where U_0 is the infinitely dilute suspension, particle fall speed, which for the case of rigid spheres is given

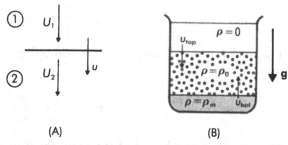

(A)   (B)

Figure 5.4.2 (A) Kinematic shock; (B) boundary conditions for kinematic shocks in batch sedimentation.

by Eq. (5.4.6). If we denote the downward speed of the discontinuity at the top of the container by u_{top}, with the boundary condition $\rho_1 = 0$ for the clarified layer at the top of the container (Fig. 5.4.2B) and with $\rho_2 = \rho_0$, where ρ_0 is the dilute suspension concentration, it follows from Eq. (5.4.9) or (5.4.10) that

$$u_{top} = U_0 \qquad (5.4.11)$$

Thus the discontinuity speed is the fall speed of the particles, which is physically evident, since all the particles at the topmost layer will be sedimenting with the speed U_0.

The corresponding upward speed of the discontinuity front from the bottom is obtained from the conditions that $\rho_1 = \rho_0$ and $\rho_2 = \rho_m$, where ρ_m is the maximum concentration of the particles in the sedimented layer. The boundary condition at the bottom is that $j_2 = 0$, since there is no flux of the sedimented layer, whence

$$-u_{bot} = \frac{0 - \rho_0 U_0}{\rho_m - \rho_0} \qquad (5.4.12)$$

The negative sign on u_{bot} indicates upward movement, the convention having been adopted that velocities are positive in the direction of **g**. We may therefore write

$$u_{bot} = \frac{\rho_0 U_0}{\rho_m - \rho_0} = \left(\frac{\rho_0}{\rho_m - \rho_0}\right) u_{top} \qquad (5.4.13)$$

In other words, the speed with which the front moves up from the bottom is approximately the density ratio of the sedimenting to sedimented particles (ρ_0/ρ_m) multiplied by the dilute particle fall speed.

We can conveniently diagram the interface positions in the position-time $(x\text{-}t)$ diagram of Fig. 5.4.3, since

$$x_{top} = u_{top} t \qquad (5.4.14a)$$

and

$$x_{bot} = H - u_{bot} t \qquad (5.4.14b)$$

The time τ at which equilibrium is reached is given by equating Eqs. (5.4.14), from which

$$\tau = \frac{H}{U_0} \left(1 - \frac{\rho_0}{\rho_m}\right) \qquad (5.4.15)$$

The corresponding height of the sedimented layer is

$$h = H - u_{top} \tau = H \frac{\rho_0}{\rho_m} \qquad (5.4.16)$$

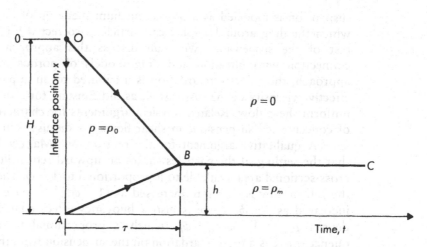

Figure 5.4.3 Interface positions as a function of time in batch sedimentation of an infinitely dilute suspension.

In the x-t diagram of Fig. 5.4.3, below ABC the concentration has its maximum value ρ_m, above OBC it is zero, and in the region OBA it is ρ_0.

The above equations can be used to deduce the properties of the suspension from observations of the front speeds, typically the one separating the clarified layer from the suspension. For example, knowing the fall speed (Eq. 5.4.6), we can determine the effective particle size if the particle density has been found independently. The extension of the results to infinitely dilute systems containing particles of two or more sizes (*polydisperse systems*) is straightforward and will not be discussed further here. It may only be mentioned that with different fall speeds there will be as many distinct downward-moving fronts as there are particle sizes. From measurements of these front speeds the particle sizes can be determined as for the monodisperse system.

Of interest in gravity sedimentation is what takes place when the concentration of particles is large enough that the particles no longer settle as individual entities. This is termed *hindered settling*, and for monodisperse spherical particles the hindered settling regime usually obtains for volume fractions $\phi > 0.15$. In hindered settling the particles tend to descend as a whole, with fall velocity

$$U = U_0 G(\phi) \tag{5.4.17}$$

where U_0 is the terminal fall velocity of an isolated particle (Eq. 5.4.6) and the function $G(\phi)$ is a positive quantity less than unity, termed the *hindered settling factor*. Physically this relation states that a sedimenting particle experiences a retardation in velocity arising from the presence of the other particles and that this retardation is dependent only on the local particle concentration.

Two approaches have been offered to explain this hindered settling behavior (Mandersloot et al. 1986, Davis & Acrivos 1985). In one, the suspension is modeled as a porous medium made up of individual particles in which the drag around a spherical particle is corrected for the presence of the rest of the suspension. We shall discuss this approach in Section 8.5 in connection with filtration and drag models of porous media. In the other approach, the velocity retardation is attributed to an apparent increase in the effective viscosity of the suspension, as did Einstein for a dilute suspension in a uniform shear flow. Related physical arguments for characterizing the viscosity of concentrated suspensions in shear flows are discussed in Chapter 9.

A qualitative argument for the retardation (Mandersloot et al. 1986) is that the settling of the particles causes an upward return flow that has a mean cross-sectional area available to it proportional to $1 - \phi$. The relative velocity of the return flow is therefore increased as $(1 - \phi)^{-1}$, whence the particle drag is increased as $1 - \phi$. There is also a buoyancy effect related to the suspension density $\rho_{sus} = (1 - \phi)\rho_{fl} + \phi\rho$, which is proportional to $1 - \phi$. As a consequence, there is a total retardation on the suspension from the two effects that is proportional to $(1 - \phi)^2$. This argument does not pretend to take into account the suspension microstructure, which is known to be important.

A commonly used empirical function for $G(\phi)$, representative of the behavior of hindered settling data over a wide range of concentrations, is generally expressed in the form

$$G(\phi) = (1 - \phi)^n \qquad (5.4.18)$$

where $n = 4.7$, when wall effects may be neglected. This type of relation is often referred to as the *Richardson-Zaki correlation*.

Kynch (1952), in his treatment of sedimenting systems, employed the general expression of Eq. (5.4.17). The consequence of this assumption is that the settling process is determined entirely from a continuity equation without the need to know the details of the forces acting on the particles; that is, the problem is purely kinematic.

Following Kynch, we ask, what would be the consequences of a concentration-dependent fall speed on the batch settling of a suspension? With hindered settling the downward movement of particles will cause the particle density to increase at the bottom of the container, and this concentration change must propagate itself upward because the particles entering the higher-density region settle slower. The adjustment in concentration can be described as a series of small discontinuities in density propagated through the fluid. These discontinuities are termed *kinematic waves*. We have previously introduced the concept of a kinematic shock wave, the speed of which is given by Eq. (5.4.10). However, if instead of a finite change in particle concentration we consider only an infinitesimal change appropriate to a continuously varying density function, then the speed u of a kinematic discontinuity propagating *upward* is given from the limit of Eq. (5.4.10) as

$$u(\rho) = -\frac{dj}{d\rho} \qquad (5.4.19)$$

with j the downward particle flux. This is then the velocity of a discontinuity between concentrations ρ and $\rho + d\rho$. Since $j(\rho)$ is usually nonlinear, concentration waves can propagate at different speeds and can interact to form kinematic shocks.

Equation (5.4.19) can be interpreted by noting that in an x-t diagram a line of constant density will describe the motion of a boundary between a suspension of density ρ and one of $\rho + d\rho$. Now the line of constant density is defined by

$$d\rho = \frac{\partial \rho}{\partial x}\, dx + \frac{\partial \rho}{\partial t}\, dt = 0 \qquad (5.4.20)$$

This line is termed a *characteristic*, which is a line in the x-t plane on which an ordinary differential equation may be written, in this case $d\rho = 0$. From Eqs. (5.4.19), (5.4.20), and the mass conservation relation $\partial \rho / \partial t = -\partial j / \partial x$, it follows that the slope of the upward-propagating characteristic curve, or *characteristic speed*, is

$$\frac{dx}{dt} = -u(\rho) \qquad (5.4.21)$$

Because ρ is constant along the characteristic curve (the kinematic wave), u is also constant along the curve, and the characteristic must be a straight line. Therefore, on an x-t diagram one such line passes through every point in the diagram below the top of the dispersion, and in a region where the density is continuous the lines (kinematic waves) do not intersect.

For batch sedimentation with an initially uniform suspension and hindered settling, a *typical but not unique behavior* is shown in the x-t diagram of Fig. 5.4.4. The characteristic lines along which ρ and u are constant are drawn dashed. In the region OBA the concentration is ρ_0 just as in Fig. 5.4.3, and the fall curve is a straight line as there. However, at point B the changes propagated from the bottom have just reached the discontinuity surface and the concentration begins to increase. In the region ABC the concentration varies from ρ_0 to the maximum settled value ρ_m, which is its value in the region below ACD. Above $OBCD$, which is the clarified region, the concentration is everywhere zero.

A problem of some practical importance that arises in connection with *clarifier-thickener systems*, in which continuously sedimented material is continuously withdrawn, is to deduce the behavior of the flux j as a function of ρ with hindered settling. For example, with reference to Fig. 5.4.4, the concentration change that is shown there as being brought about by upward-propagating waves (characteristics) is only true if the flux-density dependence is such that $u = -dj/d\rho > 0$.

To illustrate how the j-ρ curve is deduced from a batch settling experiment, suppose, for example, that a fall curve of the form of OBC in Fig. 5.4.4 is measured. Now the equation of the characteristics in the region ABC is

$$x = H - u(\rho)t \qquad (5.4.22)$$

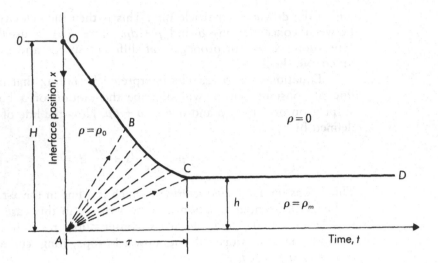

Figure 5.4.4 Interface positions as a function of time in batch sedimentation with hindered settling of the suspension.

The terminal points P of these characteristics are at the portion of the fall curve BC. The speed of fall of the upper discontinuity surface is that of the particles in the topmost layer, whence along BC,

$$\frac{dx}{dt} = U(\rho) \qquad (5.4.23)$$

Now if m_0 is the original mass of the particles per unit cross-sectional area, then after a time t the amount of mass that will have crossed the discontinuity surface will be

$$m_0(H) = \rho(U - u)t \qquad (5.4.24)$$

If U and u are eliminated by using Eqs. (5.4.22) and (5.4.23),

$$\frac{m_0(H)}{\rho} + H = \left(x + t\,\frac{dx}{dt} \right)_P \qquad (5.4.25)$$

From the x-t observations of the fall curve and knowing the initial mass of particles enables us to calculate $\rho = \rho(x, t)$ and, from Eq. (5.4.23), $U(\rho)$. With the data so obtained, we can plot a curve of $j = \rho U(\rho)$ versus ρ. A typical behavior for hindered settling is shown in Fig. 5.4.5. Note that the initially dilute system settles most rapidly, but in the hindered settling regime, as the concentration increases, the settling velocity decreases and approaches zero at very high concentrations. Thus the flux is zero at high and low concentrations and is highest at intermediate concentrations; that is, since the gravitational settling velocity declines with ρ, then $\rho U(\rho)$ passes through a local maximum.

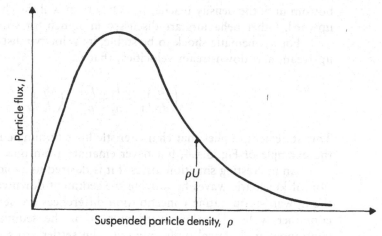

Figure 5.4.5   Flux curve for batch sedimentation with hindered settling.

As we have observed, and as discussed in Kynch's paper, for a constant-density characteristic to propagate upward, $u = -dj/d\rho$ has to be greater than zero, a condition that will not obtain for all $G(\phi)$ and all ρ. Putting $j = U_0\rho G(\phi)$ and with ρ proportional to ϕ, to have $u > 0$ requires

$$G(\phi) + \phi G'(\phi) < 0 \qquad (5.4.26)$$

Consider the simplest $j(\rho)$ curve that is everywhere concave downward (Fig. 5.4.6); that is, the curve has a maximum and no point of inflection. Since Eq. (5.4.26) is equivalent to $dj/d\rho < 0$, there will be no upward propagation of density increase until the density reaches ρ_m. Thus if the density on the bottom starts out at, say, ρ_1 (Fig. 5.4.6), a concentrated layer will build up on the

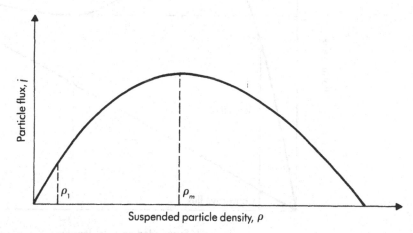

Figure 5.4.6   Simple concave downward flux curve with a maximum and no inflection point for batch sedimentation with hindered settling.

bottom until the density reaches ρ_m. Only then will the characteristics propagate upward. Other behaviors are discussed in Kynch for various j-ρ curves.

For a kinematic shock to be stable, its velocity must be intermediate to the upstream and downstream velocities; that is,

$$\left(\frac{dj}{d\rho}\right)_1 > \frac{j_2 - j_1}{\rho_2 - \rho_1} > \left(\frac{dj}{d\rho}\right)_2 \tag{5.4.27}$$

This statement implies that characteristic lines terminate in a stable shock, as in the example of Fig. 5.4.4, but never emanate from one.

An interesting situation arises if it is desired to avoid the upward propagation of kinematic waves by moving the sediment downward at a rate such that the upward-propagating concentration differences are stationary relative to the container walls. The downward motion of the sediment is obtained by its continuous withdrawal uniformly over the settler cross section. The process is termed *continuous thickening*. The continuous sedimentation process is thus composed of the batch gravitational flux and solid convective flux ρu_s. This is illustrated in Fig. 5.4.7, where the total solids flux curve j_{tot} is the sum of the batch flux and the convective flux; the shape of the curve of Fig. 5.4.5 illustrates the batch flux (Petty 1975).

The total flux-density curve is seen to have a local maximum denoted by the subscript "max" and a local minimum denoted by the subscript "lim."

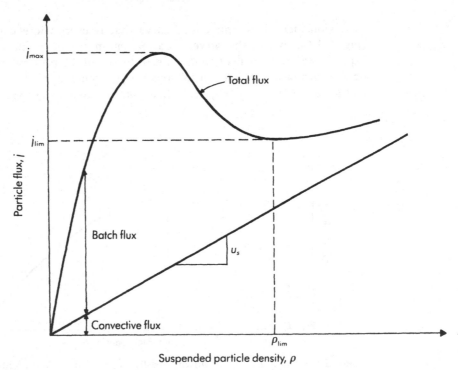

Figure 5.4.7 Flux curves in continuous sedimentation.

Detailed analysis shows that there are two possible types of steady-state solutions. The first is where $j_{tot} < j_{lim}$. The second is where $j_{max} > j_{tot} > j_{lim}$, for which there appear to be three possible solutions for ρ. However, the only steady one is that for which $j_{tot} = j_{lim}$, $\rho = \rho_{lim}$ corresponding to a characteristic speed $\partial j_{tot} / \partial \rho = 0$. As with everything said here and in connection with the Kynch analysis, compressibility of the sediment is not taken into account (Lev et al. 1986).

Finally, we make mention of a settling method that employs the fact that particles when settling in an inclined tube do so faster than if the tube is vertical. This effect is termed the *Boycott effect*, after the physician A.E. Boycott, who in 1920 made this observation in connection with the sedimentation of blood cells in a tube. A picture of batch sedimentation in an inclined channel is shown in Fig. 5.4.8. The clarified liquid, suspension, and sediment are modeled as three distinct regions separated by kinematic shocks, as in vertical settling.

Most early observations of sedimentation in inclined channels indicated that a quasi-steady interface shape between the clarified layer and suspension was formed rapidly in times short compared with the characteristic suspension settling time. Moreover, the clarified liquid layer thickness below the upper channel wall was observed to be much thinner than the channel width b. Most of the clarified fluid accumulated above the horizontal interface at the top of the suspension. It was this kinematic shock interface that was observed to fall with a vertical velocity larger than the hindered settling velocity U measured in vertical settlers.

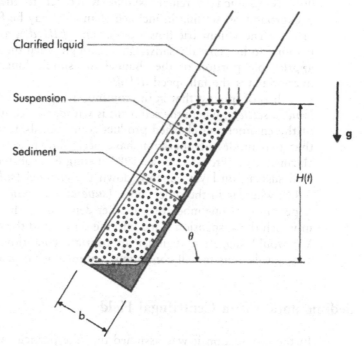

Figure 5.4.8   Batch sedimentation in inclined channel.

Kinematic models of this phenomenon supposed the rise due to buoyancy of the less dense clarified liquid to be balanced by the fall of the more dense suspension. With most of the clarified liquid assumed to accumulate above the suspension, it is a simple matter to estimate the rate of production of clarified fluid. This rate is obtained by setting the interface fall rate multiplied by the horizontal projection of the channel cross-sectional area equal to the vertical settling velocity of the particles multiplied by the sum of the horizontal projection of the area of the channel end plus the horizontal projection of the area of the lower channel wall. The result is

$$\frac{dH}{dt} = -U\left(1 + \frac{H}{b}\cos\theta\right) \tag{5.4.28}$$

where θ is the inclination angle of the channel with respect to the horizontal, and $H(t)$ is the interface height diagrammed in Fig. 5.4.8. The enhanced sedimentation rate was thus seen as a consequence of the fact that the particles could settle not only on the bottom but also on the lower channel wall of the inclined channel. The term $(H/b)\cos\theta$ is the increase in the settling rate over that for vertical settling and represents the contribution to the clarified liquid of the ascending liquid under the upper channel wall. The given relation clearly shows the augmentation in settling rate that can be achieved by decreasing the channel spacing b.

More recently a number of detailed laminar dynamic analyses have been made of the flow in inclined channels with sedimentation. In these studies the flow is assumed to remain stable. Reference to them, along with historical background on settling in inclined channels, may be found in Davis & Acrivos (1985). The volumetric flow rate $(b/\sin\theta)(dH/dt)$, as given from Eq. (5.4.28), is found to be generally satisfactory, though when the clarified layer occupies an appreciable portion of the channel the simple kinematically derived relation overestimates the fall speed dH/dt.

It is of interest that in the continuous operation of inclined settlers, termed *lamella settlers*, where the sediment is withdrawn continuously from the bottom of the channel and clarified product continuously from the top of the channel, that two modes of operation have been shown to exist for the same rate of clarified flow (Probstein et al. 1981, Leung & Probstein 1983). In one mode the feed suspension layer expands down the channel (*subcritical mode*), as in Fig. 5.4.8, whereas in the other mode (*supercritical mode*) the layer contracts. The appearance of one mode or the other depends on the geometry and the manner in which the suspension is fed into the settler and the clarified liquid withdrawn. We would suggest that under appropriate conditions there exist two types of steady solutions for all continuous sedimentation processes.

5.5   Sedimentation in a Centrifugal Field

In the last section it was assumed that the particles were sufficiently large that diffusion was negligible and sedimentation occurred with a sharp boundary. If the particles were very small, diffusion, which is inversely proportional to the particle mass, would dominate and no distinct boundary would be formed. Moreover, since the settling speed decreases as the square of the particle dimension for particles less than about 1 μm, it is not realistic to employ only gravitational force to bring about sedimentation because of the slow settling rate.

For the separation of macromolecules, such as proteins and nucleic aids, the forces can be increased appreciably by subjecting the particles to an angular rotation to produce a large radial force. The magnitude of this centrifugal force is $mr\omega^2$, with r the distance from the axis of rotation, m the particle or molecule mass, and ω the angular speed of rotation. Because the particle is subjected to a continuously increasing force as its distance from the axis of rotation increases, it never reaches a true "terminal speed." This results in a constant drift velocity. In centrifugation the term "sedimentation" refers to this migration of the particles radially from (or toward) the axis of rotation. The resulting diffusional flux is the *pressure diffusion* mentioned in Section 2.4.

Under the assumption that the fluid is in solid body rotation, then in a reference frame rotating at the speed of rotation, the fluid is at rest. If the particles move circumferentially with the fluid, and gravity neglected, then in the rotating frame the only force acting on the particles is the centrifugal force, which acts as an effective centrifugal "gravity." Therefore a particle is pulled radially outward if it is denser than the fluid or radially inward if it is less dense. For the small particles of interest in dilute solution, we may to good approximation apply the Stokes-type drag relation at any given radial distance. The same argument as applied to the derivation of the terminal settling speed for gravitational acceleration can be used for centrifugal acceleration. If we assume a mean friction coefficient $\bar{f}$, the drift velocity of the particles in the direction of increasing radial distance, U_r, is given from a balance between the force due to centrifugal acceleration and the drag force as

$$U_r = \frac{m}{\bar{f}} \omega^2 r \left(1 - \frac{\rho_{fl}}{\rho}\right) \tag{5.5.1}$$

where ρ_{fl} is the fluid solvent density and ρ is the solute particle density.

In the biological literature the particle or molecular solute density is often approximated by the reciprocal of the solute particle partial specific volume $\bar{v} = \partial V/\partial m$, which is exact for very dilute systems. In addition, because the velocity of the molecule is proportional to the magnitude of the centrifugal field, it is common to discuss sedimentation properties in terms of the velocity per unit field (essentially mobility):

$$s = \frac{U_r}{\omega^2 r} = \frac{m(1 - \bar{v}\rho_{fl})}{\bar{f}} \tag{5.5.2}$$

where the *sedimentation coefficient* s is independent of the rotation frequency of the centrifuge. The units of s are seconds, although measurements of s are frequently expressed in svedbergs ($1\,S = 10^{-13}$ s) named after the Nobel prize winner T. Svedberg, the inventor of the ultracentrifuge. This unit is not accepted in SI. Typical values of s in molecular ultracentrifugations range from 1 to 100 S.

Our interest here is largely in molecular and small particle separations, so the discussion that follows will center on ultracentrifuges as distinct from ordinary bowl or disk centrifuges used for continuous large particle separation. The ultracentrifuge is distinguished by very high accelerations. Moreover, they are almost always used in a batch mode, so our considerations will emphasize this type of operation. However, the importance of understanding the hydrodynamically more complex continuous mode should be borne in mind, particularly as it relates to future design. Modern ultracentrifuges operate at speeds up to 70,000 rpm (7300 rad s^{-1}) with accelerations up to 600,000 g's.

There are two principal types of ultracentrifuges, *analytical* and *preparative*. The former are equipped with optical systems that send light through the sample parallel to the rotation axis to determine concentration or concentration gradient distributions at any time during the measurement. In preparative centrifuges the contents are spun for a fixed time period and then removed from the centrifuge (fractionated), the purpose being to prepare or purify biological cells and macromolecules. Samples are held in tubes or sector-shaped "cells." Because of the high rotor speeds, the chamber in which the rotor spins is under high vacuum to minimize frictional heating. For balance the motor drive shaft is flexible so that the rotor can spin about an axis through its center. Figure 5.5.1

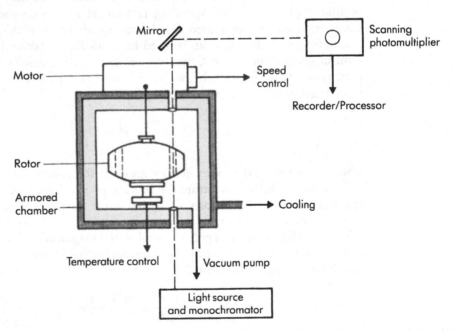

Figure 5.5.1   Schematic of a modern analytical ultracentrifuge with scanning absorption optical system. [After Cantor, C.R. & Schimmel, P.R. 1980. *Biophysical Chemistry. Part II: Techniques for the Study of Biological Structure and Function.* San Francisco: W.H. Freeman. Copyright © 1980 W.H. Freeman and Company. With permission.]

is a schematic of a modern analytical ultracentrifuge with a scanning absorption optical system.

In preparative centrifuges sample volumes of 5×10^{-6} to $10^{-4} \, m^3$ (5 to 100 mL) are used, whereas in analytical centrifuges the sample that is spun in the rotor and monitored is quite small (10^{-7} to $10^{-6} \, m^3$ or 0.1 to 1 mL). In preparative centrifuges the sample cells are cylindrical tubes. On the other hand, in analytical centrifuges the cell is sector-shaped when viewed parallel to the axis of rotation (Fig. 5.5.2). The reason for this design is that in a rectangular cell (equivalent to the cylindrical tubes used in preparative centrifuges), many particles would interact with the side walls since the motion of the sedimenting particles is principally in radial paths. The particles could remain on the side walls or ultimately pile up and settle downward, resulting in mixing. In the sector-shaped cell the solute moves unhindered along radial paths in the direction of the local centrifugal force. Were the cell to diverge more sharply than a sector, the compensating solvent flow from right to left, caused by the sedimenting solute movement from left right, will interact with the cell walls

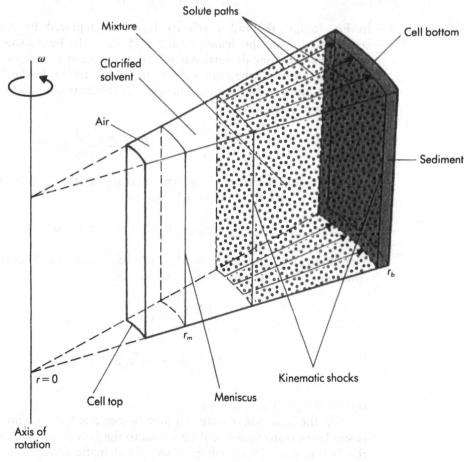

Figure 5.5.2 Sector-shaped ultracentrifuge cell.

and cause stirring. The interaction of either particles or flow in the region of the side walls is further complicated by the presence of boundary layer shear regions there (Greenspan & Ungarish 1985).

Because we are considering sedimentation of macromolecules as well as colloidal particles, in analyzing the kinematic wave fronts that will develop in the cell, we must also consider the broadening of these fronts that takes place due to diffusion. The geometry for analysis of a sector-shaped ultracentrifuge cell is shown in Fig. 5.5.2. The solvent-air interface (the meniscus) is located at r_m, and the bottom of the cell at r_b.

The solvent is assumed to be in solid body rotation at an angular speed ω, and the solute is assumed to move circumferentially with the solvent. A single solute is considered, that is, a binary mixture, and a cylindrical coordinate system rotating with the angular speed ω is adopted. The solute concentration is then a function only of the time t and radial distance r from the rotation axis. The continuity (diffusion) equation (Eq. 3.3.15) can therefore be written

$$\frac{\partial \rho}{\partial t} = \frac{1}{r} \frac{\partial}{\partial r} \left(r \left[D \frac{\partial \rho}{\partial r} - \rho s \omega^2 r \right] \right) \tag{5.5.3}$$

In Eq. (5.5.3) the radial velocity has been replaced by the sedimentation coefficient s, from the definition of Eq. (5.5.2). The fluid dynamicist should be aware that this one-dimensional diffusion equation is known in the ultracentrifuge literature as the *Lamm equation* (Fujita 1975). In the limit of infinitely dilute solutions D and s are independent of concentration and may be taken out of the derivative to give

$$\frac{\partial \rho}{\partial t} = \frac{D}{r} \frac{\partial}{\partial r} \left(r \frac{\partial \rho}{\partial r} \right) - \frac{s \omega^2}{r} \frac{\partial}{\partial r} \left(\rho r^2 \right) \tag{5.5.4}$$

Assuming initially a uniform mixture in the cell with a solute concentration ρ_0, we have the initial condition

$$\rho = \rho_0 \quad r_m < r < r_b \quad t = 0 \tag{5.5.5}$$

The boundary conditions at the meniscus and cell bottom are from the condition of no flux across these interfaces:

$$D \left(\frac{\partial \rho}{\partial r} \right)_{r_m} = s \omega^2 r_m \rho_m \quad t > 0 \tag{5.5.6a}$$

$$D \left(\frac{\partial \rho}{\partial r} \right)_{r_b} = s \omega^2 r_b \rho_b \quad t > 0 \tag{5.5.6b}$$

provided $r_m \neq 0$ and $r_b \neq \infty$.

In the limit where diffusion may be neglected, the sedimentation behavior is similar in many (but not all) respects to the gravity sedimentation discussed in the last section. There will be a sharp kinematic shock wave that propagates through the fluid, separating the clarified liquid (pure solvent) from the mixture of initially uniform particle (solute) concentration. There also will be a wave moving back into the mixture, which separates the sedimented material that accumulates at the bottom of the cell from the suspension layer. An important difference from gravity sedimentation is that the particle concentration in the mixture layer does not remain constant at its initial value but decreases with time. This happens because of a volume increase that results from the area of the cell cross section increasing linearly with the radius and also because the mixture zone thickness increases due to the axial acceleration increasing linearly with radius. The resultant decrease in solute concentration with time is termed *radial dilution*. This effect considerably complicates the simple one-dimensional wave picture presented for gravity sedimentation.

There is another complicating factor pointed out by Greenspan (1983). It has been assumed that the solute particles move circumferentially with the fluid in solid body rotation. However, this is not true in a settling process where there is a lighter clarified solvent and a heavier dense suspension phase separated by a kinematic shock wave. In the separation process the heavier particles move outward, increasing their radial distance so that by conservation of angular momentum they must have a reduced rotational speed. Through drag interaction the reduced rotational speed is transmitted to the suspending solvent, with the consequence that there is a retrograde motion of the heavier phase with respect to the rotating wall.

Conversely, there is a compensating solvent flow inward, and again, since angular momentum must be conserved, the angular speed of this fluid must be increased. This increase will be transmitted to the clarified solvent, resulting in a prograde motion of the lighter phase with respect to the rotating wall. The retrograde motion generally dominates. We mention that by using vorticity arguments instead of momentum considerations, Greenspan (1988) has shown that a negative relative vorticity, equivalent to the retrograde rotation, is always produced in the separation process.

The "slip" effect we have described is important in continuous centrifuge operation, but will not be considered further here. The interested reader is referred to Greenspan & Ungarish (1985).

Let us first consider particles sufficiently large that diffusion may be neglected. For hindered settling the assumption corresponding to Kynch's model that the fall speed depends only on the particle concentration is that $s = s(\rho)$. With these approximations Eq. (5.5.3) reduces to a first-order partial differential equation, which may be written

$$\frac{\partial \rho}{\partial t} + r\omega^2 \frac{ds\rho}{d\rho} \frac{\partial \rho}{\partial r} = -2\omega^2 s\rho \tag{5.5.7}$$

Since the equation is of first order in r, both boundary conditions of Eq. (5.5.6) cannot be satisfied. However, for dilute solutions we may neglect the accumulated material at the bottom of the cell and apply the initial condition (Eq. 5.5.5) and only the boundary condition Eq. (5.5.6a) to the description of the solvent-mixture interface motion and time dependence of the concentration distribution.

As in the gravity sedimentation problem, characteristic solutions also exist for nondiffusive sedimentation in an ultracentrifuge. We recall the definition of a characteristic as a line in the distance-time plane, here the r-t plane, on which an ordinary differential equation may be written. Such an equation must be expressed as a relation connecting total differentials in which partial derivatives do not appear. Since we wish to obtain relations involving total differentials, we write

$$\frac{\partial \rho}{\partial t}\, dt + \frac{\partial \rho}{\partial r}\, dr = d\rho \qquad (5.5.8)$$

Equation (5.5.7) is the second equation connecting the partial derivatives $\partial \rho / \partial t$ and $\partial \rho / \partial r$. Since the characteristic equation must be independent of these derivatives, the determinant of the coefficients of the partial derivatives in the two equations must vanish; that is,

$$\begin{vmatrix} dt & dr \\ 1 & r\omega^2 \dfrac{ds\rho}{d\rho} \end{vmatrix} = 0 \qquad (5.5.9)$$

Expanding the determinant gives

$$\frac{dr}{dt} = r\omega^2 \frac{ds\rho}{d\rho} \qquad (5.5.10)$$

This equation defines the characteristic speed, that is, the kinematic wave speed. Note that for $s = $ constant the characteristic speed $dr/dt = U_r$, which is analogous to the wave speed in the gravity sedimentation problem except that here U_r is *not constant* with distance. The analogy also holds with $s = s(\rho)$ if $s^{-1}d(s\rho)/d\rho$ is defined as a hindered settling factor $G(\phi)$ (Eq. 5.4.18).

Since Eqs. (5.5.7) and (5.5.8) are not homogeneous in the partial derivatives, another condition must be satisfied in order for the solutions to exist. This condition is that the partial derivatives be indeterminate (that is, on a characteristic $\partial \rho / \partial r$ and $\partial \rho / \partial t$ may be discontinuous) so that

$$\begin{vmatrix} dt & d\rho \\ 1 & -2\omega^2 s\rho \end{vmatrix} = 0 \qquad (5.5.11)$$

We therefore may write along the characteristic that

$$d\rho + 2\omega^2 s\rho\, dt = 0 \qquad (5.5.12)$$

where the solute concentration is not constant but is a decreasing function of time.

An explicit integral can be obtained for the characteristic equations (Fujita 1975). We shall write it only for an infinitely dilute mixture where $s = s_0 = $ constant. The solutions of Eqs. (5.5.10) and (5.5.12) are

$$\rho = ce^{-2s_0\omega^2 t} \qquad (5.5.13a)$$

$$\rho r^2 = d \qquad (5.5.13b)$$

where c and d are integration constants. The general solution is therefore

$$\rho = [g(r^2 e^{-2s_0\omega^2 t})]e^{-2s_0\omega^2 t} \qquad (5.5.14)$$

where g is a function of the argument indicated, which is defined by the initial and boundary conditions.

Applying the initial condition Eq. (5.5.5) and boundary condition, Eq. (5.5.6a) with $D = 0$, we obtain the solution

$$\rho = 0 \qquad r_m \le r < r_* \qquad t > 0 \qquad (5.5.15a)$$

$$\rho = \rho_0 e^{-2s_0\omega^2 t} \quad r_* < r \le r_b \qquad t > 0 \qquad (5.5.15b)$$

The solute concentration thus decays to zero exponentially with time. These expressions define a concentration distribution that is represented by a step function, that is, a kinematic shock, where the shock interface $r_*(t)$ moves toward the cell bottom following

$$r_*(t) = r_m e^{s_0\omega^2 t} \qquad (5.5.16)$$

As in gravity sedimentation, the shock arises from the boundary condition requiring the particles at the clarified solvent-mixture interface to sediment with the speed U_r. The solution behavior is illustrated in Fig. 5.5.3, where the mixture-sediment shock interface has also been sketched in.

Equation (5.5.16) may also be written

$$\ln r_* = \ln r_m + s_0\omega^2 t \qquad (5.5.17)$$

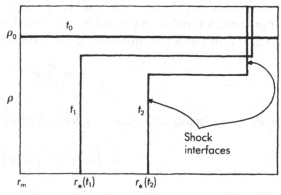

Figure 5.5.3 Solute concentration profiles during ultracentrifugation without diffusion.

which shows that a plot of $\ln r_*$ against $\omega^2 t$ gives a straight line of slope s_0. In this way s_0 can be determined by measuring the motion of the interface boundary r_* as a function of time.

Combining Eqs. (5.5.16) and (5.5.15b) shows that

$$\frac{\rho}{\rho_0} = \frac{r_m^2}{r_*^2} \tag{5.5.18}$$

That is, the solute concentration is diluted inversely as the square of the radial distance from the rotor axis of the solvent-mixture interface. In other words, $\rho r_*^2 = $ constant, which is called the *square-dilution rule* of the sector-shaped cell.

The solution behavior with a concentration-dependent sedimentation coefficient that decreases with increasing ρ (hindered settling) has a similar form as for $s = $ constant, but with Eq. (5.5.15b) replaced by

$$\int_\rho^{\rho_0} \frac{d\rho'}{\rho'[s(\rho')/s_0]} = 2s_0\omega^2 t \qquad r_*(t) < r \leq r_b \quad t > 0 \tag{5.5.19}$$

The interface radial distance $r_*(t)$ satisfies the square-dilution rule (Fujita 1975).

The discussion to this point has centered on the limiting case $D = 0$. In many instances this limit may not be realistic for macromolecules. With $D > 0$, diffusion will cause a broadening of the shock interfaces, which will increase with time, as sketched in Fig. 5.5.4. Also shown there is the concentration gradient $\partial\rho/\partial r$, which is what is commonly measured in an ultracentrifuge with a schlieren optical system. Actually, what is measured with a schlieren system is the gradient of refractive index, which can then be converted to $\partial\rho/\partial r$ when the solution is binary.

An exact analytic solution to the diffusion equation with constant D and s was given in 1938 by W.J. Archibald for the problem posed (Fujita 1975). However, it is sufficiently complex that its use in application to sedimentation experiments is difficult. To simplify the form of the solution, H. Faxén, as far back as 1929, had introduced the approximation of considering the sector cell to be infinitely extended, corresponding to $r_b \to \infty$. We shall outline the solution procedure for that case (Fujita 1975).

The diffusion equation can be thrown into a simpler form by using transformed variables suggested by the nondiffusive solution Eqs. (5.5.15b) and (5.5.18). In particular, introducing the reduced dependent variable

$$\pi = \frac{\rho}{\rho_0} e^{2s\omega^2 t} \tag{5.5.20a}$$

with $s = $ constant, and reduced independent variables

$$\tau = 2s\omega^2 t \qquad \xi = \ln\left(\frac{r}{r_m}\right)^2 \tag{5.5.20b}$$

transforms the diffusion equation (5.5.4) to

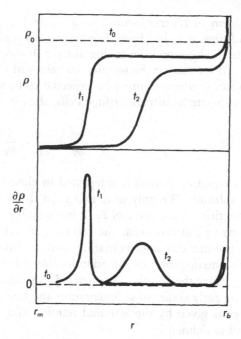

Figure 5.5.4 Solute concentration and concentration gradient profiles during ultracentrifugation of a homogeneous macromolecule with a finite diffusivity. Time scale roughly $t_1 = 3600$ s, $t_2 = 7200$ s for a large protein at 5200 rad s^{-1} (~50,000 rpm). [After Cantor, C.R. & Schimmel, P.R. 1980. *Biophysical Chemistry. Part II: Techniques for the Study of Biological Structure and Function*. San Francisco: W.H. Freeman. Copyright © 1980 W.H. Freeman and Company. With permission.]

$$\frac{\partial \pi}{\partial \tau} + \frac{\partial \pi}{\partial \xi} = \frac{e^{-\xi}}{Pe} \frac{\partial^2 \pi}{\partial \xi^2} \qquad (5.5.21)$$

Here, the Peclet number is defined by

$$Pe = \frac{s\omega^2 r_m^2}{D} = \frac{U_r r_m}{D} \qquad (5.5.22)$$

where the definition $s = U_r/\omega^2 r$ has been used.

The initial condition, Eq. (5.5.5), becomes

$$\pi = 1 \qquad 0 < \xi < \infty \quad \tau = 0 \qquad (5.5.23a)$$

where we have let $r_b \to \infty$ in accordance with Faxén's cell model. The boundary condition at the meniscus, Eq. 5.5.6a, is

$$\pi = \frac{1}{Pe} \frac{\partial \pi}{\partial \xi} \qquad \xi = 0 \quad \tau > 0 \qquad (5.5.23b)$$

This boundary condition is the same one met earlier in our analysis of the problem of reverse osmosis in a channel (Eq. 4.4.9b). Note that the boundary condition at r_b has been dropped and is replaced by the requirement that π be finite in the semi-infinite domain $\xi = 0 \rightarrow \infty$.

If we restrict the solution to values of r close to r_m, that is, to the air-liquid meniscus, where it might be expected to be most accurate and which is actually appropriate to ultracentrifuge cells, then $e^{-\xi} \approx 1$ and

$$\frac{\partial \pi}{\partial \tau} + \frac{\partial \pi}{\partial \xi} = \frac{1}{Pe} \frac{\partial^2 \pi}{\partial \xi^2} \qquad (5.5.24)$$

This equation is readily integrated in closed form, although we shall not write the solution. We only note that even though it is not necessary to impose any restriction on the value of Pe, it is physically necessary that Pe $\gg$ 1; otherwise the effects of diffusion near the cell bottom will extend to the solution region, and the infinite cell approximation will be invalidated. Typical values of Pe for ultracentrifugation of macromolecular solutes range from 10^2 to 10^3.

Since the solution is restricted to values close to the meniscus, this limits it to the early stages of sedimentation and hence small times. The measure of small times is given by the Strouhal number (Eq. 3.5.12), which for the problem at hand is defined by

$$St = \frac{1}{\omega^2 st} = \frac{r_m}{U_r t} \qquad (5.5.25)$$

Small times imply St $\gg$ 1, and for typical values of s for macromolecules of 10^{-13} to 10^{-11} s and with $\omega \sim 5200$ rad s^{-1} ($\sim$50,000 rpm) this parameter will generally be large for sedimentation times measured in hours.

The solution behavior is as shown in Fig. 5.5.4. An important result is that the maximum of the concentration gradient curve follows Eq. (5.5.17). This provides a means of evaluating the sedimentation coefficient s as in the absence of diffusion. The diffusion coefficient itself may be determined from matching the measurements to the theoretical solution.

For a dilute suspension we may use Eq. (5.5.2) and the Einstein relation $D = kT/\bar{f}$ to write

$$s = \frac{m(1 - \bar{v}\rho_{fl})}{\bar{f}} = \frac{mD(1 - \bar{v}\rho_{fl})}{kT} \qquad (5.5.26)$$

where ρ_{fl} is the solvent density. This assumes that the friction coefficients affecting diffusion and sedimentation are the same. With $m = M/N_A$, where M is the molar mass and $R = N_A k$, Eq. (5.5.26) may be rewritten as

$$M = \frac{sRT}{D(1 - \bar{v}\rho_{fl})} \qquad (5.5.27)$$

Equation (5.5.27) is known as the *Svedberg equation*. If the partial specific volume of the molecules have been measured, the Svedberg equation provides a means of determining the molar mass by using the results for s and D derived from sedimentation measurements. Of course, $\bar{f}$ could be estimated from the theoretical solutions discussed earlier, in which case it would not be necessary to determine D. Evidently a number of alternatives are possible.

Keep in mind that the solutions discussed did not consider viscous effects at the walls and interfaces or the retrograde and prograde rotations of the mixture and clarified fluid, respectively. In most instances of batch separation it is not necessary to consider these effects, although they must generally be taken into account for continuous separations (Greenspan & Ungarish 1985).

Before concluding this discussion, we mention another ultracentrifuge technique, *density-gradient centrifugation*. An advantage of this technique can be seen by considering it in conjunction with a procedure used for the separation by ultracentrifugation of a mixture of several components into discrete bands. In this procedure, referred to as *zonal sedimentation* (Cantor & Schimmel 1980), a thin band of a mixture of macromolecular components is layered on top of a large volume of solvent and ultracentrifuged (Fig. 5.5.5). This causes the components to separate into bands of pure components, each traveling at its characteristic sedimentation speed, with diffusion broadening the bands into roughly Gaussian shapes. The problem with the procedure is that in a homogeneous solution the density of the sample would be higher in the band than in the solution below it, and because of gravitational instability, which is magnified by the high g forces in the centrifuge, the bands would collapse.

To eliminate the problem of band collapse, one creates, in one version of density-gradient centrifugation, a stabilizing gradient in density in the solvent by means of an "inert" small molecule, typically a salt, which is added in dilute solution to the original mixture. At equilibrium a continuous increase of salt density is obtained in the direction of increasing centrifugal force. Now if the densities of the macromolecular components lie between the density extremes at the top and bottom of the cell, then the bands will come to equilibrium at

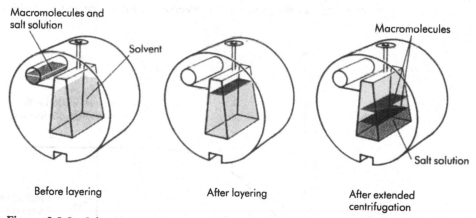

Figure 5.5.5 Schematic of zonal sedimentation in presence of stabilizing density gradient (density-gradient centrifugation). [After Cantor, C.R. & Schimmel, P.R. 1980. *Biophysical Chemistry. Part II: Techniques for the Study of Biological Structure and Function*. San Francisco: W.H. Freeman. Copyright © 1980 W.H. Freeman and Company. With permission.]

different positions along the cell (Fig. 5.5.5). The Watson-Crick hypothesis—that one strand of each daughter DNA molecule is newly synthesized, whereas the other is derived from the parent DNA molecules—was shown experimentally by Meselson & Stahl (1958) by labeling DNA molecules with two different isotopes of nitrogen and separating them with density-gradient centrifugation.

5.6   Ultrafiltration

Ultrafiltration, like reverse osmosis, is a pressure-driven membrane separation process. The applied pressures usually range from about 7×10^4 to 7×10^5 Pa, and the solvent, most often water, passes through the membrane. Material that does not pass through the membrane includes particulate matter, colloids, suspensions, and dissolved macromolecules of molecular weight generally greater than 10,000 and often greater than 2000. Rejection is usually close to complete.

From a theoretical point of view, ultrafiltration is like ordinary filtration except that very small particles are held back by the membrane. The size of the rejected particles depends on the pore size of the membrane. The most common ultrafiltration membranes are of the asymmetric type described in Section 4.4 for reverse osmosis. Practical ultrafiltration rates are in the range of 7 to 35×10^{-6} m s^{-1}, which is about 1/200 of the usual practical filtration rates.

For small applied pressures the solvent flux through the membrane is proportional to the applied pressure. However, as the pressure is increased further, the flux begins to drop below that which would result from a linear flux-pressure behavior (Fig. 5.6.1). For macromolecular solutes this nonlinear behavior may be ascribed to concentration polarization, that is, the buildup of rejected solute at the membrane surface. This buildup increases the local osmotic pressure and leads to a lower effective driving pressure and, hence, lower flux, as was shown for reverse osmosis. For macromolecular solutes the

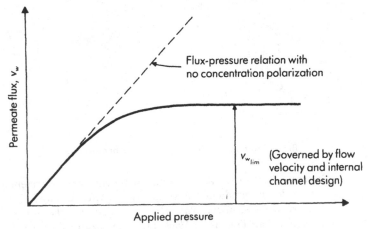

Figure 5.6.1   Schematic of an ultrafiltration flux-pressure excursion curve.

osmotic pressure may have a strong nonlinear dependence on solute concentration, so the osmotic pressure effects can be quite significant when compared with reverse osmosis where the solute diffusivities are 10 to 100 times larger. For particulates and undissolved species the observed flux behavior will be similar due to an apparent fouling of the membrane, although any analysis is considerably more complicated.

For macromolecular solutes at sufficiently high pressures, a limiting flux is reached where any further pressure increase no longer results in an increase in flux (Fig. 5.6.1). Two explanations have been offered. One put forward by Michaels (1968) is that the accumulated solute at the surface reaches a concentration where gelation begins to form a gel layer that offers limiting hydrodynamical resistance, which is large in comparison with the membrane resistance. Thereafter the limiting flux ceases to become pressure dependent and becomes flow, or mass transfer, dependent. Any increase in pressure will cause a transient increase in flux, which results in more solute transport to the membrane. However, since the gel concentration cannot increase, the back diffusive transport away from the membrane is unchanged. Hence, the accumulation of solute at the membrane will just thicken the gel layer and increase the resistance to flow until the flux is reduced to its former value. Applying Darcy's law (Eq. 4.7.7) for the limiting flux through the gel layer gives

$$v_{w_{\text{lim}}} = \frac{k_g}{\delta_g} \frac{\Delta p}{\mu} \tag{5.6.1}$$

where k_g is the gel permeability and δ_g is the gel layer thickness. The quantity δ_g/k_g is the hydrodynamic resistance of the gel layer, and, as stated, any increase in pressure merely results in a corresponding increase in gel layer thickness, so the flux remains essentially independent of pressure. Note that the membrane constant A' defined in Eq. (4.4.4) would in this case be $k_g/\mu\delta_g$. The concept of polymer gelation in which the solute reaches its solubility limit is not always appropriate since a true gel that has a sharp phase boundary and no fluidity may simply not be formed. A highly concentrated macromolecular solution exhibiting similar properties might be more representative in some cases.

A second explanation for a pressure-independent permeate rate could be a strong osmotic pressure dependence on concentration with the osmotic pressure approaching the applied pressure. This description is viable only where the osmotic pressure has meaning. In a system in which solidlike particulates coalesce, the osmotic pressure model would not be a good one.

Both the concentration polarization and osmotic pressure descriptions can be applied to polymer solutions that form well-defined gels at high concentrations. In a gel the thermodynamic osmotic pressure results from the solvent-mediated interactions between the randomly moving gel monomers, and this pressure tends to swell the gel. Both descriptions have been calculated in some detail for gelling macromolecular solutions and shown to produce similar behaviors (Probstein et al. 1979, Trettin and Doshi 1981). Actually a relatively simple argument shows that the two approaches are equivalent if the diffusion coefficient for concentration polarization is appropriately defined in terms of the gel permeability and gel osmotic pressure.

To determine the value of the diffusivity that connects the two approaches, we follow Einstein's thermodynamic arguments given in Section 5.2 for evaluating the translational Brownian diffusion coefficient. The basis for this is the random Brownian motion of the monomer units in the gel, which translates into the gel osmotic pressure. If, as above, the flow through the gel is assumed to follow Darcy's law (Eq. 4.7.7), then we may write the applied hydrodynamic force per mole of solution flowing through the gel as

$$F = -\frac{\mu}{k_g}\frac{U}{c_{soln}} \qquad (5.6.2)$$

Here, ∇p has been replaced by F/V_{soln}, where V_{soln} is the *solution* volume, and the force has been divided by the number of moles of solution, so c_{soln} is the molar concentration of the solution. Here, k_g is the gel permeability and μ is the solution viscosity, not to be confused with chemical potential. This equation is equivalent to the Stokes-type drag law used by Einstein (Eq. 5.2.13).

From the equilibrium balance of the convective and diffusive fluxes,

$$-D\nabla\pi = c\left(\frac{d\pi}{dc}\right)U \qquad (5.6.3)$$

where c is the molar concentration of *solute*, π is the osmotic pressure, and ∇c has been replaced by $\nabla\pi(dc/d\pi)$, the osmotic pressure at constant temperature and pressure being assumed to be a function of the solute concentration only. This can be shown to be a consequence of the *Gibbs-Duhem equation* $\Sigma\, n_i\mu_i = 0$, where n_i is the number of moles of species i and μ_i is the corresponding chemical potential.

Lastly, to ensure equilibrium we assume, following Einstein, that the hydrodynamic force is balanced by a "thermodynamic" force that is identified with the change in chemical potential. In terms of osmotic pressure this change in chemical potential can be shown from thermodynamic arguments to be given by

$$F_{therm} = -\nabla\mu = \bar{V}_{solv}\nabla\pi \qquad (5.6.4)$$

where $\bar{V}_{solv} = \partial V/\partial n_{solv}$ is the partial molar volume of the *solvent* and n_{solv} is the numbers of moles of solvent.

Equating the forces and assuming a moderately dilute solution with the partial molar volume of the solvent replaced by V/n_{solv}, we obtain

$$D = \frac{k_g}{\mu}\,c\left(\frac{d\pi}{dc}\right)\left(1 + \frac{c}{c_{solv}}\right) \qquad (5.6.5)$$

This relation may be interpreted as another form of the Stokes-Einstein equation. It was obtained in a different manner by Wijmans et al. (1985) in the dilute limit, where the term c/c_{solv} did not appear. The importance of the expression is that it shows the equivalency of the gel polarization and osmotic pressure approaches with the appropriate definition of the diffusivity. We note that a more general related treatment of Brownian diffusion with hydrodynamic interaction may be found in Batchelor (1976, 1983).

To illustrate the behavior of the ultrafiltration flux, we here adopt Michaels' model of gel layer formation. As was done for reverse osmosis, let us again consider the geometry of a two-dimensional parallel plate channel with fully developed flow. Moreover, to simplify the presentation, we examine only the limiting-flux problem.

In Michaels' analysis he determined the limiting flux by assuming a steady, one-dimensional thin-film mass transfer model (Nernst layer) in which streamwise convection parallel to the membrane surface is neglected. Assuming the solute to be completely rejected and balancing the limiting convective solute flux against the back diffusion from the solute concentration gradient normal to the surface, we write

$$cv_{w_{\lim}} = D \frac{dc}{dy} \tag{5.6.6}$$

For a constant diffusion coefficient Eq. (5.6.6) can be integrated over the thin film (diffusion layer) of thickness δ_D to give Michaels' result

$$v_{w_{\lim}} = \frac{D}{\delta_D} \ln \frac{c_g}{c_0} \tag{5.6.7}$$

Here, c_0 is the bulk feed concentration and c_g is the solute gelling concentration, which is identified with the wall concentration.

We can use the channel flow solution for constant wall concentration derived in Section 4.3 to estimate the diffusion layer thickness δ_D, since at the wall ($y = 0$) the concentration is constant at the gel value c_g. In particular, from Eq. (4.3.18),

$$\delta_D = 1.475 \left(\frac{2hDx}{3\bar{u}} \right)^{1/3} \tag{5.6.8}$$

where h = channel half-width
$\bar{u}$ = mean flow velocity
x = distance along channel measured from the entrance

The average limiting flux over a length of membrane L is

$$\bar{v}_{w_{\lim}} = \frac{1}{L} \int_0^L v_{w_{\lim}}(x) \, dx \tag{5.6.9}$$

Substituting δ_D into Eq. (5.6.7) and integrating according to Eq. (5.6.9) gives

$$\bar{v}_{w_{\lim}} = 1.16 \left(\frac{\bar{u}D^2}{hL} \right)^{1/3} \ln \frac{c_g}{c_0} \tag{5.6.10}$$

From the above result it can be seen that the gel concentration is a quantity that can be determined by extrapolating data for the limiting flux as a function of bulk concentration to zero values of the flux with all other parameters held fixed. The basis of the extrapolation is that complete gelation of the bulk fluid by definition reduces the flux to zero. To evaluate the gelation concentration at the membrane surface from experiments at finite fluxes, observe from Eq. (5.6.10) that

$$\bar{v}_{w_{\lim}} \left(\frac{b^2}{Q} \right)^{1/3} \sim \ln c_g - \ln c_0 \qquad (5.6.11)$$

where Q is the volume flow rate. If this functional relationship is correct, the limiting flux when plotted against bulk concentration should be linear on a semilog plot. Straight-line extrapolation of the flux data to zero flux then determines the gelation concentration.

Although the parametric dependences of Eq. (5.6.10) have been found to agree with experiment, the quantitative predictions are somewhat less satisfactory. However, these differences can in large measure be accounted for by the variable diffusivity in the concentration polarization layer that is associated with the much higher solute concentration there. From the physical nature of the problem it is evident that the critical region for mass transfer is near the surface. This has been verified by an integral solution of the convective diffusion equation with variable diffusivity, which yielded the same form of result as Eq. (5.6.10) but with the diffusivity evaluated at the gel concentration. This modification has provided much better quantitative agreement with macromolecule ultrafiltration data. Moreover, it has enabled the diffusivity at the gelling concentration to be determined from ultrafiltration flux measurements (Probstein et al. 1979).

5.7   Hydrodynamic Chromatography

Hydrodynamic chromatography is a size exclusion chromatographic procedure used for the size analysis of submicrometer colloidal particles (McHugh 1989). The method is a variation on the molecular fractionation procedure of gel chromatography in which the separation columns are packed with non-porous beads such as styrene-divinylbenzene or glass, rather than porous gels or resins. The largest application of the technique is in the sizing of polymer latexes (Fig. 1.3.4). To illustrate the discussion that follows, neutrally buoyant, force-free spherical particles are examined with nonhydrodynamic surface forces such as charge not considered.

Characterizing the porous bed by means of a capillary model of the interstitial space, the physical basis of the size separation procedure can be demonstrated through examination of the convection and Brownian diffusion of the colloidal particles in a liquid flow through a circular capillary. Figure 5.7.1 shows two freely-rotating spherical Brownian particles of different size sampling a nonuniform Poiseuille flow. The center of the larger particle in its travel

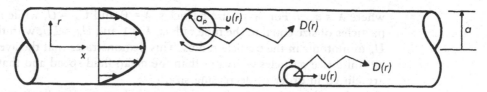

Figure 5.7.1 Convection and diffusion of spherical Brownian particles through a circular capillary.

through the tube samples a smaller volume than does that of the smaller particle because a particle center is unable to approach the tube wall closer than a distance equal to its radius. Thus, the larger a particle, the larger is the volume that is inaccessible. This argument is not modified by the fact that in actuality the particles are always separated from the wall by a thin fluid layer associated with the motion relative to the tube wall of the particles in proximity to the tube wall. The closest approach of a particle center to the wall is therefore a distance somewhat greater than the particle radius.

Because the smaller particle can approach closer to the tube wall, it is able to sample velocities lower than the larger particle and can therefore be expected to have a lower mean speed through the capillary. Moreover, because of the particles' excluded volume, both must have on average a velocity greater than the mean fluid velocity, although the velocity of the larger particle will be greater than that of the smaller one. It follows that particles will be eluted from the capillary in order of decreasing size.

The picture described is that of convective-diffusion of finite size spherical Brownian particles through a circular capillary. In consequence, this may be looked upon as a generalizaton of the Taylor problem for point size particles (Brenner & Edwards 1993). A detailed analysis of this problem based on Brenner's moment analysis method has been carried out by Brenner & Gaydos (1977), taking into account the tube wall effects on the motions of the particles. Neglecting wall interactions, the essential element of the chromatographic technique can be illustrated by a simple calculation for the average velocity of a particle.

From the Taylor–Aris formulation for times $t \gg a^2/D$, where a is the capillary radius and D the Stokes–Einstein diffusion coefficient of the particle, the particle of radius a_p will have had sufficient time to sample the full velocity profile. With the local particle velocity taken to be equal to that of the fluid (Eq. 4.2.14), the average particle velocity over the tube cross-section U_p is given by

$$U_p = \frac{2U}{\pi(a - a_p)^2} \int_0^{a-a_p} \left[1 - \left(\frac{r}{a}\right)^2\right] 2\pi r \, dr \qquad (5.7.1)$$

where U is the mean fluid velocity. On integrating, the expression for U_p can be written

$$U_p = U(1 + 2\lambda - \lambda^2) \qquad (5.7.2)$$

where $\lambda = a_p/a$. For point size particles, $\lambda = 0$ and $U_p = U$, while in the limit of particles of size equal to the tube radius, $\lambda = 1$ and $U_p = 2U$, with the increase in U_p monotonic in the particle radius. This demonstrates that the average speed of all finite size particles is greater than the mean fluid speed and that the particles are eluted in order of decreasing size.

A well known result from the theory of inertia free flows is that the effect of the tube wall on the particle motion through a capillary is to slow the particle down relative to the fluid in the neighborhood of the wall. The mean particle velocity given by Eq. (5.7.2) must therefore be too large. From their moment analysis, Brenner & Gaydos found for small values of λ that to terms of lowest order in λ

$$U_p = U[1 + 2\lambda - \lambda^2 + 3.9\lambda^2 + O(\lambda^3)] \tag{5.7.3}$$

The first three terms are the Poiseuille flow contribution, with the additional terms representing the wall effect. In agreement with the physical picture, Eq. (5.7.3) shows the wall effect not to be of leading order, at least for small λ.

The corresponding modification to the Taylor dispersion coefficient, which takes into account both the excluded volume and wall effects, is given by

$$D_{\text{eff}} = \frac{a^2 U^2}{48D} [1 - 1.9\lambda + 9.7\lambda^2 + O(\lambda^3)] \tag{5.7.4}$$

The finite particle size thus reduces the dispersivity in comparison with the Taylor value. The smaller dispersion coefficient results from the excluded volume which does not allow the particle center to sample the region of highest velocity gradient near the wall, thereby reducing the mean radial diffusion and hence dispersivity. Unlike with the mean particle velocity, the wall effects enter to first order in λ, and reduce the values of all the numerical coefficients of the λ terms in comparison with the values obtained by only accounting for the excluded volume effect.

References

BACHELIER, L. 1900. *Théorie de la Speculation. Ann. Sci. École Normale Superieure, 3.* Paris: Gauthier-Villars. English translation in Cootner, P.H. (ed.) 1964. *The Random Character of Stock Market Prices.* Cambridge, MA: MIT Press.

BATCHELOR, G.K. 1967. *An Introduction to Fluid Dynamics.* Cambridge: Cambridge Univ. Press.

BATCHELOR, G.K. 1972. Sedimentation in a dilute suspension of spheres. *J. Fluid Mech.* **52**, 245–268.

BATCHELOR, G.K. 1976. Developments in microhydrodynamics. In *Theoretical and Applied Mechanics* (ed. W.T. Koiter), pp. 33–55. Amsterdam: North-Holland.

BATCHELOR, G.K. 1976. Brownian diffusion of particles with hydrodynamic interaction. *J. Fluid Mech.* **74**, 1–29.

BATCHELOR, G.K. 1983. Diffusion in a dilute polydisperse system of interacting spheres. *J. Fluid Mech.* **131**, 155–175.

BIRD, R.B., ARMSTRONG, R.C. & HASSAGER, O. 1987. *Dynamics of Polymeric Liquids. vol. 2. Kinetic Theory*, 2nd edn. New York: Wiley.

BRENNER, H. 1974. Rheology of a dilute suspension of axisymmetric Brownian particles. *Int. J. Multiphase Flow* 1, 195–341.

BRENNER, H. & EDWARDS, D.A. 1993. *Macrotransport Processes*. Boston: Butterworth-Heinemann.

BRENNER, H. & GAYDOS, L.J. 1977. The constrained Brownian movement of spherical particles in cylindrical pores of comparable radius: Models of the diffusive and convective transport of solute molecules in membranes and porous media. *J. Colloid Interface Sci.* 58, 312–356.

CANTOR, C.R. & SCHIMMEL, P.R. 1980. *Biophysical Chemistry. Part II: Techniques for the Study of Biological Structure and Function*. San Francisco: W.H. Freeman.

CHANDRASEKHAR, S. 1943. Stochastic problems in physics and astronomy. *Rev. Mod. Phys.* 15, 1–89.

DAVIS, R.H. & ACRIVOS, A. 1985. Sedimentation of noncolloidal particles at low Reynolds numbers. *Ann. Rev. Fluid Mech.* 17, 91–118.

EINSTEIN, A. 1956. *Investigations on the Theory of the Brownian Movement* (ed. R. Fürth). New York: Dover.

FUJITA, H. 1975. *Foundations of Ultracentrifugal Analysis*. New York: Wiley.

GREENSPAN, H.P. 1983. On centrifugal separation of a mixture. *J. Fluid Mech.* 127, 91–101.

GREENSPAN, H.P. 1988. On the vorticity of a rotating mixture. *J. Fluid Mech.* 191, 517–528.

GREENSPAN, H.P. & UNGARISH, M. 1985. On the centrifugal separation of a bulk mixture. *Int. J. Multiphase Flow* 11, 825–835.

HAPPEL, J. & BRENNER, H. 1983. *Low Reynolds Number Hydrodynamics*. The Hague: Martinus Nijhoff.

HIEMENZ, P.C. 1986. *Principles of Colloid and Surface Chemistry*, 2nd edn. New York: Marcel Dekker.

KIM, S. & KARRILA, S.J. 1991. *Microhydrodynamics*. Boston: Butterworth-Heinemann.

KYNCH, G.J. 1952. A theory of sedimentation. *Trans. Faraday Soc.* 48, 166–176.

LANDAU, L.D. & LIFSHITZ, E.M. 1987. *Fluid Mechanics*, 2nd edn. Oxford: Pergamon Press.

LAVENDA, B.H. 1985. Brownian motion. *Sci. Amer.* 252(2), 70–85.

LEAL, L.G. 1992. *Laminar Flow and Convective Transport Processes*. Boston: Butterworth-Heinemann.

LEUNG, W-F. & PROBSTEIN, R.F. 1983. Lamella and tube settlers. 1. Model and operation. *I & EC Process Des. Dev.* 22, 58–67.

LEV, O., RUBIN, E. & SHEINTUCH, M. 1986. Steady state analysis of a continuous clarifier-thickener system. *AIChE J.* 32, 1516–1525.

MANDELBROT, B.B. 1982. *The Fractal Geometry of Nature*. San Francisco: W.H. Freeman.

MANDERSLOOT, W.G.B., SCOTT, K.J. & GEYER, C.P. 1986. Sedimentation in the hindered settling regime. In *Advances in Solid-Liquid Separation* (ed. H.S. Muralidhara), pp. 63–77. Columbus: Battelle.

McHUGH, A.J. 1989. Hydrodynamic chromatography. In *Size Exclusion Chromatography* (eds. B.J. Hunt & S.R. Holding), pp. 248–270. New York: Chapman and Hall.

MERTON, R.C. 1992. *Continuous-Time Finance*, Revised edn. Cambridge, MA: Blackwell.

MESELSON, M. & STAHL, F.W. 1958. The replication of DNA in *Escherichia coli. Proc. Nat. Acad. Sci.* **44**, 671–682.

MICHAELS, A.S. 1968. New separation techniques for the CPI. *Chem. Eng. Prog.* **64**(12), 31–43.

PERRIN, F. 1934. Mouvement Brownien d'un ellipsoide (I). Dispersion diélectrique pour des molécules ellipsoidales. *J. Physique et Radium* **5**, 497–511.

PERRIN, F. 1936. Mouvement Brownien d'un ellipsoide (II). Rotation libre et dépolarisation des fluorescences. Translation et diffusion de molécules ellipsoidales. *J. Physique et Radium* **7**, 1–11.

PERRIN, J.(B.) 1923. *Atoms*, 2nd English edn. London: Constable.

PETTY, C.A. 1975. Continuous sedimentation of a suspension with a nonconvex flux law. *Chem. Eng. Sci.* **30**, 1451–1458.

PROBSTEIN, R.F., LEUNG, W-F. & ALLIANCE, Y. 1979. Determination of diffusivity and gel concentration in macromolecular solutions by ultrafiltration. *J. Phys. Chem.* **83**, 1228–1232.

PROBSTEIN, R.F., YUNG, D. & HICKS, R.E. 1981. A model for lamella settlers. In *Theory, Practice and Process Principles for Physical Separations* (ed. M.P. Freeman & J.A. FitzPatrick), pp. 53–92. New York: Engineering Foundation.

RUSSELL, W.B., SAVILLE, D.A. & SCHOWALTER, W.R. 1989. *Colloidal Dispersions.* Cambridge: Cambridge Univ. Press.

SADRON, CH. 1953. Dilute solutions of impenetrable rigid particles. In *Flow Properties of Disperse Systems* (ed. J.J. Hermans), pp. 131–198. Amsterdam: North-Holland.

TAYLOR, G.I. 1932. The viscosity of a fluid containing small drops of another fluid. *Proc. Roy. Soc.* **A138**, 41–48.

TAYLOR, G.I. 1969. Motion of axisymmetric bodies in viscous fluids. In *Problems of Hydrodynamics and Continuum Mechanics*, pp. 718–724. Philadelphia: SIAM.

TRETTIN, D.R. & DOSHI, M.R. 1981. Pressure-independent ultrafiltration—Is it gel limited or osmotic pressure limited? In *Synthetic Membranes: vol. II Hyper- and Ultrafiltration Uses* (ed. A.F. Turbak), pp. 373–409. ACS Symp. Series 154. Washington: Am. Chem. Soc.

WIJMANS, J.G., NAKAO, S., VAN DEN BERG, J.W.A. & TROELSTRA, F.R. 1985. Hydrodynamic resistance of concentration polarization boundary layers in ultrafiltration. *J. Membrane Sci.* **22**, 117–135.

Problems

5.1 Show that a long slender axisymmetric body falling in a highly viscous fluid at constant speed can never fall along a path inclined at more than 19.5° and that this occurs when the axis of the body lies at an angle to the vertical given by $\cos^{-1}(1/\sqrt{3})$. Suggest how you might employ this fall behavior of long thin bodies for particle characterization when Brownian motions can be neglected.

5.2 To estimate the effect of concentration diffusion, one often uses the relation $t_{\text{diff}} \sim L^2/D$ to estimate the diffusion time, where D is the diffusion coefficient and L is a characteristic diffusion length. Note that there is no concentration term in this equation. However, if there is no concentration gradient, there should not be a diffusive flux and t_{diff} should approach infinity while D and L are finite. What is the source of this contradiction?

5.3 a. Show analytically that in a dilute solution of spherical particles being sheared, the shear stress at any plane is unaffected by the presence of the particles, so the only effect of the perturbation flow caused by the particles is to decrease the relative horizontal velocity in these planes.

b. Knowing the result that the decrease in relative horizontal velocity due to a single plane of particles is independent of the coordinates x, y, z, show what the modified velocity profile would look like if all the particles in the flow were assumed to form an infinite slab of thickness Δ somewhere in the middle of the bounded shear layer. What would the decrease in relative velocity be a function of in this simple picture?

c. Using the simple model of an altered shear velocity profile due to the introduction of an infinite slab of thickness Δ, show that for small particle volume fractions this leads to a functional form for the relative viscosity $\eta = 1 + B\phi$, where B is a constant and ϕ is the particle volume fraction, which may be taken to be proportional to the slab thickness.

5.4 The increased viscosity in a shear flow due to the introduction of particles can be associated with the increased energy dissipation due to the introduced solid boundaries. For a dilute solution the total dissipation with the particles present can be taken to be the energy dissipation per unit time without the particles plus the power dissipation associated with rotating the particles relative to the flow. From this and the fact that for a small volume fraction of spherical particles the perturbation velocity induced by a particle is $u' \sim a^3 \dot{\gamma}/r^2$, where a is the particle radius, r is the radial distance, and $\dot{\gamma}$ is the applied shear rate, show that the functional form for the relative viscosity is given by $\eta = 1 + B\phi$, where B is a constant and ϕ is the particle volume fraction.

5.5 Consider the batch sedimentation of a mixture of particles of two different characteristic sizes but the same density and configuration (concentrations ρ_A and ρ_B). Assume that both size particles pack to a concentration ρ_m.

a. Derive the speeds of the kinematic shock fronts if the solution is very dilute.

b. Sketch the sedimentation x-t diagram.

c. Determine the times and sedimentation layer heights at which the speed of the sedimentation layer front is discontinuous, in terms of the initial mixture height H and the front speeds found in part a.

5.6 A dilute binary mixture is ultracentrifuged at an angular speed ω in a sector-shaped cell. The mixture in the cell is initially uniform with a solute concentration ρ_0.

a. Neglecting all particle interactions and wall effects, obtain an expression for the steady-state solute concentration distribution in terms of the sedimentation coefficient s, diffusivity D, meniscus radius r_m, cell bottom radius r_b, and ρ_0 and ω. This state is referred to as sedimentation equilibrium.

b. What does the expression for the concentration distribution reduce to if $r_b - r_m \ll r_m$?

c.   Rigorously speaking, it takes an infinite time for a mixture to reach sedimentation equilibrium; however, estimate the times in practice at which equilibrium would be attained if the column length $r_b - r_m$ relative to r_m is 10^{-2}, $s = 10^{-13}$ s, and $\omega = 1000$ and 5000 rad s^{-1}.

5.7   The osmotic pressure difference of macromolecular solutions relative to the pure solvent can be represented approximately by $\Delta\pi = ac^n$, where c is the molar concentration, $n > 1$, and a is a constant. This solution is ultrafiltered across a membrane with complete solute rejection. The permeate flux in terms of the applied pressure difference and osmotic pressure difference is assumed to be given by Eq. (4.4.2).

a.   Show that the macromolecular solution exhibits a "limiting-flux-type" behavior at high osmotic pressure by assuming the applicability of a steady, thin-film model in which streamwise convection is neglected and by obtaining an expression for the change in solvent flux with respect to the change in applied pressure difference across the membrane.

b.   Obtain an expression for the limiting permeate flux as a function of the applied pressure difference and the bulk solute concentration.

5.8   The problem is to determine the effect of the excluded volume of particles of radius a_p on the Taylor dispersion coefficient (Eq. 4.6.27). In so doing, note that in the dispersion coefficient, a should be replaced by $a - a_p$, and U should not represent the average translational speed but rather the magnitude of the difference between the largest and smallest velocities in the flow field. The largest particle velocity remains equal to the centerline solvent velocity u_{max}, but due to the excluded volume, the smallest velocity is not zero but is the value of the solvent velocity u' at a distance a_p from the wall. This represents a reduction in the velocity difference and, therefore, in the dispersion coefficient. Accordingly, U should be multiplied by $(u_{max} - u')/u_{max}$. Based on these arguments, show that the excluded volume correction to the Taylor dispersion coefficient is $1 - 6\lambda + 15\lambda^2 + O(\lambda^3)$, where $\lambda = a_p/a$.

6 Solutions of Electrolytes

6.1 The Electrolytic Cell

The fluid mechanical and electrical equations governing the distribution of ion concentration and potential in flowing electrolyte solutions were set down in Section 3.4. Recall that for dilute solutions the ion flow is due to migration in the electric field, diffusion, and convection. For simplicity of presentation the following discussion will be restricted to a dilute binary electrolyte, that is, an unionized solvent and a dilute fully ionized salt.

It was shown in Section 3.4 that if the bulk of a dilute binary electrolyte solution may be assumed electrically neutral, then the distribution of reduced ion concentration is governed by the same convective diffusion equation as for a neutral species with an effective diffusion coefficient related to the difference in charge and diffusion coefficients of the positive and negative ions. Once the concentration distribution has been found, the potential distribution in the solution can be obtained by integrating the equation for current continuity (Eq. 3.4.16) to give

$$\Delta\phi = -\frac{RT}{F^2 z_+ \nu_+ (z_+ D_+ - z_- D_-)} \int \frac{1}{c}\mathbf{i}\cdot d\mathbf{x}$$

$$-\frac{RT(D_+ - D_-)}{F(z_+ D_+ - z_- D_-)} \int \nabla \ln c \cdot d\mathbf{x} \qquad (6.1.1)$$

The potential difference can be seen to be made up of two terms. The first term represents the *ohmic potential* drop due to the flow of current through a medium of given electrical conductance. The second term, called the *diffusion potential* drop, is associated with a region in which there is a concentration gradient (concentration polarization region). This term does not disappear in the absence of a current and is due to unequal rates of diffusion of the charged particles, thus giving rise to a diffusional electric field.

To define a unique solution, we must specify the corresponding boundary and initial conditions. Normally electrolyte solutions are in contact with or bounded by electrodes. An *electrode* in its simplest form is a metal immersed in an electrolyte solution so that it makes contact with it. For example, copper in a solution of cupric sulfate is an example of an electrode. A system consisting of two electrodes forms an *electrochemical cell*. If the cell generates an emf by chemical reactions at the electrodes, it is termed a *galvanic cell*, whereas if an emf is imposed across the electrodes it is an *electrolytic cell* (Fig. 6.1.1). If a current is generated by the imposed emf, the electrochemical or electrolytic process that occurs is known as *electrolysis*. Now whether or not a current flows, the electrolyte can be considered to be neutral except at the solution-electrode interface. There a thin layer, termed a *Debye sheath* or *electric double layer*, forms that is composed predominately of ions of charge opposite to that of the metal electrode. We shall examine this double layer in Section 6.4, but for our purposes here it may be neglected.

By way of example consider the electrolytic cell of Fig. 6.1.1 with two copper electrodes in a solution of cupric sulfate. The cupric sulfate will dissociate into charged cupric ions Cu^{2+} and sulfate ions SO_4^{2-}. When a potential difference is applied between the electrodes, there will be a current flow and reactions at the electrode. The electric field drives the cupric ions (cations) toward the negative electrode (cathode) and the sulfate ions (anions) toward the positive electrode (anode). At the anode there will be a dissolution of copper,

$$Cu \rightarrow Cu^{2+} + 2e^- \qquad (6.1.2)$$

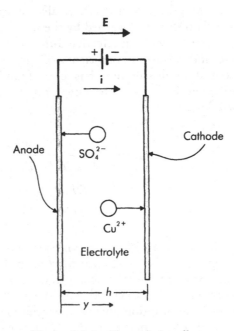

Figure 6.1.1   Electrolytic cell.

while at the cathode there will be a deposition of copper. In general, the electrode at which an electron-producing reaction, such as given by Eq. (6.1.2), takes place is the *anode*, and the electrode at which electrons are consumed is the *cathode*. The cathode is negative in an electrolytic cell, as shown in Fig. 6.1.1, and is positive in a galvanic cell. The current (flow of electrons) in the external circuit is from anode to cathode in both types of electrochemical cell.

In the electrolytic cell of Fig. 6.1.1 the cupric ions and sulfate ions both contribute to the conduction mechanisms, but only the cupric ions enter into the electrode reaction and pass through the electrode-solution interface. The electrode therefore acts like a semipermeable membrane which is permeable to the Cu^{2+} ions but impermeable to the SO_4^{2-} ions. Anions accumulate near the anode and become depleted near the cathode, resulting in concentration gradients in the solution near the electrodes of both ions. This is termed *concentration polarization*, in accord with the meaning of the phrase when applied to neutral species.

As discussed above, concentration gradients will produce a potential drop. Because of the electrode reaction even at equilibrium, that is, with no current flow, there will be a potential drop. The formation of this metal-electrolyte potential difference, which is based on a metal-ion potential, arises from the transfer of metal ions from the metal into the electrolyte, and vice versa. This transfer of metal ions through the electric double layer (here assumed infinitely thin) takes place simultaneously in both directions. The amount of this transference is generally not equal in both directions and gives rise to the metal-electrolyte potential difference. This *electrode potential* is the potential difference that forms at the boundaries of the two phases.

When no current flows in the outer circuit and the metal dissolution is fast in comparison with metal deposition, the metal is charged negatively with respect to the electrolyte. The potential of the metal becomes more negative with respect to the electrolyte. In this way the rate of metal dissolution is retarded, and the rate of metal deposition is accelerated. The potential will become more negative until an *equilibrium potential* $\mathscr{E}$ is reached. This is equivalent to chemical equilibrium with a chemical reaction. In this case the rates of metal dissolution and deposition are equal.

When the potential of the metal is more negative than the equilibrium potential, metal deposition is more rapid than metal dissolution. The entering positive charge from the metal ions shifts the potential to more positive values until the equilibrium potential is reached. At this point both processes occur at equal rates. Each process involves an *exchange current density* which is equal in magnitude and opposite in sign to that of the other, so that the net current is zero. In the absence of an external current, the equilibrium potential $\mathscr{E}$ is the stable limiting value.

In place of the reaction of Eq. (6.1.2), let us consider the general equilibrium electrode reaction

$$Me \rightarrow Me^{z_{+}+} + z_{+}e^{-} \tag{6.1.3}$$

where Me denotes the metal considered.

Now the electrode potential can only be measured against a reference electrode. The standard hydrogen electrode is usually used as the reference electrode, and it is arbitrarily taken to have a potential of zero. The *standard hydrogen electrode* is "platinized" (high surface area) platinum immersed in an acid containing H^+ ions at an activity of $1 \, mol \, dm^{-3}$ and dissolved hydrogen gas at atmospheric pressure (Vetter 1967). The overall electrode reaction is

$$H_2 \rightarrow 2H^+ + 2e^-$$ (6.1.4)

The equilibrium potential of a metal-ion electrode in conjunction with a standard hydrogen electrode to form a complete cell can be shown to be given by (Castellan 1983, Koryta & Dvorak 1987)

$$\mathscr{E} = \mathscr{E}^\circ + \frac{RT}{z_+ F} \ln \frac{a_{Me^{z+}}}{a_{Me}}$$ (6.1.5)

where $a_{Me^{z+}}$ = activity of metal ions in solution
a_{Me} = activity of pure metal
$\mathscr{E}^\circ$ = standard electrode potential of metal-ion electrode

This equation is known as the *Nernst equation*. By convention the activity ratio for both anode and cathode is the ratio of the activity of the oxidized species to that of the reduced species, where

oxidized species + z electrons $\rightleftharpoons$ reduced species

Here, the number of electrons that enters into the reaction is equal to the charge number z_+ or $|z_-|$. Note that what is relevant is the number of electrons that enter into the reaction, and this number may not always be equal to the charge number. For example, the oxidation of ferrous to ferric is represented by $Fe^{2+} \rightarrow Fe^{3+} + e^-$.

When there is a net current through the electrolytic cell, the rates of deposition and dissolution are not equal. As a result, the potential drop at the electrode surface is different from the equilibrium potential. The difference is called the *overpotential*. If the magnitude of the external current density is small compared to the exchange current density, the departure from equilibrium is also small. In this case, the electrode potential is close to the equilibrium value given by the Nernst equation and for practical purposes, the overpotential can be neglected. It should be noted that two or more electrode processes that have different equilibrium potentials may occur independently of each other at the same metal surface (Newman 1991).

The *standard electrode potential* $\mathscr{E}^\circ$ is a characteristic value of a metal and its valence, the solvent, the temperature, and the pressure. Table 6.1.1 lists the standard electrode potentials of some metal-ion electrodes. The least noble metals have the most negative values, and the noblest metals have the most positive values.

Table 6.1.1
Standard Electrode Potentials of Metal-Ion Electrodes at
25°C and Atmospheric Pressure in Aqueous Solution[a]

Electrode	Potential, V
Li/Li^+	−3.04
Na/Na^+	−2.71
Al/Al^{3+}	−1.66
Zn/Zn^{2+}	−0.762
Ni/Ni^{2+}	−0.250[b]
H_2/H^+	0
Cu/Cu^{2+}	0.342
Ag/Ag^+	0.799
Au/Au^{3+}	1.50[b]

[a]Koryta & Dvorak (1987).
[b]Vetter (1967).

The activity of the pure metal may be assumed to be constant, and incorporated in $\mathscr{E}°$. For dilute solutions, where activity may be replaced by concentration, we have

$$\mathscr{E} = \mathscr{E}° + \frac{RT}{z_+ F} \ln(c_+)_{el} \qquad (6.1.6)$$

where the subscript el denotes the ion concentration at the electrode surface. The last term depends on the concentration of metal ions at the interface between the electrolyte and electrode. It arises from the equilibrium requirement of a balance between ions crossing the interface in both directions. Because the standard electrode potentials tabulated in the literature are effectively for a concentration of 1 mol dm^{-3}, the concentration appearing in Eq. (6.1.6) must also be in these units. The characteristic potential RT/F, sometimes referred to as the *Nernst potential*, appears frequently in electrochemical studies and has a value of 25.7 mV at 25°C.

Our considerations have centered on the ion transfer from the bulk of the solution to the electrode surface. As in any heterogeneous reaction, however, it sometimes is necessary to consider the rate of the electrolysis reaction and the rate of product deposition or dissolution at the electrode surface. Generally, the ion transfer step is the slowest, and in what follows we shall assume this to be the case. Discussion of the potential drops associated with finite-rate electrochemical reactions and finite electrode reaction rates may be found in Levich (1962) and Newman (1991).

By way of example, let us consider the simple electrolytic cell of Fig. 6.1.1, containing a motionless dilute binary metal electrolyte. We wish to determine the current-voltage characteristic of the cell, that is, the concentration polarization. To do this, we must calculate the flux of metal ions (cations) arriving at the cathode and depositing on it. As noted above, we assume that the overall rate of the electrode reaction is determined by this flux. Once the cation distribution is known, the potential drop can then be calculated. Note that the anions are effectively motionless and do not produce a current.

The electrodes of the electrolytic cell are taken to be infinite planes at the anode ($y = 0$) and cathode ($y = h$). Since the electrolyte velocity is zero, it follows from the definitions of the current densities $i_{\pm} = Fz_{\pm} j_{\pm}^{*}$ and Eq. (3.4.1) for the molar fluxes that

$$i_+ = -D_+ F z_+ \frac{dc_+}{dy} - \frac{F^2 z_+^2 D_+ c_+}{RT} \frac{d\phi}{dy} \qquad (6.1.7)$$

$$i_- = -D_- F z_- \frac{dc_-}{dy} - \frac{F^2 z_-^2 D_- c_-}{RT} \frac{d\phi}{dy} \qquad (6.1.8)$$

where the mobilities have been replaced by the diffusion coefficients through the relations $v_{\pm} = D_{\pm}/RT$. The current has only a y component, and z_+ and z_- are, respectively, the charge numbers (valences) of the positive and negative ions, with z_+ positive and z_- negative. The condition of electroneutrality is assumed to hold within the fluid; that is,

$$z_+ c_+ + z_- c_- = 0 \qquad (6.1.9)$$

Let us now consider the solution of the current density equations subject to the electroneutrality condition. From the solution the current-voltage characteristic is determined. Our approach will be to first find the concentration distribution and then express the potential drop in terms of this distribution.

In the simple electrolytic cell only the cations deposit on the cathode. The flux of anions, and therefore the anion current at the cathode, must be zero. However, since the electrolyte is motionless, the anion current must be everywhere zero; that is,

$$i_- = 0 \qquad (6.1.10)$$

The current flowing in the cell due to electromigration and diffusion is thus due only to the cation transport. It is the magnitude of this current which we wish to determine for a given value of the applied voltage (or vice versa).

As in Section 3.4, it is convenient to express the basic equations in terms of the reduced ion concentration

$$c = \frac{c_+}{v_+} = \frac{c_-}{v_-} \qquad (6.1.11)$$

where v_+ and v_- are the number of positive and negative ions produced by the dissociation of one molecule of electrolyte. In terms of v the electroneutrality condition becomes

$$z_+ v_+ = -z_- v_- \qquad (6.1.12)$$

while Eqs. (6.1.7) and (6.1.8) may be written

$$i_+ = -D_+ F z_+ \nu_+ \frac{dc}{dy} - \frac{F^2 z_+^2 D_+ \nu_+ c}{RT} \frac{d\phi}{dy} \qquad (6.1.13)$$

$$0 = -D_- F z_- \nu_- \frac{dc}{dy} - \frac{F^2 z_-^2 D_- \nu_- c}{RT} \frac{d\phi}{dy} \qquad (6.1.14)$$

where the anion current has been set to zero.

From Eq. (6.1.14),

$$\frac{d\phi}{dy} = - \frac{RT}{F z_- c} \frac{dc}{dy} \qquad (6.1.15)$$

This equation states that the migration of anions under the prevailing potential gradient is balanced by the diffusion of the anions in the concentration gradient. On using this relation to eliminate $d\phi/dy$ in Eq. (6.1.13) and replacing $z_+ \nu_+ / \nu_-$ by $-z_-$, we find

$$i_+ = -D_+ F \nu_- (z_+ - z_-) \frac{dc}{dy} \qquad (6.1.16)$$

Since there is a fixed current for a given applied voltage, it follows that c must be linear in distance across the cell, y. Now $i_+ > 0$ and $z_+ - z_- > 0$, so $dc/dy < 0$; that is, the concentration of cations decreases from anode to cathode. The potential drop is also in the same direction. From charge neutrality $c_+/c_- =$ constant, so the anion distribution behaves similarly to the cation distribution, which was noted earlier in the discussion of the characteristics of the electrolytic cell.

The concentration distribution is fixed by specifying the reduced ion concentration at the anode:

$$c = c_a \qquad \text{at } y = 0 \qquad (6.1.17)$$

The concentration c_a is not known in advance, but must be determined from the overall conservation of species if we know the initial uniform species concentration and the concentration drop across the cell, which for a given voltage or current is specified by the solution. Integrating Eq. (6.1.16) and applying this boundary condition gives

$$c = c_a - \frac{i_+ y}{D_+ F \nu_- (z_+ - z_-)} \qquad (6.1.18)$$

The concentration at the cathode ($y = h$), denoted by c_c, is

$$c_c = c_a - \frac{i_+ h}{D_+ F \nu_- (z_+ - z_-)} \qquad (6.1.19)$$

When $c_c = 0$, the current density approaches a limiting value given by

$$i_{\text{lim}} = \frac{D_+ F \nu_- (z_+ - z_-) 2c_0}{h} \tag{6.1.20}$$

In this equation, from conservation of species we have used the condition $c_c + c_a = 2c_0$, where c_0 is the initial uniform reduced species concentration. This diffusion limiting current is reached when the cations at the cathode have been completely depleted by the electrode reaction. As we show below, it is attained at sufficiently large values of applied voltage.

To obtain the potential distribution, we substitute the concentration distribution of Eq. (6.1.18) into the integral form of Eq. (6.1.13), which on integration yields

$$\Delta\phi = \frac{RT}{z_+ F} \frac{z_+ - z_-}{z_-} \ln\left(\frac{i_{\text{lim}} - i_+}{i_{\text{lim}} + i_+}\right) - \frac{RT}{z_+ F} \ln \frac{c_a}{c_c} \tag{6.1.21}$$

Here, $\Delta\phi = \phi(y = 0) - \phi(y = h)$, and i_{lim} is defined by Eq. (6.1.20). As discussed earlier, the potential difference between the electrodes is made up of two terms. The first term is the usual ohmic drop due to the flow of current through the electrolyte whose electrical conductivity varies because of the variation in ion concentration across the cell. The second part of the drop, which arises from the concentration gradient term, is associated with the presence of the background immobile anions in equilibrium. It represents a counteracting force to compensate for the gradient in osmotic pressure.

The total difference in potential between the anode and cathode is composed of three parts (Levich 1962). The first part is the potential drop in the fluid where charge neutrality is assumed, which is here given by Eq. (6.1.21). The second part is the difference in equilibrium potential between the anode and cathode, given by the Nernst equation (Eq. 6.1.6). Since at equilibrium the concentration difference term across the cell vanishes, this second potential drop is equal to $\Delta\mathscr{E}°$. However, since for the example considered both electrodes are copper, we also have $\Delta\mathscr{E}° = 0$.

The last part of the total potential difference requires some discussion. This part represents an additional potential drop at the electrodes, called the *concentration overpotential*. It is a consequence of current flow which leads to a lower ion concentration in the solution at the cathode and a higher concentration at the anode. Now at an electrode there is a change in concentration of the ions in solution to the concentration of the adsorbed ions at the surface, a change which is here assumed to be discontinuous but which in actuality takes place over the thin double layer adjacent to the electrode in which charge neutrality does not hold and in which a potential gradient exists. The concentration overpotential is a consequence of the electrochemical reaction, that is, the charge transfer reaction, which occurs at a metal-ion electrode. In physical terms, because of the higher concentration at the anode, in comparison with the equilibrium concentration, it is harder to dissolve a metal ion there, and because of the decreased concentration at the cathode it is harder to plate out an ion there. The value of the concentration overpotential is

$$\eta_{conc} = \frac{RT}{z_+F} \ln \frac{c_a}{c_c} \tag{6.1.22}$$

This formula represents the difference between the logarithm of the anode and cathode concentrations, each of which are referenced to the equilibrium value c_0. The result derives from the Nernst equation under the assumption that the current in the fluid may be considered small in comparison with the equilibrium charge transfer rate that takes place at the electrodes.

The potential boundary condition for the electrolytic cell is therefore

$$V = \Delta\phi + \frac{RT}{z_+F} \ln \frac{c_a}{c_c} \tag{6.1.23}$$

where V is the applied voltage. Substituting the solution for the potential difference $\Delta\phi$, Eq. (6.1.21), and solving for the current density, we find the following result defining the current-voltage characteristic:

$$\frac{i_+}{i_{lim}} = \frac{1 - \exp[z_+z_-FV/(z_+ - z_-)RT]}{1 + \exp[z_+z_-FV/(z_+ - z_-)RT]} \tag{6.1.24}$$

From the current-voltage characteristic it is seen that the current is inversely proportional to the electrode spacing, since $i_{lim} \sim h^{-1}$. At low values of FV/RT the current is linear in the applied voltage, and at sufficiently high values it approaches the limiting current exponentially. This behavior is sketched in Fig. 6.1.2. The ideal electrolytic cell behavior will be modified with a real electrolyte as a consequence of dissociation of the solvent, say water, at sufficiently high voltages. This will result in a plateau and then a subsequent current increase, as sketched in Fig. 6.1.2.

The overall *cell reaction* for water dissociation (*electrolysis of water*) is a "split" one with H^+ ions and O_2 produced at the anode, and OH^- ions and H_2 produced at the cathode

$$H_2O \rightarrow 2H^+ + \frac{1}{2} O_2(g) + 2e^- \tag{6.1.25a}$$

$$2H_2O + 2e^- \rightarrow 2OH^- + H_2(g) \tag{6.1.25b}$$

The cell reaction is the sum of the two electrode reactions. It will take place with the current carried by the H^+ and OH^- ions if the applied voltage exceeds the equilibrium potential difference for the cell given by the Nernst equation. In an electrolytic cell with an aqueous electrolyte, it is usual to employ *inert* electrodes for which the equilibrium potential is such that water electrolysis takes place for the voltages applied rather than metal dissolution.

The example we have considered is without flow, and, as may be recognized from our earlier study of concentration polarization in nonionic solutions, the limiting flux (current density in this case) can be increased by

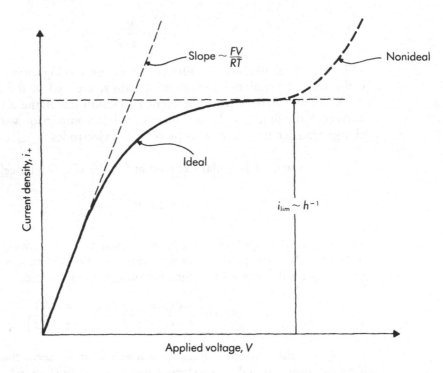

Figure 6.1.2   Current-voltage characteristic for an electrolytic cell ($\Delta \mathscr{E}° = 0$).

having the solution flow parallel to the walls (electrodes). With flow the ions are carried with the solution, and the polarization at the electrodes is thereby reduced. We shall consider hydrodynamic effects in the following section in connection with electrodialysis.

6.2   Electrodialysis

Electrodialysis is a membrane process in which dissolved ions are removed from solution through membranes under the driving force of a dc electric field. Electrodialysis membranes are *ion exchange membranes* and are of two types: cation exchange membranes that essentially allow only cations to pass through, and anion exchange membranes that allow only anions to pass through (Shaffer & Mintz 1980).

In an electrodialysis "stack," a common form of which is shown schematically in Fig. 6.2.1, flat membrane sheets are arranged to form parallel channels. The membranes are arranged so that cation exchange membranes and anion exchange membranes alternate, with electrodes at each end. The ion-containing solution, which for discussion purposes we may take to be a simple salt like sodium chloride in water, flows through the channels. When an electric field is applied transverse to the membranes, cations such as Na^+ pass through the cation exchange membranes and anions such as Cl^- pass through the anion exchange membranes. With reference to Fig. 6.2.1, this reduces the salt content

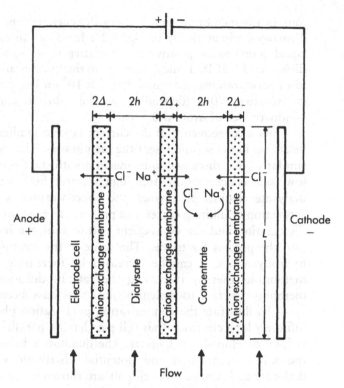

Figure 6.2.1   Electrodialysis cell pair.

in the channel formed by the left pair of membranes, termed the *dialysate channel*, and increases it in the channel formed by the right pair of membranes, termed the *concentrate channel*.

The salt solution is pumped through the dialysate and concentrate channels, with salt removed continuously along the length from the dialysate channel and transferred to the concentrate channel. A dialysate and concentrate channel with the associated membranes are termed a *cell pair*. A typical electrodialysis stack may have 50 to 300 cell pairs between a single pair of electrodes, and a number of stacks may be used in series to achieve the desired level of salt removal.

As the salt content is reduced in the dialysate channels, the fluid conductivity decreases. The resulting potential drop for a given current (salt removal) is minimized by making the channel spacing small. Typical channel widths are about 1 mm with channel lengths of from about 0.25 to 1 m.

The membranes themselves are essentially ion exchange resins which have been made in sheet form. The ion exchange material is an organic polymer in which there is a fixed charge of one sign and a mobile charge of the opposite sign that is free to move in and out of the polymer matrix when in solution. A membrane permeable to cations will contain a high concentration of mobile cations relative to the external concentration. A membrane permeable to anions will contain a relatively high concentration of mobile anions. It is these mobile ions in the membrane which carry the current. The relative concentrations are illustrated schematically in Fig. 6.2.2 for a cation exchange membrane and for equal numbers of positive and negative ions ($\nu_+ = \nu_-$) so that $c_+ = c_-$ (say, dissociated NaCl). Typical membrane thicknesses are around 0.5 mm with fixed ion concentrations of around $2|z|^{-1} \times 10^3$ mol m^{-3} (2 equiv dm^{-3}), selectivities greater than 90% (depending on the solution concentration), and electrical conductivities of around 0.5 S m^{-1}.

As a consequence of the current flow, a gradient in ion concentration will be set up in the solution near the membranes. This concentration polarization is similar to that discussed in connection with the electrolytic cell. It gives rise to a low salt concentration and high electric field near the membranes in the dialysate channel and, when the concentration is low enough, to a current saturation or limiting current. In practice, as mentioned in the last section, water dissociation and the consequent entrance of the hydrogen and hydroxyl ions into the process limits this. The value of the limiting current is affected by the hydrodynamics. It can be increased by increasing the flow velocity past the membrane, thereby increasing the rate of diffusion of salt ions toward the membrane surface to replenish those that have been depleted.

To illustrate the concentration polarization phenomenon, we consider an infinitely long electrodialysis cell pair having parallel channels in which the flow is fully developed and laminar. The qualitative behavior of the development of the salt concentration and potential distributions along the channels of a dialysate and concentrate cell pair are shown schematically in Fig. 6.2.3 for the case where the inlet salt concentrations are the same in both channels (Probstein 1972).

At the point the fluid enters the region where current is allowed to flow, the velocity profile is already fully developed and the ion concentration is uniform. As a result of the nearly uniform concentration close to the inlet, the fluid responds to the applied electric field simply like a medium with constant electrical conductivity; that is, there is a linear potential drop in the fluid as well as the membranes.

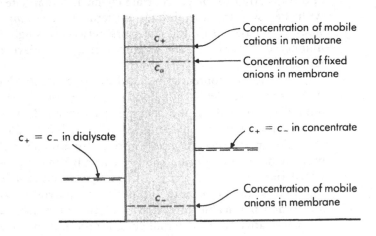

Figure 6.2.2 Cation exchange membrane.

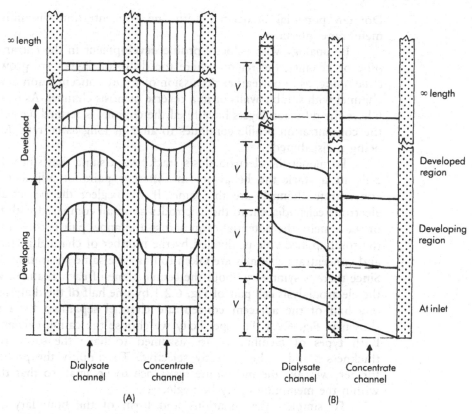

∞ length

Developed

Developing

V

V

V

V

∞ length

Developed region

Developing region

At inlet

Dialysate channel

Concentrate channel

(A)

Dialysate channel

Concentrate channel

(B)

Figure 6.2.3 Development in electrodialysis cell pair of (A) salt concentration, and (B) potential distribution. [After Probstein, R.F. 1972. Desalination: Some fluid mechanical problems. *Trans. ASME J. Basic Eng.* **94**, 286–313. With permission.]

Further downstream, concentration gradients develop adjacent to the membranes as a result of the ion loss or gain (depending on the channel), and these concentration boundary layers grow in thickness. The ion concentration decreases at the membrane surfaces in the dialysate channel and increases in the concentrate channel. As a result of the large concentration gradients, the potential drop in the dialysate channel is larger than that associated with the average conductivity in the channel, with the drop being sharpest at the membrane boundaries. In the concentrate channel the opposite situation prevails. We note also that there is a *Donnan potential drop* at the fluid-membrane boundaries resulting from the concentration discontinuities across the membranes. The Donnan potential drop has the same origin and form as the electrode concentration overpotential.

Eventually, the diffusion layers fill the channels, and thereafter the ion concentration begins to decrease in the center of the dialysate channel and to increase in the concentrate channel. At infinite channel lengths the concentrations in the dialysate and concentrate channels would tend to limiting values corresponding to the total applied potential drop being taken up by the

Donnan potential associated with the concentration discontinuities at the membrane interfaces.

By analogy with velocity profile development in the entrance region of a pipe or channel, the region where the diffusion layers are growing is termed *developing*, as in our earlier discussion of solute concentration development in a channel with soluble walls or in a reverse osmosis channel. As before, the region where the diffusion layers have filled the channel is termed *developed*, although the concentration profile continues to alter as long as current flows and salt is being redistributed.

To model the electrodialysis stack, we assume that since there are many cells in a stack the behaviors in different pairs of adjacent dialysate and concentrate channels are the same. If we neglect the potential drop in the electrode cells adjacent to the electrodes as small compared with that in the rest of the system, the potential drop across a channel pair is constant and equal to the total applied voltage divided by the number of channel pairs. The dialysate and concentrate channels are taken to have the same separation $2h$ (Fig. 6.2.1). Since there is symmetry about the center plane of each channel, we may model the electrodialysis cell pair of Fig. 6.2.1 by one half of the dialysate channel and one half of the adjacent concentrate channel separated by a membrane, as shown in Fig. 6.2.4. For specificity we choose the cation exchange membrane. Both types of membranes are assumed to have the same resistances and thicknesses and to be perfectly selective. To simplify the problem somewhat further, we take the membrane resistance to be small so that the ohmic drop within the membranes may be neglected.

To simplify the equations and form of the boundary conditions, we

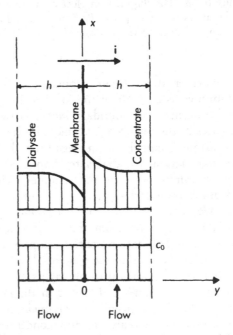

Figure 6.2.4 Electrodialysis cell pair model.

assume that the dilute electrolyte consists of one positive and one negative charged species with equal numbers of positive and negative ions,

$$\nu_+ = \nu_- = \nu \tag{6.2.1}$$

and equal diffusion coefficients,

$$D_+ = D_- = D \tag{6.2.2}$$

These assumptions are introduced simply to minimize the algebra.

A rectangular Cartesian coordinate system is chosen with orientation as shown in Fig. 6.2.4. As in our previous channel flow analyses, the flow is taken to be laminar and fully developed with the same mean velocity U in each channel. The salt concentration at the entrance to each channel is assumed uniform with value

$$c = c_0 \qquad \text{at } x = 0 \tag{6.2.3}$$

Finally, the length in the direction of the flow is large with respect to the channel half-width.

The basic equations for this problem have already been set out in Section 3.4. In particular with

$$c_+ = c_- = c \tag{6.2.4}$$

and with the previously noted assumptions, from Eq. (3.4.14) the equation governing the concentration distribution is

$$u(y) \frac{\partial c}{\partial x} = D \frac{\partial^2 c}{\partial y^2} \tag{6.2.5}$$

Here, $u(y)$ is given by the Poiseuille profile (Eq. 4.2.14). The equation governing the potential distribution follows from current continuity (Eq. 3.4.16), which, with $z_+ = -z_-$, is

$$\frac{\partial \phi}{\partial y} = -\frac{i(y)RT}{2F^2 z_+^2 Dc} \tag{6.2.6}$$

Note that c is the actual molar ion concentration. As with the electrolytic cell, our approach will be to first determine the concentration distribution from Eq. (6.2.5) and then express the potential drop in terms of this distribution through Eq. (6.2.6).

The boundary condition on the concentration at the cation exchange membrane is defined by the condition $i_- = 0$, which is the same condition that applied at the cathode in the electrolytic cell examined in the last section. For the present problem from Eq. (6.1.16),

$$D \frac{\partial c}{\partial y} = - \frac{i}{2 z_+ F} \qquad \text{at } y = 0 \qquad\qquad (6.2.7)$$

At the anion exchange membrane the sign of Eq. (6.2.7) is reversed. Finally, the condition of symmetry at the channel centers is

$$\frac{\partial c}{\partial y} = 0 \qquad \text{at } y = \pm h \qquad\qquad (6.2.8)$$

The system comprising the differential equation for the concentration distribution (Eq. 6.2.5), and the initial and boundary conditions (Eqs. 6.2.3, 6.2.7, and 6.2.8) is essentially the same as that for the concentration profile development in a reverse osmosis channel treated in Section 4.4. There, $i/2z_+F$ of Eq. (6.2.7) is replaced by $v_w c_w$. The current density i is simply $Fz_+ j_+$, so the boundary conditions are indeed essentially the same, whence we may expect the solution behavior to be the same. For the present problem, however, the current is coupled to the equation for the potential, and it is therefore defined by the applied voltage.

For small distances from the channel entrance ($x/h \to 0$), the concentration layer is developing, and, as with the reverse osmosis problem, the solution is self-similar. We therefore choose the same similarity variable as defined by Eq. (4.4.18), namely,

$$\eta = y \left(\frac{3U}{hDx} \right)^{1/3} \qquad\qquad (6.2.9)$$

The choice of the reduced streamwise coordinate ξ is somewhat modified from Eq. (4.4.16) and is here

$$\xi = \left(\frac{z_+ FV}{2RT} \right)^3 \left(\frac{D}{3Uh} \right) \left(\frac{x}{h} \right) \qquad\qquad (6.2.10)$$

where V is the constant applied voltage drop across the channel half-pair. The replacement of the Peclet number $v_w h/D$ by the dimensionless applied voltage drop $V^*/2$, where

$$V^* = \frac{z_+ FV}{RT} \qquad\qquad (6.2.11)$$

is readily derived in the same manner as Eq. (4.4.16).

From the convective diffusion equation the diffusion layer thickness is given in order of magnitude by

$$\delta_D^3 \sim h^3 \left(\frac{D}{3Uh} \right) \left(\frac{x}{h} \right) \qquad\qquad (6.2.12)$$

where we have set $u \sim 3U\delta_D/h$. Now the ion concentration at the membrane (wall) may be estimated from the boundary condition Eq. (6.2.7), with $c_w < c_0$, as

$$\frac{c_0 - c_w}{c_0} \sim \frac{i\delta_D}{2z_+ Fc_0 D} \tag{6.2.13}$$

But the current density i is determined from Eq. (6.2.6) as

$$i \sim \frac{V}{2h} \frac{2F^2 z_+^2 Dc_0}{RT} \tag{6.2.14}$$

Eliminating i and δ_D in Eq. (6.2.13) by using the estimates of Eqs. (6.2.12) and (6.2.14), we find that

$$\frac{c_0 - c_w}{c_0} \sim \xi^{1/3} \tag{6.2.15}$$

with ξ defined by Eq. (6.2.10). As with the reverse osmosis problem, this parameter measures the extent of the concentration polarization. Note that reference here is to the dialysate channel with a positive sign on $\xi^{1/3}$.

The analysis for the ion concentration distribution now follows exactly as in Section 4.4. A function $f(\eta)$ is defined as in Eq. (4.4.17), and the resulting ordinary differential equation is solved exactly as for the reverse osmosis problem. The result for the wall concentration from Eq. (4.4.28) is

$$\frac{c_w}{c_0} = 1 - 1.536 \xi^{1/3} \tag{6.2.16}$$

With the ion concentration so determined the current-voltage characteristic can be obtained by integrating the equation for the potential distribution. Again, as in the case of the electrolytic cell, some care must be exercised with respect to the boundary conditions. In particular, the total potential drop must equal that in the dialysate half-channel, plus that in the concentrate half-channel, plus the Donnan potential drop across the membrane. The Donnan potential drop arises from the discontinuities in concentration at the boundaries of the membranes (in this case, the cation exchange membrane for the half-cell as considered). The origin and expression for the Donnan potential are the same as for the electrode concentration overpotential. For the cation exchange membrane the Donnan potential drop is

$$\eta_{\text{Don}} = -\frac{RT}{z_+ F} \ln \frac{(c_w)_{\text{conc}}}{(c_w)_{\text{dial}}} \tag{6.2.17}$$

It follows that the applied voltage V is therefore

$$V = \Delta\phi - \frac{RT}{z_+ F} \ln \frac{(c_w)_{\text{conc}}}{(c_w)_{\text{dial}}} \tag{6.2.18}$$

where $\Delta\phi$ is the potential drop in the fluid. This drop is given by

$$\Delta\phi = -\frac{iRT}{2F^2 z_+^2 D}\left[\int_{-h}^{0}\frac{dy}{c} + \int_{0}^{+h}\frac{dy}{c}\right] \qquad (6.2.19)$$

which is obtained from integrating Eq. (6.2.6).

Employing the solution for the ion concentration distribution and carrying out the quadratures in Eq. (6.2.19), we find, to consistent order, that

$$i^*\frac{2}{V^*} = 1 - 3.072\frac{\xi^{1/3}}{V^*} \qquad (6.2.20)$$

where

$$i^* = \frac{ih}{2Fz_+ Dc_0} \qquad (6.2.21)$$

The solution as given by Eqs. (6.2.16) and (6.2.20) requires that both $\xi^{1/3}$ and $\xi^{1/3}/V^*$ be small.

It can be seen that there are two limits: (a) $\xi \to 0$ and (b) $V^* \to \infty$. For V^* large the current approaches the limiting value

$$i_{\text{lim}}^* = \frac{V^*}{2} \qquad (6.2.22)$$

and the point at which the concentration at the membrane becomes zero is

$$\xi_{\text{lim}} = (1.536)^{-3} = 0.276 \qquad (6.2.23)$$

Both cases (a) and (b) correspond to x/h small. However, whereas in case (a) the similarity solution covers only a small part of the region of interest, in case (b) the solution is valid up to ξ_{lim} (Solan & Winograd 1969).

6.3 Ion Exchange

Ion exchange is the reversible exchange of ions in a solution with an equivalent amount of ions of the same charge in a solid phase or ion exchanger. There is no exchange of ions of the opposite charge between the solution and the exchanger. The solid ion exchangers may be spherical or irregular granules, are relatively insoluble in water, contain charges fixed within their solid structure and mobile charges which can be exchanged, and are always in a state of electroneutrality. They are normally highly porous synthetic polymeric cation and anion exchange resins. In the last section the ion exchange resin was introduced as the basic element of the electrodialysis membrane. We recall that in a cation exchanger the mobile ions are positive ions (cations), and in an anion exchanger they are negative ions (anions).

A principal use of the process of ion exchange is to remove or concentrate ions in solution, usually inorganic ions. An example is the *softening* of water which involves the removal of calcium ions, which form an insoluble salt with soap (water *hardness*), and replacing them with innocuous ions such as sodium. Another example is the *deionization* of water in which both cations and anions are exchanged, usually by employing a cation exchange resin saturated with hydrogen ions and an anion exchange resin saturated with hydroxyl ions. Ion exchange also has many preparative and analytical uses, one example of which is *ion exchange chromatography* in which charged molecules are adsorbed by the exchanger and then eluted by changing the ionic environment. This is analogous to the size exclusion chromatography discussed in Section 4.7. Because we have already illustrated the chromatographic separation principle, our treatment of ion exchange will focus on its use for the removal (concentration) of some "contaminant" ion and its replacement by an "innocuous" ion.

The conventional ion exchange process for the softening of water may be represented by the chemical equation

$$2\overline{Na^+R^-} + Ca^{2+}Cl_2^- \rightleftharpoons \overline{Ca^{2+}R_2^-} + 2Na^+Cl^- \tag{6.3.1}$$

or simply

$$2\overline{Na^+} + Ca^{2+} \rightleftharpoons \overline{Ca^{2+}} + 2Na^+ \tag{6.3.2}$$

Here, the unbarred quantities are ions in solution and the barred quantities are ions in the resin phase, with R denoting the negative immobile ions. During calcium removal the reaction is to the right. To regenerate "used up" resin, a concentrated salt solution with at least an equivalent amount of sodium must be provided to replace the calcium in the resin. In this case the reaction is to the left. As mentioned, the exchange conserves charge and maintains bulk electrical neutrality within the solution as well as within the exchanger. An electric double layer in which neutrality is not maintained forms around the exchanger particles, but it is small and will be neglected here.

More generally, if A and B are the exchangeable ions, the reversible exchange reaction may be represented by the equation

$$\overline{A} + B \rightleftharpoons \overline{B} + A \tag{6.3.3}$$

where the ion exchanger is initially in the A form and the *counterion* in solution is B. Ion exchange occurs, and the ion A in the exchanger is partially replaced by B. At equilibrium, both the ion exchanger and the solution contain both competing ion species A and B.

The concentration ratio of the two competing species in the ion exchanger is usually different from that in the solution, with the ion exchanger selecting one species in preference to the other. Ion exchange equilibrium is characterized by the *ion exchange isotherm*, which shows at a given temperature the equilibrium ionic composition of the ion exchanger as a function of the ionic composition and concentration of the exterior solution. Usually the *equivalent ionic fraction* of the counterion A (or B) in the ion exchanger is plotted as a function of the equivalent ionic fraction of A (or B) in the solution for a given total concentration of the external solution, different curves corresponding to different concentrations (Helfferich 1962).

The equivalent ionic fraction of A in solution is

$$x_A = \frac{z_A c_A}{z_A c_A + z_B c_B} \tag{6.3.4}$$

where $z_i c_i$ is the concentration in equiv m^{-3} (charge per unit volume in units of F). We recall from Section 2.5 that equivalents are not recognized in SI. However, because equivalents are used in the ion exchange literature and because we are concerned principally with equivalent ionic fractions, which are dimensionless, we retain the usage here. The equivalent ionic fraction may be interpreted as the ratio of the number of electronic charges contributed by A to the total number of electronic charges contributed by the exchangeable ion. The corresponding equivalent ionic fraction of A in the exchanger is

$$\bar{x}_A = \frac{z_A \bar{c}_A}{z_A \bar{c}_A + z_B \bar{c}_B} \tag{6.3.5}$$

We may also write

$$x_B = 1 - x_A \tag{6.3.6}$$

$$\bar{x}_B = 1 - \bar{x}_A \tag{6.3.7}$$

Note that the total concentration of the exterior solution in equiv m^{-3} is

$$C = z_A c_A + z_B c_B \tag{6.3.8}$$

where here and in what follows C is used to denote an equivalent concentration. The *capacity* of the ion exchanger is correspondingly

$$\bar{C} = z_A \bar{c}_A + z_B \bar{c}_B \tag{6.3.9}$$

Since the nonexchangeable neutralizing ions cannot move from one phase to another, then, by charge neutrality during exchange, C and $\bar{C}$ must be constant for a given solution and given exchanger.

In Fig. 6.3.1 are sketched ion exchange equilibrium isotherms for a given exchanger and different external solution concentrations. In a hypothetical system in which the ion exchanger shows no preference for A or B, the equivalent ionic fractions in the ion exchanger are the same as those in the solution and the isotherm is linear, as shown by the dashed diagonal line in Fig. 6.3.1. If the exchange is *favorable* for A—that is, there is a preference for the exchanger to absorb A—the isotherm is concave downward and lies above the diagonal. If the exchange is *unfavorable* for A—that is, there is a preference for the exchanger to absorb B—the isotherm is concave upward and lies below the diagonal.

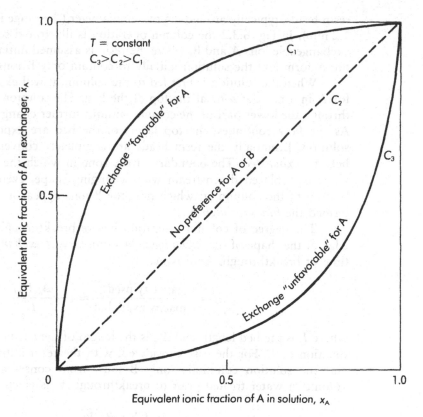

Figure 6.3.1 Ion exchange equilibrium isotherms.

The origin and end point of any isotherm must lie at $(0, 0)$ and $(1, 1)$, respectively, since at equilibrium the absence of one of the ions in solution requires the absence of this ion in the exchanger. It follows that at $(0, 0)$ there is no species A in solution and therefore none in the resin. At $(1, 1)$ there is no species B in solution and therefore none in the resin. In general, the preference of a given exchanger for a specific ion increases with dilution, as indicated by the isotherm behavior sketched.

We may illustrate the use of the isotherm by asking what the equilibrium exchanger concentrations $\bar{c}_A$ and $\bar{c}_B$ would be, given z_A, z_B, c_A, and c_B for the solution. From the formulas $x_A = z_A c_A / C$ and $C = z_A c_A + z_B c_B$, the quantities x_A and C are calculated, whence $\bar{x}_A$ is determined from the isotherm and $\bar{x}_B$ from the definition $\bar{x}_B = 1 - \bar{x}_A$. Finally, from

$$z_A \bar{c}_A = \bar{x}_A \bar{C} \qquad z_B \bar{c}_B = \bar{x}_B \bar{C} \qquad (6.3.10)$$

and the relation $\bar{C} = z_A \bar{c}_A + z_B \bar{c}_B$, two equations are obtained for $\bar{c}_A$ and $\bar{c}_B$.

Normally, the ion exchange process is carried out in columnar operation with the fluid from which the contaminant ion is to be removed introduced at the top of the column and allowed to flow downward under gravity through a packed bed of exchanger. The exchanger is usually in the form of fine polymeric resin beads, typically around 0.5 mm in diameter (the range is from about 0.3 to 1.2 mm). In Fig. 6.3.2 the column operation is illustrated schematically for the exchangeable ions A and B, where the resin is assumed initially to be entirely in the A form and the solution initially to contain only B ions.

When the solution is first fed to the column, it will exchange all its ions B for A in a narrow zone at the top of the bed. The solution with A ions passes through the lower part of the column without further changing its composition. As the flow continues, the top layers of the bed are exposed to fresh B-ion solution. Eventually the resin beads are completely converted to B form and become *exhausted*. The boundary of the zone in which the ion exchange takes place is displaced downstream while changing shape, eventually reaching the bottom of the column, at which time the B ions appear in the effluent. This is termed the *breakthrough* of B.

The degree of column utilization before breakthrough requires a knowledge of the shape of the *exchange zone boundary* or *exchange zone front* at the time of breakthrough. Approximately,

$$\beta = \frac{\text{capacity used}}{\text{capacity available}} \approx 1 - \frac{\Delta_{ex}(L)}{L} \qquad (6.3.11)$$

where L is the bed length and Δ_{ex} is the length of the exchange zone, which is a function of L. For the case considered, with the resin initially in pure A form and the solution containing only B ions, from conservation of species the volume of water treated prior to breakthrough V, is given by

$$CV = \beta \bar{C} \bar{V} \qquad (6.3.12)$$

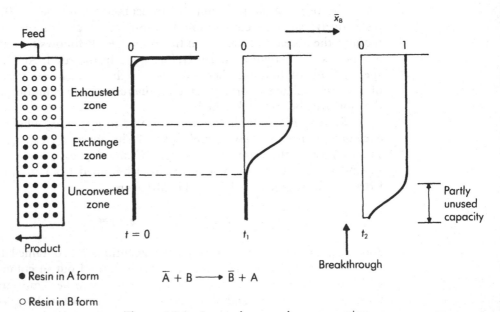

Figure 6.3.2 Ion exchange column operation.

Here, $\bar{V}$ is the resin volume and $\bar{C}$ is its capacity. *Regeneration* of the exhausted resin, by passing a regenerant fluid through the column, is necessary to bring the resin back to its original composition.

The shape of the isotherm and whether the equilibrium is favorable or unfavorable alter the shape of the S-shaped exchange zone front sketched in Fig. 6.3.2. An important problem in ion exchange column operations, both for contaminant ion removal and elution, is the determination of the shape of the front as a function of the isotherm shape and distance along the column. The exchange zone front represents a moving continuity wave or kinematic wave across which there is a change in ion concentration. This wave is analogous to the kinematic waves and shocks studied in connection with gravity sedimentation and centrifugal sedimentation.

In the particular case of an infinitely fast exchange rate with diffusional effects neglected, the exchange zone front is discontinuous; that is, it is a kinematic shock in the sense of Section 5.4 with, for example, the ionic fraction x_B changing discontinuously across the front, which moves downward with speed u_{ex} (Fig. 6.3.3).

The mean interstitial speed of the fluid is

$$U_e = \frac{Q}{\varepsilon A} = \frac{U}{\varepsilon} \qquad (6.3.13)$$

where Q = volume flow rate
A = column cross-sectional area
ε = void fraction
U = superficial speed of the fluid

Here, the solute is assumed to be excluded by the resin. The front speed can be related to the fluid speed by species conservation, assuming that the cross-

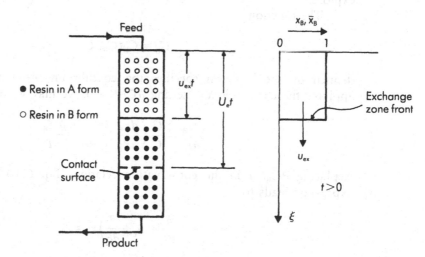

Figure 6.3.3 Discontinuous exchange zone front.

sectional area available to the fluid is εA and that to the resin is $(1 - \varepsilon)A$ (cf. Eq. 4.7.5).

With respect to a coordinate system moving down with the front speed u_{ex}, the solution moves down with speed $U_e - u_{ex}$ and the resin moves up with speed u_{ex}. In this reference frame the flow is steady, and the ion transfer down by the solution must equal the ion transfer up by the resin, whence

$$\varepsilon C(U_e - u_{ex}) = (1 - \varepsilon)\bar{C}u_{ex} \tag{6.3.14}$$

or

$$u_{ex} = \frac{U}{\varepsilon + (1 - \varepsilon)\bar{C}/C} \tag{6.3.15}$$

with $U = \varepsilon U_e$ the superficial velocity.

In the more general case where the exchange front is not discontinuous but diffusion is neglected, from Eq. (3.3.18) we may write the species conservation relation as

$$\varepsilon\left(\frac{\partial C_B}{\partial t} + U_e \frac{\partial C_B}{\partial \xi}\right) = R_B = -(1 - \varepsilon)\frac{\partial \bar{C}_B}{\partial t} \tag{6.3.16}$$

Here, ξ is the coordinate in the direction of flow, and R_B is the rate at which B ions are removed from solution and taken up by the resin. The factor ε on the left side of the equation characterizes the average portion of the fixed control surface through which the fluid flows because of the presence of the resin, and the factor $1 - \varepsilon$ on the right side characterizes the volume of resin contributing to the decrease in B ions from adsorption. Although Eq. (6.3.16) is written here and in the adsorption literature in a cavalier fashion, it should be recognized that there are many subtleties involved which the interested reader may wish to explore.

The function

$$\bar{C}_B = \bar{C}_B(C_B, C) \tag{6.3.17}$$

defined by the isotherm specifies the equilibrium production rate R_B. On applying the chain rule and setting $\partial C/\partial t = 0$, we may write

$$\frac{\partial \bar{C}_B}{\partial t} = \frac{\bar{C}}{C}\left(\frac{d\bar{x}_B}{dx_B}\right)_{C = \text{const}} \frac{\partial C_B}{\partial t} \tag{6.3.18}$$

Replacing $\partial \bar{C}_B/\partial t$ in the conservation relation (Eq. 6.3.16) with the above expression leads to

$$\left(\frac{\partial}{\partial t} + u_{eff}\frac{\partial}{\partial \xi}\right)x_B = 0 \tag{6.3.19}$$

where

$$u_{\text{eff}}(x_B) = \frac{U}{\varepsilon + (1 - \varepsilon)(\bar{C}/C)(d\bar{x}_B/dx_B)_{C=\text{const}}} \tag{6.3.20}$$

The speed u_{eff} is the speed with which a point with ionic fraction x_B in solution moves. In general, Eq. (6.3.19) is a nonlinear hyperbolic equation possessing real characteristic solutions. Note that the characteristic direction is

$$\frac{d\xi}{dt} = u_{\text{eff}} \tag{6.3.21}$$

From Eq. (6.3.20) it can be seen how favorable and unfavorable equilibria result in different exchange front patterns. With reference to Fig. 6.3.4 and given an initial S-shaped front near $\xi = 0$ and two points on the front a and b, if the equilibrium is favorable, then $(d\bar{x}_B/dx_B)_a > (d\bar{x}_B/dx_B)_b$. Thus, from Eq. (6.3.20), $u_{\text{eff}}^a < u_{\text{eff}}^b$. This means that point b catches up with point a and the front tends to steepen. Physically, any ions A behind the front are displaced preferentially by B ions and catch up with the front. The B ions ahead of the front are preferentially retained and have opposite behavior to the A ions. In this case, the front is said to be *self-sharpening* (Helfferich 1962). At large distances down the column the front will approach the discontinuous form of Fig. 6.3.3 with a speed given by Eq. (6.3.15), assuming no diffusion and an infinite exchange rate. In a ξ-t diagram the characteristics would be seen to coalesce to form a discontinuity.

If the isotherm is linear, $d\bar{x}_B/dx_B = 1$; that is, there is no preference for A or B, and the front propagates unchanged. Finally, if the isotherm is unfavorable, $(d\bar{x}_B/dx_B)_a < (d\bar{x}_B/dx_B)_b$ and $u_{\text{eff}}^a > u_{\text{eff}}^b$. The situation is opposite to that described above. Here the A ions behind the front are preferentially retained and fall further behind. The B ions ahead of the front have the opposite behavior. This type of front spreads out and is termed *nonsharpening*. It is the form characteristic of most elution techniques. Here, in a ξ-t diagram the characteristics would appear as a spreading fan emanating from the origin.

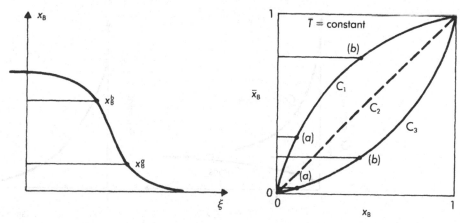

Figure 6.3.4 Effect of equilibrium isotherm shape on shape of exchange zone front.

From the arguments given it can be seen that at large distances the width of the front for the favorable isotherm is governed by diffusion-dispersion and/or the reaction rate (Helfferich 1962, Liberti & Helfferich 1983). On the other hand, for unfavorable equilibria the front spreads indefinitely, and at large distances its width and shape are governed essentially by the isotherm shape alone, diffusion and reaction rate having little effect.

6.4   The Electric Double Layer and Electrokinetic Phenomena

In the last several sections we have discussed the effect of charge transfer in neutral electrolytes (the electrode, electrodialysis membrane, and ion exchange resin particles). Generally, most substances will acquire a surface electric charge when brought into contact with an aqueous (polar) medium. Some of the charging mechanisms include ionization, ion adsorption, and ion dissolution. The effect of any charged surface in an electrolyte solution will be to influence the distribution of nearby ions in the solution. Ions of opposite charge to that of the surface (*counterions*) are attracted toward the surface while ions of like charge (*coions*) are repelled from the surface (Fig. 6.4.1). This attraction and

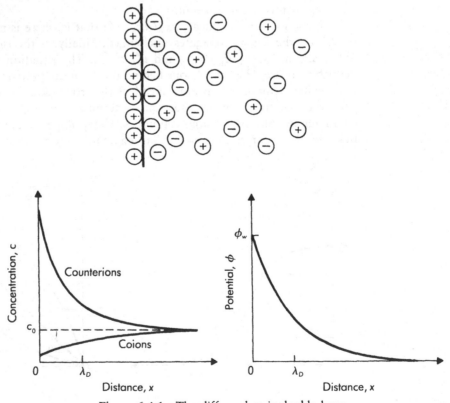

Figure 6.4.1 The diffuse electric double layer.

repulsion, when combined with the mixing tendency resulting from the random thermal motion of the ions, leads to the formation of an *electric double layer*. The electric double layer is a region close to the charged surface in which there is an excess of counterions over coions to neutralize the surface charge, and these ions are distributed in a "diffuse" manner. Evidently there is no charge neutrality within the double layer because the number of counterions will be large compared with the number of coions.

We may observe that if there were no thermal motion there would be just as many counterions in the electric double layer as needed to balance the charge on the surface. This is termed *perfect shielding*, since all of the other ions are shielded from the surface charge. However, because of the finite temperature and associated random thermal motion of the ions, those ions at the edge of the "cloud" where the electric field is weak have enough thermal energy to escape from the electrostatic potential well. Therefore the "edge" of the double layer is at a position where the potential energy is approximately equal to the thermal energy of the counterions ($RT/2$ per mole per degree of freedom), and the shielding is not complete.

We may estimate the approximate thickness of the double layer in the one-dimensional picture of Fig. 6.4.1. There, the electric field is taken to be parallel to the x axis, that is, everywhere perpendicular to the plane charged surface. We consider a simple fully dissociated symmetrical salt in solution for which the number of positive and negative ions are equal, so

$$z_+ = -z_- = z \tag{6.4.1}$$

For specificity the surface is taken to be positively charged as in Fig. 6.4.1.

Let us first make a rough estimate of the thickness by assuming that there are no positive ions (coions) present. Then the electric potential from the Poisson equation (Eqs. 3.4.5 and 3.4.7) is defined by

$$\frac{d^2\phi}{dx^2} = \frac{Fzc}{\epsilon} \tag{6.4.2}$$

where c is taken to be the average molar negative ion (counterion) concentration. The electrical potential energy per mole of negative ion is

$$W = -Fz\phi \tag{6.4.3}$$

The change in W across a plane layer of width x is obtained by integrating Eq. (6.4.2):

$$\Delta W = -\frac{F^2 z^2 c x^2}{2\epsilon} \tag{6.4.4}$$

This result assumes that the electric field vanishes on one side of the plane layer. If we assume only planar translational motion, the value of x for which the absolute value of ΔW equals RT is

$$x \equiv \lambda_D = \left(\frac{\epsilon RT}{2F^2 z^2 c} \right)^{1/2} \tag{6.4.5}$$

The quantity λ_D is termed the *Debye shielding distance* or, more frequently, the *Debye length*.

For an aqueous solution of a symmetrical electrolyte at 25°C,

$$\lambda_D = \frac{9.61 \times 10^{-9}}{(z^2 c)^{1/2}} \tag{6.4.6}$$

with λ_D in meters and c in mol m^{-3}. For a univalent electrolyte the Debye length is thus about 1 nm for a concentration of 10^2 mol m^{-3} and 10 nm for 1 mol m^{-3}.

From Eq. (6.4.5) or (6.4.6) it can be seen that λ_D decreases inversely as the square root of the concentration. Physically this is a result of the fact that there are more counterions per unit of depth. The Debye length also decreases with increasing valency because fewer ions are required to equilibrate the surface charge. More importantly, λ_D increases as the square root of RT. That is, without thermal agitation the double layer would collapse to an infinitely thin layer.

From the above considerations we can now define what is meant by electrically neutral solutions. If the dimensions of the system L are much larger than λ_D, then whenever local charge concentrations arise or external potentials are introduced into the solution they are shielded out in a distance short compared with L, leaving the bulk of the solution free of large electric potentials or fields. Based on a Debye length of 1 to 10 nm, the assumption of electrical neutrality is generally justified for the problems so far considered. However, as we shall discuss in the next section, in the case of very small charged microscopic capillaries, such as are characteristic of membranes and finely porous media, the double layer is central to the calculation of the solute and ion fluxes.

A more detailed calculation than that given above of the one-dimensional diffuse double layer recognizes that the concentration of ions in the sheath has the Boltzmann distribution

$$c_{\pm} = c_0 \exp\left(\frac{\mp zF\phi}{RT} \right) \tag{6.4.7}$$

where the concentration far from the surface $c \to c_0$ as $\phi \to 0$. The charge density from $\rho_E = F \sum z_i c_i$ is therefore

$$\rho_E = zFc_0 \left[\exp\left(\frac{-zF\phi}{RT} \right) - \exp\left(\frac{zF\phi}{RT} \right) \right]$$

$$= -2Fzc_0 \sinh\left(\frac{zF\phi}{RT} \right) \tag{6.4.8}$$

whence from Poisson's equation in place of Eq. (6.4.2) we have

$$\frac{d^2\phi}{dx^2} = \frac{2zFc_0}{\epsilon} \sinh\left(\frac{zF\phi}{RT} \right) \tag{6.4.9}$$

The above form of the Poisson equation can be integrated explicitly. However, note that for small potentials $zF\phi \ll RT$ (recall that $RT/F = 25.7 \, \text{mV}$ at 25°C) we may expand $\sinh(zF\phi/RT)$ with the result

$$\frac{d^2\phi}{dx^2} = \frac{\phi}{\lambda_D^2} \tag{6.4.10}$$

This approximation is termed the *Debye-Hückel approximation*. Integrating Eq. (6.4.10) subject to the conditions that $\phi = \phi_w$ at $x = 0$ and $\phi = 0$, and $d\phi/dx = 0$ as $x \to \infty$ gives

$$\phi = \phi_w \exp\left(-\frac{x}{\lambda_D}\right) \tag{6.4.11}$$

The Debye length is thus seen to be the $1/e$ decay distance for the potential and electric field at low potentials. Close to the charged surface where the potential is relatively high and the Debye-Hückel approximation inapplicable, the potential decreases faster than the exponential fall-off indicated.

The potential ϕ_w can be related to the charge density at the surface by equating the surface charge with the net space charge in the diffuse part of the double layer.

The treatment given above of the diffuse double layer is based on the assumption that the ions in the electrolyte are treated as point charges. The ions are, however, of finite size, and this limits the inner boundary of the diffuse part of the double layer, since the center of an ion can only approach the surface to within its hydrated radius without becoming specifically adsorbed (Fig. 6.4.2). To take this effect into account, we introduce an inner part of the double layer next to the surface, the outer boundary of which is approximately a hydrated ion radius from the surface. This inner layer is called the *Stern layer*, and the plane separating the inner layer and outer diffuse layer is called the *Stern plane* (Fig. 6.4.2). As indicated in Fig. 6.4.2, the potential at this plane is close to the *electrokinetic potential* or zeta (ζ) *potential*, which is defined as the potential at the *shear surface* between the charge surface and the electrolyte solution. The shear surface itself is somewhat arbitrary but characterized as the plane at which the mobile portion of the diffuse layer can "slip" or flow past the charged surface.

The above discussion leads to what we mean by *electrokinetic phenomena*, which is the term applied to four phenomena that arise when the mobile portion of the diffuse double layer and an external electric field interact in the viscous shear layer near the charged surface. With reference to Fig. 6.4.3, if an electric field is applied tangentially along a charged surface then the electric field will exert a force on the charge in the diffuse layer. This layer is part of the electrolyte solution, and the migration of the mobile ions will carry the solvent with them and cause it to flow. On the other hand, an electric field is created if the charged surface and diffuse part of the double layer are made to move relative to each other. In the example of Fig. 6.4.3 the value of the velocity U outside the diffuse layer is a constant, and for a sufficiently thin diffuse layer U may be regarded as a "slip velocity" relative to the surface. As shown in Fig.

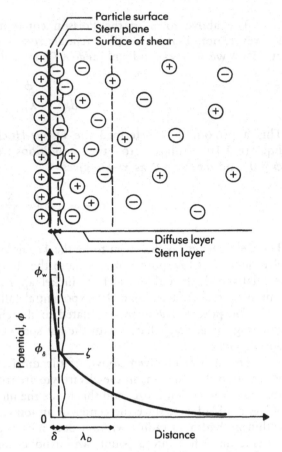

Figure 6.4.2   Structure of electric double layer with inner Stern layer. [After Shaw, D.J. 1980. *Introduction to Colloid and Surface Chemistry*, 3rd edn. London: Butterworths. With permission.]

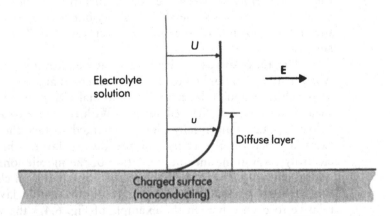

Figure 6.4.3   Flow produced by electric field acting on diffuse double layer.

6.4.3, the true velocity of the fluid at the surface must be zero from the viscous flow condition of no slip.

The four electrokinetic phenomena following the description of Shaw (1980) are

1.   *Electrophoresis*—the movement of a charged surface plus attached material (i.e., dissolved or suspended material) relative to stationary liquid by an applied electric field.
2.   *Electroosmosis*—the movement of liquid relative to a stationary charged surface (e.g., a capillary or porous plug) by an applied electric field (i.e., the complement of electrophoresis). The pressure necessary to counterbalance electroosmotic flow is termed the *electroosmotic pressure*.
3.   *Streaming potential*—the electric field created when liquid is made to flow along a stationary charged surface (i.e., the opposite of electroosmosis).
4.   *Sedimentation potential*—the electric field created when charged particles move relative to stationary liquid (i.e., the opposite of electrophoresis).

Both electroosmosis and streaming potential relate to the motion of electrolyte solutions and are therefore considered in the following section. However, we shall reserve the detailed discussion of streaming potential for the next chapter in connection with the treatment of sedimentation potential, which together with electrophoresis deals with the motion of dissolved or suspended charged particles.

6.5   Electroosmosis

The discovery of electrokinetic phenomena may be credited to F.F. Reuss, whose experiments on electroosmosis and electrophoresis were described in 1809 in the *Proceedings of the Imperial Society of Naturalists of Moscow*. Reuss demonstrated that under the influence of an applied electric field water migrated through porous clay diaphragms toward the cathode. This is understood today to be a consequence of the fact, illustrated schematically in Fig. 6.5.1, that clay, sand, and other mineral particles usually carry negative surface charges when in contact with water; the water normally containing small quantities of dissociated salts. As described in the last section, the charged surface will attract positive ions present in the water and repel negative ions. The positive ions will therefore predominate in the Debye sheath next to the charged surface, so application of an external electric field results in a net migration toward the cathode of ions in the surface water layer. Due to viscous drag, the water in the pores is drawn by the ions and therefore flows through the porous medium.

Electroosmosis has been used in a variety of applications, including the dewatering of soils for construction purposes and the dewatering of mine tailings and waste sludges. It has also been used to characterize and design the salt rejection properties of reverse osmosis membranes and to help understand the behavior of biological membranes. Electroosmosis is also being investigated as a means of removing contaminants from soils.

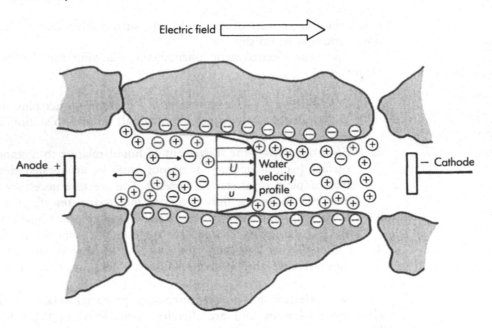

Figure 6.5.1 Electroosmotic flow of water in a porous charged medium.

To analyze the flow through a porous medium, we can, as before, model the medium as a collection of parallel cylindrical microcapillaries. As noted in Section 4.7, the actual sinuous nature of the capillaries may be accounted for by the introduction of an empirical tortuosity factor. The results for electroosmotic flow through a capillary are then readily carried over to the porous medium by using Darcy's law (Eq. 4.7.7) and, for example, the Kozeny-Carman permeability (Eq. 4.7.16).

Let us estimate the electroosmotic velocity produced in a fine circular capillary by a uniform electric field applied along the axis as in Fig. 6.5.1. If the surface is assumed to be negatively charged, then the flow will be in the direction of the cathode, as shown. With the electric body force per unit volume given by $\mathbf{f}_E = \rho_E \mathbf{E}$, the momentum equation (Eq. 3.1.8) may be written

$$\rho \, \frac{D\mathbf{u}}{Dt} = \nabla \cdot \mathfrak{T} + \rho \mathbf{g} + \rho_E \mathbf{E} \qquad (6.5.1)$$

Neglecting gravitational forces and supposing the flow to be an inertia free capillary flow, with no pressure gradient Eq. (6.5.1) simply reduces to a balance between viscous and electrical forces:

$$\mu \nabla^2 \mathbf{u} = -\rho_E \mathbf{E} \qquad (6.5.2)$$

With x the coordinate directed along the axis toward the cathode and u the velocity component in that direction, we have for a long capillary that the derivatives of u with respect to x are zero and u is a function only of the transverse coordinate. If, in addition, the diffuse layer thickness λ_D is small compared with the tube radius a, then the curvature terms can be neglected and Eq. (6.5.2) reduces to the one-dimensional form appropriate to a long plane channel or infinite plane surface:

$$\mu \frac{\partial^2 u}{\partial y^2} = -\rho_E E_x = \epsilon \frac{\partial^2 \phi}{\partial y^2} E_x \tag{6.5.3}$$

Here, y is the Cartesian coordinate normal to x with origin at the lower channel wall (or plate surface) and directed into the flow. The component of the electric field E_x is parallel to the surface in the positive x direction. Poisson's equation has been used to eliminate the charge density ρ_E.

Integrating Eq. (6.5.3) gives

$$\mu \frac{\partial u}{\partial y} = \epsilon \frac{\partial \phi}{\partial y} E_x \tag{6.5.4}$$

where at the edge of the diffuse layer ($y \to \infty$) we have set $\partial u/\partial y = \partial \phi/\partial y = 0$. Integrating once more and setting $\phi = \zeta$ (the zeta potential) at $u = 0$, we find

$$U = -\frac{\epsilon \zeta E_x}{\mu} \tag{6.5.5}$$

This formula for the electroosmotic velocity past a plane charged surface is known as the *Helmholtz-Smoluchowski equation*. Note that within this picture, where the double layer thickness is very small compared with the characteristic length, say $a/\lambda_D \gg 100$, the fluid moves as in plug flow. Thus the velocity "slips" at the wall; that is, it goes from U to zero discontinuously. For a finite-thickness diffuse layer the actual velocity profile has a behavior similar to that shown in Fig. 6.5.1, where the velocity drops continuously across the layer to zero at the wall. The constant electroosmotic velocity therefore represents the velocity at the "edge" of the diffuse layer. A typical zeta potential is about 0.1 V. Thus for $E_x = 10^3$ V m^{-1}, with viscosity that of water, the electroosmotic velocity $U \sim 10^{-4}$ m s^{-1}, a very small value.

We observe here that in a capillary the volume flow rate due to a fixed pressure gradient is proportional to a^4 ($\pi a^4/8\mu(dp/dx)$ for a circular capillary). The electroosmotic flow rate is proportional to U multiplied by the cross-sectional area πa^2. Therefore, the ratio of electroosmotic to hydraulic flow rate will be proportional to a^{-2}. Thus, for example, if we employ a capillary model for a porous medium, it is evident that as the average pore size decreases electroosmosis will become increasingly effective in driving a flow through the medium, compared with pressure, *provided* $\lambda_D/a \ll 1$.

We next calculate how the electroosmotic flow and potential change in a long capillary for different ratios of Debye length to radius. We shall allow for pressure gradients, but the mean velocity and tube radius are assumed sufficiently small that inertia effects can be neglected. The dilute electrolyte solution in the capillary is taken to be binary.

Under the assumptions noted, the momentum equation reduces to

$$0 = -\nabla p + \mu \nabla^2 \mathbf{u} - F(z_+ c_+ + z_- c_-)\nabla \phi \tag{6.5.6}$$

where $\mathbf{E}$ has been replaced by $-\nabla \phi$, and the electric charge density has been written in terms of the ion concentration, using $\rho_E = F \sum z_i c_i$. To simplify the calculation, we assume an ideal solution of a fully dissociated symmetrical salt so that $z_+ = -z_- = z$. For the axially symmetric circular capillary we adopt a cylindrical coordinate system (x, r) with x positive in the direction of flow and r the radial coordinate with origin at the axis of symmetry:

$$\frac{\mu}{r} \frac{\partial}{\partial r}\left(r \frac{\partial u}{\partial r}\right) = \frac{\partial p}{\partial x} + Fz(c_+ - c_-) \frac{\partial \phi}{\partial x} \tag{6.5.7}$$

To define the interaction between the electric field and ion concentration, we invoke Poisson's equation $\nabla^2 \phi = -\rho_E/\epsilon$. For a capillary of length L large compared with its radius a, the term $\partial^2 \phi/\partial x^2$ may be neglected, to give

$$\frac{1}{r} \frac{\partial}{\partial r}\left(r \frac{\partial \phi}{\partial r}\right) = -\frac{Fz}{\epsilon}(c_+ - c_-) \tag{6.5.8}$$

In other words, at any small segment of the capillary the ion concentrations are in a local quasi equilibrium determined solely by the radial variation in ϕ.

Because of the behavior indicated, it is convenient to divide the potential into two parts,

$$\phi(x, r) = \Phi(x) + \psi(x, r) \tag{6.5.9}$$

and write Poisson's equation as

$$\frac{1}{r} \frac{\partial}{\partial r}\left(r \frac{\partial \psi}{\partial r}\right) = -\frac{Fz}{\epsilon}(c_+ - c_-) \tag{6.5.10}$$

Since there is no radial flux of ions or radial flow, we can integrate the radial component of the Nernst-Planck equations for the ion fluxes and obtain the Boltzmann distribution of Eq. (6.4.7):

$$c_\pm(x, r) = c_0(x) \exp\left(\mp \frac{zF\psi}{RT}\right) \tag{6.5.11}$$

Here, we have set $c_+^0 = c_-^0 \approx c_0$ an approximation that we show is generally justified even when the Debye length is not small, a condition for which the bulk of the solution is electrically neutral.

Recasting the right side of Eq. (6.5.10) in the same manner as Eq. (6.4.8), we have

$$\lambda^{*2} \frac{1}{r^*} \frac{\partial}{\partial r^*}\left(r^* \frac{\partial \psi^*}{\partial r^*}\right) = \sinh \psi^* \tag{6.5.12}$$

where $r^* = r/a$, $\lambda^* = \lambda_D/a$, and $\psi^* = zF\psi/RT$. The boundary conditions from symmetry are that

$$\frac{\partial \psi^*}{\partial r^*} = 0 \qquad \text{at } r^* = 0 \qquad (6.5.13a)$$

and, at the surface,

$$\psi^* = \psi_w^* = \zeta^* \qquad \text{at } r^* = 1 \qquad (6.5.13b)$$

In writing the condition at the surface, we assume that the solid wall is charged and that its electrical nature affects the electroosmosis only through an effective surface potential, which we identify with the ζ potential.

Equation (6.5.12), subject to the boundary conditions indicated, can be solved analytically only under limiting conditions on the Debye length described below. In general, it must be solved numerically. Figure 6.5.2 gives numerical solutions obtained by Gross & Osterle (1968) for a constant surface potential $\psi_w^* = \zeta^* = 2.79$ for various values of the Debye length ratio λ^*. It can be seen that for $\lambda^* < 0.1$ the potential is zero over most of the capillary cross section, whereas for $\lambda^* > 10$ the potential is nearly constant over the cross section.

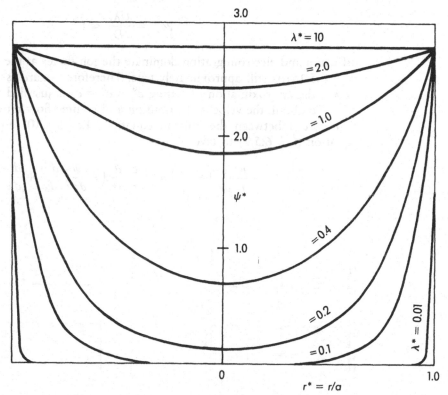

Figure 6.5.2 Dimensionless potential distribution across a cylindrical capillary for different values of the Debye length ratio λ^* and a constant surface potential $\psi_w^* = \zeta^* = 2.79$ (after Gross & Osterle 1968).

For the limiting case of small Debye length ratio, shown schematically in Fig. 6.5.3A, $\psi = 0$ at the center and the solution is electrically neutral there, so the approximation $c_+^0 = c_-^0 = c_0$ employed in Eq. (6.5.12) is, in this case, exact. Moreover,

$$\frac{\partial \Phi}{\partial x} = \text{constant} \qquad (6.5.14)$$

For the limiting case of large Debye length, the entire capillary (pore) is within the double layer. If 1 and 4 designate locations in the external solution just outside of the double layer and 2 and 3 designate locations inside the pore entrance, then at equilibrium with no flow ($u = 0$) and no flux ($j_i^* = 0$) $\psi_1 = \psi_4 = 0$ and $\psi_2 = \psi_3 = \zeta$. Hence, from the Boltzmann distribution,

$$c_{\pm 2} = c_{\pm 3} = c_0 \exp\left(\mp \frac{zF\zeta}{RT}\right) \qquad (6.5.15)$$

But this is exactly the Donnan potential, which was written down without proof as Eq. (6.2.17) in connection with the electrodialysis analysis. At small Peclet number based on λ_D, that is,

$$\text{Pe}_{\lambda_D} = \frac{U\lambda_D}{D} \ll 1 \qquad (6.5.16)$$

diffusion and electromigration dominate the ion fluxes at the pore entrance and Eq. (6.5.15) is still approximately true. Therefore, regardless of the magnitude of λ^*, the approximation of setting $c_+^0 = c_-^0 = c_0$ is justified (Liang 1976).

To obtain the velocity distribution and volume flow rate, we can eliminate $Fz(c_+ - c_-)$ between the Poisson equation (Eq. 6.5.10) and the momentum equation (Eq. 6.5.7) to give

$$\frac{\mu}{r} \frac{\partial}{\partial r}\left(r \frac{\partial u}{\partial r}\right) = -\frac{\epsilon}{r} \frac{\partial}{\partial r}\left(r \frac{\partial \psi}{\partial r}\right) \frac{d\Phi}{dx} + \frac{dp}{dx} \qquad (6.5.17)$$

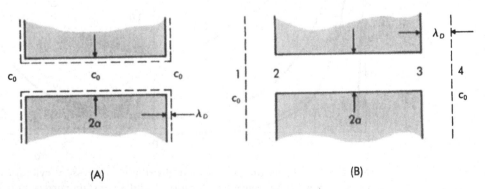

Figure 6.5.3 Debye layer location in a cylindrical pore: (A) λ small; (B) λ large.

This equation can be integrated subject to the electrokinetic and flow boundary conditions

$$\frac{\partial u}{\partial r} = 0 \quad \frac{\partial \psi}{\partial r} = 0 \quad \text{at } r = 0 \qquad (6.5.18a)$$

$$u = 0 \quad \psi = \zeta \quad \text{at } r = a \qquad (6.5.18b)$$

The result is

$$u(r) = -\frac{\epsilon(\psi - \zeta)}{\mu}\frac{d\Phi}{dx} + \frac{r^2 - a^2}{4\mu}\frac{dp}{dx} \qquad (6.5.19)$$

From the velocity profile the volume flow rate is obtained by integrating across the capillary cross section to get

$$Q = -\int_0^a \frac{\epsilon(\psi - \zeta)}{\mu}\frac{d\Phi}{dx} 2\pi r\, dr - \frac{\pi a^4}{8\mu}\frac{dp}{dx} \qquad (6.5.20)$$

One limit for which an analytic solution is readily obtained is that for small Debye length, since here $\psi = 0$ over most of the capillary cross section. Therefore, it is necessary to solve for ψ only near the pore wall, where $r \sim a \gg \lambda_D$. That is, we can neglect the curvature effect, and the equation for ψ is exactly the same as that for ϕ in the one-dimensional problem of the last section (Eq. 6.4.9):

$$\frac{\partial^2 \psi^*}{\partial y^{*2}} = \sinh \psi^* \qquad (6.5.21)$$

Here, $y^* = y/\lambda_D$.

With the Debye-Hückel linearization the solution is simply

$$\psi = \zeta e^{-(a-r)/\lambda_D} \qquad (6.5.22)$$

Substituting this result into the general expressions for velocity and volume flow rate, we obtain (Liang 1976)

$$u = \frac{\epsilon\zeta}{\mu}\left(1 - e^{-(a-r)/\lambda_D}\right)\frac{d\Phi}{dx} + \frac{r^2 - a^2}{4\mu}\frac{dp}{dx} \qquad (6.5.23)$$

$$Q = \frac{\epsilon\zeta}{\mu}\frac{d\Phi}{dx}\pi a^2\left(1 - 2\frac{\lambda_D}{a}\right) - \frac{\pi a^4}{8\mu}\frac{dp}{dx} \qquad (6.5.24)$$

For a thin diffuse layer, that is, for $a/\lambda_D \to \infty$, the expression for the electroosmotic velocity reduces identically to the Helmholtz-Smoluchowski equation (Eq. 6.5.5) for zero pressure gradient.

Another limit for which an analytic solution is readily obtained is that of large Debye length. In this case the ion concentrations are uniform across the pore and are given by the Boltzmann distribution of Eq. (6.5.15); that is,

$$c_{\pm} = c_0 \exp\left(\mp \frac{zF\zeta}{RT}\right) \tag{6.5.25}$$

The result for u is then obtained simply by integrating the momentum equation (Eq. 6.5.7) subject to the no-slip boundary condition $u = 0$ at $r = a$ and the symmetry condition $\partial u/\partial r = 0$ at $r = 0$. Note that the Poisson equation is not used here; it is replaced by the condition of equilibrium, that is, overall charge neutrality. Carrying out the integration, we get

$$u = -\frac{a^2 - r^2}{4\mu}\frac{d}{dx}\left(p - 2zFc_0\Phi\sinh\left(\frac{zF\zeta}{RT}\right)\right) \tag{6.5.26}$$

$$Q = -\frac{\pi a^4}{8\mu}\frac{d}{dx}\left(p - 2zFc_0\Phi\sinh\left(\frac{zF\zeta}{RT}\right)\right) \tag{6.5.27}$$

We had earlier observed, as may be seen from Eq. (6.5.24), that the ratio of the electroosmotic flow to hydraulic flow for $\lambda_D/a \ll 1$ goes as a^{-2}. It may be seen from Eq. (6.5.27) that for $\lambda_D/a \gg 1$ this ratio is independent of the capillary radius, so in terms of flow rates achievable there is no particular advantage in using an electric field rather than a pressure gradient.

The application of the above results for a capillary to a porous medium is straightforward and may be found in Liang (1976), Jacazio et al. (1972), and Shapiro & Probstein (1993). Although we have not presented it here, the current density can also be calculated by using Eq. (3.4.3), and the total current can be obtained by integrating across the capillary cross section.

The simple capillary model presented has also been used to model salt rejection in flow through charged porous materials. The physical mechanism of the salt rejection is just that the surface charge gives rise to a potential field that extends a distance approximately equal to the Debye length into the liquid within the pore, as we have shown above. If a pressure gradient is applied to make the salt solution flow through the pore, then, because of an excess of charge of one sign within the pore liquid, there results a net transport of charge and the buildup of the streaming potential defined in the last section. The effect of the potential is to set up an electric field parallel to the surface that will increase the transport of the coions and reduce that of the counterions until there is an equal steady-state transport of positive and negative ions, the feed and effluent solutions being insulated from each other so that there is no current flow. In this steady state, salt and water, but not charge, are transported through the capillary. However, the net effect of the coion exclusion and the axial field is such as to cause the ratio of the molar salt flux to the volume flux of water to be less than the molar salt concentration on the upstream side of the membrane. In other words, the membrane tends to reject salt. This model has been used to describe the salt rejection characteristics of reverse osmosis membranes (Jacazio et al. 1972, Sonin 1976).

6.6   Effects of Chemical Reactions

We have to this point neglected any effects of chemical reactions that are frequently associated with electrolyte solutions in the presence of an electric field. One ubiquitous example is the dissociation of water (Section 6.1), while another is reactions brought about by the presence of dissolved impurities.

An application of electrokinetics that serves to illustrate a number of chemical effects is that of electroosmotic purging of contaminated liquids from soils (Shapiro & Probstein 1993). A dc electric field is applied across electrode pairs in the ground and under the action of the field, the contaminants in the liquid phase in the soil are moved by electroosmosis to one set of electrodes, typically the cathodes. The electric field simultaneously draws in a noncontaminating liquid to help wash and treat the soil to enhance the restoration process. The contaminated effluent is then removed by pumping from wells surrounding the cathodes.

The problem is a complex one for not only is there convection of the liquid by electroosmosis but if the dissolved contaminants are themselves charged, then an additional electromigration velocity will be imposed on them. Moreover, to the extent that concentration gradients are set up, there will also be transport of dissolved species by diffusion. In addition, there are chemical reactions in the bulk fluid and at the electrodes, together with adsorption or desorption at the soil surface.

Typically electrolysis of water takes place at the electrodes (Eqs. 6.1.25). The electrolysis reaction lowers the pH at the anode and raises it at the cathode, accompanied by the propagation of an acid front into the soil from the anode and a base front from the cathode. This process can have a significant effect on the soil zeta potential, which is strongly dependent upon pH, as well as on the solubility, ionic state and charge, and level of adsorption of the contaminants. Most importantly, nearly all soils become less negatively charged with decreasing pH because adsorbed hydrogen ions neutralize the negative charge.

The phenomena and processes described can be modeled by convective diffusion equations with chemical reactions. In the simplest model, we may apply these equations in a cylindrical capillary and by means of a capillary model to a porous medium. Assuming dilute solutions, rapid chemical reactions, the double-layer thickness to the soil pore radius and the Peclet number based on the pore radius both small, the overall transport rate for the ith species in a straight cylindrical capillary is

$$\frac{\partial \bar{c}_i}{\partial t} + \frac{\partial}{\partial x}\left[\bar{c}_i(\bar{u}_{e,i} + \bar{u}_c)\right] = D_i \frac{\partial^2 \bar{c}_i}{\partial x^2} + \bar{R}_i + \bar{R}_i^a \tag{6.6.1}$$

This equation follows from continuity (Eq. 3.3.2) and the Nernst–Planck relation (Eq. 3.4.1). The overbars indicate that the variable has been averaged over the tube cross-section (Eq. 4.6.19), and $\bar{R}_i$ and $\bar{R}_i^a$ are the molar rates of production due to chemical reactions and sorption, respectively.

Here, we have defined two velocities that appear in the brackets. The first is the electromigration velocity

$$\bar{u}_{e,i} = -v_i z_i F \frac{\partial \bar{\phi}}{\partial x} \tag{6.6.2}$$

and the second is the convection velocity, that is, the bulk electroosmotic velocity

$$\bar{u}_c = \frac{\epsilon}{\mu} \left\langle \zeta \frac{\partial \bar{\phi}}{\partial x} \right\rangle \tag{6.6.3}$$

The brackets denote the volume average of the scalar product of the local ζ potential and the electric field in the x-direction. This result can be seen from Eq. (6.5.24) by dividing by the capillary length and cross-sectional area and in the limit of a thin diffuse layer and integrating over this length. Although there is an induced pore pressure gradient, this term drops out in the integration because the pressures at the capillary ends are equal. Thus $\bar{u}_c$ is just the average interstitial fluid velocity in the x-direction and, owing to the incompressibility of the system, is independent of position and is the same for each species. The electromigration velocity $u_{e,i}$ on the other hand depends on the local electric field and differs for each species.

The electric field from Eq. (3.4.4) is given by

$$\bar{i} = -\sigma(x) \frac{\partial \bar{\phi}}{\partial x} - \sum z_i D_i \frac{\partial \bar{c}_i}{\partial x} \tag{6.6.4}$$

The current remains constant along the length of the cell, so the electric field varies with position to compensate for variations in concentration. Because the total applied voltage is known, Eq. (6.6.4) provides a relationship between $\partial \bar{\phi}/\partial x$ and $\bar{c}_i$.

The sorption rate $\bar{R}_i^a$ is characterized by a particular isotherm, and it relates the concentration in the adsorbed phase to that in the bulk solution. The reaction rate $\bar{R}_i$ is eliminated from Eq. (6.6.1) by using the equilibrium constant for the reaction and the conservation of mass of the elements to determine the equilibrium concentration of the species. In the special case of hydrogen and hydroxyl ions, mass conservation cannot be used because the total amount of water present is not known (unit activity is assumed). However, the electroneutrality condition, Eq. (3.4.8) provides the extra equation required and, with the dissociation constant for water, fixes the concentrations of hydrogen and hydroxyl ions. Using the transport equations to track the movement of these ions is therefore not necessary and in fact would overspecify the problem.

The solution of these equations subject to the boundary conditions must generally be carried out numerically although the procedures are not described here. Two limiting cases are one in which electroosmosis is the dominant removal mechanism for the dissolved contaminant and one where electromigration is the principal removal mechanism. The removal of aqueous soluble organics would be principally by electroosmosis, and the removal of soluble metal ions would be mainly by electromigration.

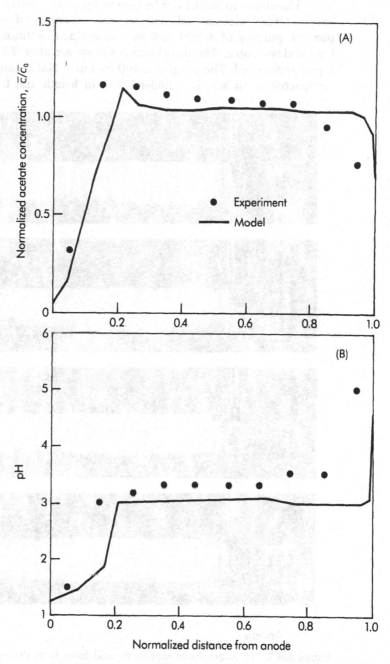

Figure 6.6.1 Comparison after removing 0.12 pore volumes of liquid between finite element calculations and experiment for electroosmotic purging of acetic acid from a cylindrical clay sample 0.5 m long, initially saturated with a 100 mol m^{-3} acetic acid solution ($\bar{c}_0$), across which 25 V is applied: (A) normalized acetate concentration along sample; (B) pH profile along sample (after Shapiro & Probstein 1993).

The ability to model can be seen in Fig. 6.6.1 where comparisons are given between finite element calculations and experimental results for the electro-osmotic purging of a 100 mol m^{-3} acetic acid solution initially saturating a kaolin clay sample. The distributions shown are after 0.12 of a pore volume of liquid is removed. The purge is a 100 mol m^{-3} NaCl solution and the sample is compacted in an acrylic cylinder 0.5 m in length and 0.1 m in diameter with

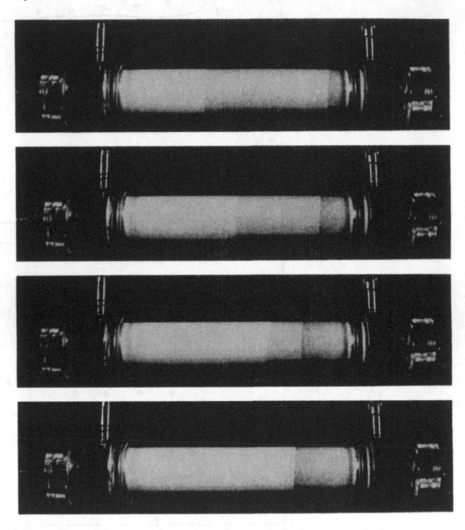

Anode   Cathode

Figure 6.6.2 Photographs of motion of acid front from the anode and base front from the cathode for removal by electromigration of zinc from a cylindrical clay sample 0.2 m long, initially saturated with a 7.7 mol m^{-3} zinc solution, across which 8 V is applied. The frame times from top to bottom are 6, 8, 10, and 11.3 h, respectively. [Courtesy of Dr. Sebastian Tondorf. From Probstein & Hicks 1993. Removal of contaminants from soils by electric fields. *Science* **260**, 498–503. Copyright 1993 by the AAAS. With permission.]

porous carbon fiberboard electrodes at each end across which a constant voltage of 25 V is applied. The only parameter chosen is a volume-averaged zeta potential that is matched to give the mean experimental flow rate. Such a procedure is dictated by the fact that the ζ potential generally is not known as a function of the pH, ion concentration, and soil type although it can be determined semi-empirically. In this instance, the deeper penetration of the acid front from the anode than the base front from the cathode results primarily from the electroosmotic flow.

The physics and chemistry of removal of a dissolved metal is quite different from that of a dissolved organic in that the metals are initially present in solution as positively charged ions and the principal removal mechanism is electromigration (Probstein & Hicks 1993). Diffusion also plays a significant role, particularly in regions where steep gradients in the species concentration develop.

As the metal ion moves under the action of the electric field, it enters the region of high pH near the cathode, where it may adsorb onto the soil, precipitate, or form hydroxo complexes. Typically, an intermediate pH of minimum solubility exists at which virtually all of the metal precipitates. This "focusing" of the metal at an intermediate point between the electrodes, which has been observed experimentally, generally takes place at the point where there is a sharp jump in the pH. The formation of such a pH jump is shown in Fig. 6.6.2 for a zinc removal test. In the experiment, a clay sample packed into an acrylic cylinder 200 mm in length and 32 mm in diameter is saturated initially with a 7.7 mol m^{-3} zinc solution. A voltage of 8 V is applied across the sample and pH indicators are used to define the acid front from the anode and base front from the cathode which move into the clay at constant speeds. The sharp front forming at about 11 h remains relatively stationary during the test period of over a week. If ionic contaminants are to be removed, the increase in pH in the cathode region must be prevented. A number of simple procedures exist, typically involving rinsing the cathode with a solution to neutralize or wash away the hydroxyl ions.

References

CASTELLAN, G.W. 1983. *Physical Chemistry*, 3rd edn. Reading, Mass.: Addison-Wesley.

GROSS, R.J. & OSTERLE, J.F. 1968. Membrane transport characteristics of ultrafine capillaries. *J. Chem. Phys.* **49**, 228–234.

HELFFERICH, F.(G.) 1962. *Ion Exchange*. New York: McGraw-Hill.

JACAZIO, G., PROBSTEIN, R.F., SONIN, A.A. & YUNG, D. 1972. Electrokinetic salt rejection in hyperfiltration through porous materials. Theory and experiment. *J. Phys. Chem.* **76**, 4015–4023.

KORYTA, J. & DVORAK, J. 1987. *Principles of Electrochemistry*. New York: Wiley.

LEVICH, V.G. 1962. *Physicochemical Hydrodynamics*. Englewood Cliffs, N.J.: Prentice-Hall.

LIANG, L. 1976. Electroosmotic dewatering of wastewater sludges. Ph.D. dissertation, M.I.T., Cambridge, Mass.

LIBERTI, L. & HELFFERICH, F.G. (ed.) 1983. *Mass Transfer and Kinetics of Ion Exchange*. The Hague: Martinus Nijhoff.

NEWMAN, J.S. 1991. *Electrochemical Systems*, 2nd edn. Englewood Cliffs, N.J.: Prentice-Hall.

PROBSTEIN, R.F. 1972. Desalination: Some fluid mechanical problems. *Trans. ASME J. Basic Eng.* **94**, 286–313.

PROBSTEIN, R.F. & HICKS, R.E. 1993. Removal of contaminants from soils by electric fields. *Science* **260**, 498–503.

SHAFFER, L.H. & MINTZ, M.S. 1980. Electrodialysis. In *Principles of Desalination*, 2nd edn., Part A (ed. K.S. Spiegler & A.D.K. Laird), pp. 258–357. New York: Academic.

SHAPIRO, A.P. & PROBSTEIN, R.F. 1993. Removal of contaminants from saturated clay by electroosmosis. *Environ. Sci. Technol.* **27**, 283–291.

SHAW, D.J. 1980. *Introduction to Colloid and Surface Chemistry*, 3rd edn. London: Butterworths.

SOLAN, A. & WINOGRAD, Y. 1969. Boundary-layer analysis of polarization in electrodialysis in a two-dimensional laminar flow. *Phys. Fluids* **12**, 1372–1377.

SONIN, A.A. 1976. Osmosis and ion transport in charged porous membranes: A macroscopic mechanistic model. In *Charged Gels and Membranes I* (ed. E. Sélégny), pp. 255–265. Dordrecht-Holland: D. Reidel.

VETTER, K.J. 1967. *Electrochemical Kinetics. Theoretical and Experimental Aspects*. New York: Academic.

Problems

6.1 The current-voltage characteristic of an electrolytic cell was analyzed in Section 6.1, where the solution contained between the copper electrodes was cupric sulfate. Suppose that the cupric sulfate solution is replaced by another electrolyte which is indifferent to the electrodes; that is, no chemical reactions take place at the electrode surfaces. A constant potential difference is applied across the electrodes. Determine the potential and concentration distributions in the solution between the electrodes.

6.2 A practical problem in electrodialysis is surface fouling of the membranes caused by deposits of macromolecules or colloidal particles. If the fouling material is uncharged, then its effect may be modeled by considering the addition of a thin film of thickness Δ_f that acts as an added resistive layer through which the ions must pass before reaching the membrane. It is desired to estimate the effect of such a fouling film using the simple Nernst diffusion layer model in which the salt concentration is taken to drop linearly from the bulk value c_0 over a diffusion layer thickness δ_D to a value c_w at the edge of the film. It is then assumed to drop linearly from c_w to c_w' at the edge of the membrane proper. For specificity consider the dialysate side of a cation exchange membrane and take the dilute salt solution to consist of equal numbers of anions and cations having equal diffusion coefficients.

    a. Will the value of the limiting current be lower or higher with the fouling film present? Why?

b. If D_f is the diffusion coefficient in the film, obtain an expression for the limiting current in terms of D_f, Δ_f, and the other parameters in Eq. (6.2.13).

c. What is the effect on the limiting current of a film on the concentrate side of the membrane?

6.3 Consider an unfavorable ion exchange process carried out in columnar operation. Initially, the equivalent ionic fraction of the exchangeable ion B in the resin is $\bar{x}_B = 0$ while in the solution $x_B = 1$. If u_a is the velocity of the leading edge of the exchange front and u_b is the velocity of the trailing edge, estimate the front width Δ divided by the variable $z = (u_a + u_b)t/2$, where t is time measured from the start of the operation. Discuss the behavior of the front thickness where the capacity of the ion exchanger is large compared with the total equivalent concentration of the solution; that is, $\bar{C} \gg C$.

6.4 It is desired to determine the response of a very thin impermeable membrane to a cyclic load of frequency $\omega = 100\ \text{s}^{-1}$, where the magnitude of the pressure difference across the membrane is required to be very small, about 10 Pa. To develop the cyclic load, a cylindrical tube with no surface charge and containing a dilute aqueous solution of a fully dissociated singly charged symmetrical binary salt of concentration $c_0 = 1\ \text{mol m}^{-3}$ at 300 K is closed at both ends by electrodes. The distance between the electrodes is $L = 0.2\ \text{m}$, and located symmetrically between the ends is a porous plug of axial thickness $t = 0.1\ \text{m}$, which extends across the tube, whose diameter is about 0.05 m. The porous plug is made up of finely drilled thin capillaries parallel to the tube axis and of uniform radii $a = 10^{-5}\ \text{m}$. When in contact with the electrolyte, the walls of these fine capillaries acquire a constant surface (zeta) potential of $10^{-2}\ \text{V}$. An alternating voltage drop of 2 V is applied across the electrodes at the desired frequency ω. In this manner a reversing pressure difference is created across the plug due to electroosmosis. By means of closed fluid connections from both sides of the plug to both sides of the membrane, it is then possible to transmit the resulting cyclic load reversal across the membrane.

a. Calculate the Debye length λ_D and the Debye length ratio λ_D/a.

b. What is the pressure gradient developed across the porous plug at any instant of time?

c. Show that the inertial forces developed are small compared with the pressure forces.

d. Show that the Peclet number, based on the double layer thickness ($\sim\lambda_D$), and the magnitude of the velocity in the double layer are small. Take the diffusion coefficient D to be $10^{-9}\ \text{m}^2\ \text{s}^{-1}$. What would happen if the Peclet number were large?

6.5 Rederive the small Debye length result for the electroosmotic velocity in a long circular capillary for a symmetrical dilute binary electrolyte (Eq. 6.5.23) with a constant surface charge density in place of constant surface (zeta) potential. Show that the appropriate boundary condition at the tube wall that replaces the one of constant potential is $\partial \psi / \partial r = q_s / \varepsilon$, where q_s is the constant surface charge density.

6.6 Consider a long circular capillary containing a symmetrical dilute binary electrolyte where the Debye length is large compared with the capillary radius. Calculate the electroosmotic volume flow rate Q for a constant surface charge density q_s.

6.7 It was suggested that in the simplest capillary model of a porous medium, the basic electrokinetic equations (6.6.1)–(6.6.4) for a straight cylindrical capillary could be applied. However, in an actual porous medium, the capillaries are sinuous so that the flow length is actually longer than the straight-through distance. This is frequently accounted for by introducing a tortuosity factor, τ, which is a constant determined empirically. One way to model the effect of the longer flow path is to assume that the flow takes place in a tilted path (x-direction) between two surfaces of constant potential, while the straight-through or shortest path between the surfaces is in the z-direction. This is like crossing a street at an angle. In this case $dz/dx = 1/\tau$. Using this definition, what forms do Eqs. (6.6.1)–(6.6.4) take in terms of z and τ.

6.8 If Eq. (6.6.1) is applied to a porous medium by a simple capillary model, then the superficial velocity is given by $U = \varepsilon \bar{u}_c$, where ε is the porosity and $\bar{u}_c$ is the convection velocity in Eq. (6.6.1). Locally, the interstitial convection velocity is made up of hydraulic and electroosmotic contributions and is given by

$$\bar{u}_c = k \frac{\partial p(x)}{\partial x} + \frac{\varepsilon}{\mu} \zeta(x) \frac{\partial \bar{\phi}(x)}{\partial x}$$

where p is the pressure and k is the hydraulic permeability of the porous medium. Show that Eq. (6.6.3) for the convection velocity is true between two points of equal pressure.

7   Solutions of Charged Macromolecules and Particles

7.1   The Charge of Macromolecules and Particles

In this chapter we consider the motion of charged macromolecules or particles through a solvent medium, with the aim of utilizing the motional characteristics for particle characterization, fractionation, or preparation. Such motion of a charged surface in a liquid under the action an electric field was termed electrophoresis. It is the mirror image of the phenomena of electroosmosis considered in the last section. We previously noted that most substances will acquire a surface charge when brought into contact with a polar medium. The origin of the charge for various organic and inorganic compounds lies in the acid-base equilibria of the solution.

Electrophoresis is commonly used to differentiate proteins of different sizes. The origin of the protein charge can be related to the charge on amino acids, which are the basic structural units of protein molecules. We recall that an amino acid consists of an amino group, a carboxyl group, a hydrogen atom, and a "side chain" or R group bonded to a carbon atom. Amino acids in solution at neutral pH are predominantly dipolar ions rather than un-ionized molecules (Fig. 7.1.1B). In the dipolar form of an amino acid, the amino group is protonated ($-NH_3^+$), and the carboxyl group is dissociated ($-COO^-$). The ionization state of an amino acid varies with pH. In a strongly acidic solution (Fig. 7.1.1C), where the pH is low, the carboxyl group is un-ionized ($-COOH$), and the amino group is ionized ($-NH_3^+$). In a strongly alkaline solution (Fig. 7.1.1A), where the pH is high, the carboxyl group is ionized ($-COO^-$), and the amino group is un-ionized ($-NH_2$). The surface charge on microorganisms can also be explained in terms of charged groups similar to the amino and carboxyl groups above.

The mechanism responsible for charge on crystalline inorganic materials such as clays is associated with the lattice imperfections. The substitution of Al^{3+} for Si^{4+} and Mg^{2+}, Li^+, etc., for Al^{3+} in the clay lattice results in net

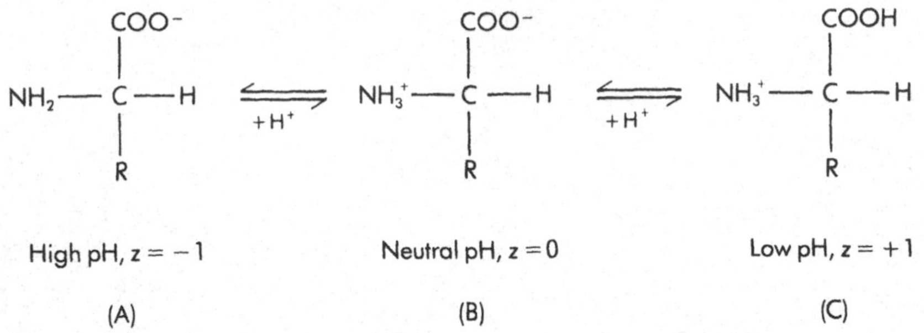

$$\text{NH}_2\text{—}\overset{\displaystyle \text{COO}^-}{\underset{\displaystyle R}{\text{C}}}\text{—H} \;\underset{+\text{H}^+}{\overset{}{\rightleftharpoons}}\; \text{NH}_3^+\text{—}\overset{\displaystyle \text{COO}^-}{\underset{\displaystyle R}{\text{C}}}\text{—H} \;\underset{+\text{H}^+}{\overset{}{\rightleftharpoons}}\; \text{NH}_3^+\text{—}\overset{\displaystyle \text{COOH}}{\underset{\displaystyle R}{\text{C}}}\text{—H}$$

High pH, z = −1   Neutral pH, z = 0   Low pH, z = +1

(A)   (B)   (C)

Figure 7.1.1   Ionized forms of an amino acid as a function of solution pH.

negative charges that are relatively constant and independent of the ion concentration and pH of the fluid. Physical adsorption is due to van der Waals attraction, and hydrogen bonding is yet another charge mechanism. The occurrence of negative charges on hydrocarbons, for example, is a consequence of the preferential adsorption of simple anions (e.g., Cl^-) over simple cations, enhanced by the hydrophobic nature of the interface with the water. The anions are polarizable and have smaller hydrated radii than cations do. Specific chemical adsorption due to chemical bonds between charged groups and the surface of the solids is another mechanism for some materials. Finally, the charge may be caused by the adsorption of ions that are identical to those in the solid lattice. For example, the charge on aluminum hydroxide sols depends on the relative concentrations of Al^{3+} and OH^- ions, which in turn depend on the pH. The mechanisms cited are among the more important physical-chemical ones that give rise to the charge on macromolecules and particles when in solution.

In our discussion of electroosmosis the charged state of the surface was described in terms of the surface potential, which was in turn identified with the ζ potential. Of course, the surface potential is related to the charge density at the surface. We may calculate this relation from charge conservation and Poisson's equation by noting that the net charge of a particle q must be equal and opposite to the total charge in the double layer. As an example, consider a spherically symmetric diffuse double layer arising from a stationary spherical, nonconducting charged particle of radius a. The amount of charge in a layer dr of the spherical double layer shell is

$$dq = 4\pi r^2 \rho_E \, dr \tag{7.1.1}$$

where ρ_E is the volume electric charge density (C m^{-3}), and r is the radial spherical coordinate measured from the sphere center. Substituting the Laplacian of the potential for the charge density and using the Poisson equation, we get the total charge in the double layer:

$$q = -\int_a^\infty 4\pi\epsilon \, \frac{\partial}{\partial r}\left(r^2 \frac{\partial \phi}{\partial r}\right) dr \tag{7.1.2}$$

where spherical symmetry has been assumed. Integrating and letting $\partial\phi/\partial r \to 0$ as $r \to \infty$, we obtain

$$q = 4\pi\epsilon a^2 \left(\frac{\partial\phi}{\partial r}\right)_{r=a} \tag{7.1.3}$$

We recall here the definition of surface charge distribution as the limit of a charge distribution in a surface layer of finite thickness δ when $\delta \to 0$ but the charge per unit area of surface remains finite. This charge per unit area of surface is the surface charge density q_s, which is opposite in sign to the charge in the double layer and has units of $C\,m^{-2}$. From Eq. (7.1.3),

$$q_s = -\epsilon\left(\frac{\partial\phi}{\partial r}\right)_{r=a} \tag{7.1.4}$$

This boundary condition relates q_s to the potential distribution. It is generally valid with r replaced by the appropriate normal coordinate, for example, x in the one-dimensional case of Fig. 6.4.1 or $-r$ for the cylindrical electroosmosis problem of Section 6.5.

The net charge can be calculated from the Poisson equation, which for spherical symmetry may be written

$$\frac{1}{r^2}\frac{d}{dr}\left(r^2\frac{d\phi}{dr}\right) = \frac{\phi}{\lambda_D^2} \tag{7.1.5}$$

where we have here introduced the Debye-Hückel approximation of small potential. With $\xi = r\phi$, the equation transforms to

$$\frac{d^2\xi}{dr^2} = \frac{\xi}{\lambda_D^2} \tag{7.1.6}$$

which is exactly the same form as for the plane wall problem (Eq. 6.4.9). Integrating and applying the boundary condition $\phi \to 0$ as $r \to \infty$ gives

$$\phi = \frac{A}{r}\exp\left(-\frac{r}{\lambda_D}\right) \tag{7.1.7}$$

where A is a constant of integration.

At the particle surface, $\phi = \zeta$, whence

$$\zeta = \frac{A}{a}\exp\left(-\frac{a}{\lambda_D}\right) \tag{7.1.8}$$

Eliminating the integration constant between Eqs. (7.1.7) and (7.1.8), we have

$$\left(\frac{\partial\phi}{\partial r}\right)_{r=a} = -\zeta\left(\frac{1}{a} + \frac{1}{\lambda_D}\right) \tag{7.1.9}$$

With this result, surface charge can now be related to surface potential by means of Eq. (7.1.4), to give

$$q_s = \epsilon \zeta \left(\frac{1}{a} + \frac{1}{\lambda_D} \right) \tag{7.1.10}$$

From the definition of surface charge, we can write

$$\zeta = \frac{q}{4\pi\epsilon a(1 + a/\lambda_D)} \tag{7.1.11a}$$

or

$$\zeta = \frac{q}{4\pi\epsilon a} - \frac{q}{4\pi\epsilon(a + \lambda_D)} \tag{7.1.11b}$$

In other words, the ζ potential is the sum of two superposed potentials: one arising from a charge q on the surface of radius a, and a second arising from a charge $-q$ on a sphere of radius $a + \lambda_D$. But the sum is just the net potential between two concentric spheres carrying equal and opposite charges and differing in radius by λ_D.

In the small Debye length limit, from Eq. (7.1.10),

$$q_s = \frac{\epsilon\zeta}{\lambda_D} \tag{7.1.12}$$

When treating the small Debye length case, we shall frequently use surface charge density and ζ potential interchangeably, relating them through Eq. (7.1.12).

The use of constant potential and constant surface charge will generally bracket the actual conditions. In the case of charged polyelectrolytes, for example, where there are fixed charge sites, the surface charge is relevant. For colloids where the surface charge density varies with the solution state, the surface potential or ζ potential would appear to be the more appropriate variable.

7.2   Electrophoresis

In this section we consider the electrophoretic motion of a charged spherical particle in an electrolyte solution under the influence of an applied electric field (Fig. 7.2.1). The particle is assumed rigid and generally considered to be nonconducting with a uniformly distributed surface charge. The assumption of a nonconducting particle is usually appropriate inasmuch as most conducting particles become polarized by the applied field, preventing the passage of current through the particle and so causing it to behave like a nonconductor.

As shown in Fig. 7.2.1, a negatively charged particle will move in the direction opposite to the electric field. In general, because of the motion, the

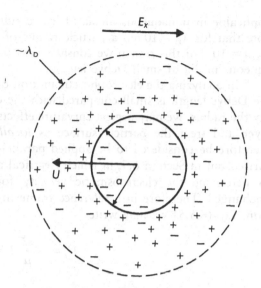

Figure 7.2.1 Electrophoretic motion of spherical particle.

diffuse double layer will not remain concentric with the sphere but will become distorted. This results in the particle being retarded. However, for both small and large Debye lengths, the distortion is small and may be neglected. We therefore examine these two limits first.

We begin by first considering large Debye length, where λ_D is large compared with the radius a. In this case the particle may be treated as a point charge in an unperturbed electric field E_x. Equating the electrical force with the Stokes drag on the particle, we obtain

$$qE_x = 6\pi\mu Ua \tag{7.2.1}$$

where q is the net charge between the charged sphere and the concentric spherical double layer of predominantly opposite charge. We note that the use of Eq. (7.2.1), is appropriate to an isolated charge and hence large λ_D, since the isolated charge corresponds to the condition of concentration approaching zero, that is, λ_D tending to infinite.

From the above equation and the formula relating ζ potential and charge (Eq. 7.1.11a), the electrophoretic velocity is

$$U = \frac{2}{3}\frac{\zeta\epsilon(1 + a/\lambda_D)E_x}{\mu} \approx \frac{2}{3}\frac{\zeta\epsilon E_x}{\mu} \tag{7.2.2}$$

Now, in the electrophoresis literature reference is often made to the electrophoretic mobility U/E_x or velocity per unit field, with units $C\,s\,kg^{-1}$ (cf. Eq. 2.5.3) in place of electrophoretic velocity. Here, we shall generally refer to the electrophoretic velocity.

Equation (7.2.2) with the term in a/λ_D neglected is known as the *Hückel equation*. It is not generally applicable in aqueous media, although it may be applicable in nonaqueous media of low conductivity. For example, we would note that for $\lambda_D = 10$ nm a particle radius of 1 nm would be required to give $\lambda_D/a = 10$. For this reason we consider next the more relevant limiting case for aqueous media of small Debye length.

In analyzing the electrophoretic motion of a nonconducting particle where the Debye length is small compared with the characteristic particle dimension, say the radius, we may neglect curvature effects in the diffuse part of the double layer and treat the particle surface as *locally plane*. The electric field may therefore be considered to be applied parallel to the surface, and the analysis carried out in Section 6.5, in which electrical and viscous forces were balanced to determine the electroosmotic velocity for a fixed surface, applies here unchanged. Therefore in a reference frame in which the particle is stationary, from Eq. (6.5.5) we may write

$$U_t = -\frac{\epsilon\zeta}{\mu} E_t \qquad (7.2.3)$$

Here, u_t is the velocity of the liquid tangential to the surface, and E_t is the applied electric field tangential to the surface. The tangential velocity and electric field are not constant but vary along the surface and are coupled through this equation.

To specify the velocity distribution the electric field distribution must be determined and our procedure follows in outline that of Morrison (1970). With $\mathbf{E} = -\nabla\phi$, the electric field is defined by the Laplace equation

$$\nabla^2\phi = 0 \qquad (7.2.4)$$

subject to the boundary conditions that the normal component of the current density vanishes at the surface, $\mathbf{n}\cdot\nabla\phi = 0$, while far from the particle, the potential approaches the value corresponding to the uniform applied field E_x.

Recall that within the small double layer approximation, the velocity "slips" at the wall going from u_t to zero discontinuously. Now irrotational flow defined by the Laplace equation is a solution for the velocity field that admits of a "slip" condition at the surface but also satisfies the full Navier–Stokes equations, although not of course the usual no-slip boundary condition. Such an irrotational flow exerts no force or moment on the particle and the velocity is derivable from a potential, that is, $u = -\nabla\Phi$ where Φ is the velocity potential. The velocity must also satisfy the boundary conditions of no normal flow through the surface, whence $\mathbf{n}\cdot\nabla\Phi = 0$, while far from the particle the velocity potential approaches the value corresponding to the uniform velocity U.

The slip flow condition at the surface is given by Eq. (7.2.3) which can also be written in terms of the potentials Φ and ϕ. As we have seen, however, both the differential equations and boundary conditions are identical for the electric potential and velocity potential so that from Eq. (7.2.3)

$$\Phi = -\frac{\epsilon\zeta}{\mu}\phi \qquad (7.2.5)$$

This equation can be expressed in terms of the velocity and electric field and on changing to a reference frame in which the particle is moving relative to a stationary fluid, we have

$$U = \frac{\epsilon \zeta E_x}{\mu} \qquad (7.2.6)$$

This is just the Helmholtz-Smoluchowski equation, as might have been expected, since electrophoresis is just the complement of electroosmosis. Its derivation shows that the electrophoretic velocity of a nonconducting particle is independent of the particle size and shape for a constant surface potential when the Debye length is everywhere small compared with the characteristic body dimension. Note that Eq. (7.2.6) differs from the Hückel large Debye length result (Eq. 7.2.2) only by the factor $\frac{2}{3}$.

With a finite-thickness double layer we may distinguish three effects that will alter the electrophoretic velocity from that given by the Helmholtz-Smoluchowski or Hückel relations. These effects, which in general are not mutually exclusive, are termed *electrophoretic retardation, surface conductance,* and *relaxation* (Shaw 1969).

We first consider *electrophoretic retardation*, which results from the fact the ions in the double layer will have a net movement opposite to that of the particle. Through drag interaction this creates a local electroosmotic flow of the solvent that opposes the motion of the particle. Henry (1931) accounted for this effect and calculated the electrophoretic velocity for spherical particles with arbitrary values of λ_D and particle conductivity. Henry's results for conducting particles do not take into account that the particles will become polarized at the surface and therefore are of limited applicability (Levich 1962). For nonconducting particles he showed that within the assumptions made, the limits for small and large Debye lengths were the velocities given by the Hückel and Helmholtz-Smoluchowski relations, respectively. Here, we shall restrict the discussion to nonconducting particles.

The three principal assumptions made by Henry in his analysis were:

1. The double layer is undistorted, and the potential field in the double layer arising from the particle charge can be superposed with the potential field about the spherical particle that results from the application of an electric field parallel to the direction of motion.
2. The surface potential on the particle is low enough that the Debye-Hückel approximation is applicable.
3. The viscous flow generated is inertia free.

Note that the mutual distortion of the fields resulting from their interaction, which Henry did not take into account in his 1931 paper, is what gives rise to the relaxation and surface conductance effects discussed later.

The details of Henry's solution are somewhat lengthy, so we only outline it here. The particle is taken to have a constant electrophoretic velocity U as a consequence of applying a uniform electric field E_x, as in Fig. 7.2.1. It is convenient to examine the problem in a reference frame in which the particle is stationary and the fluid has the steady velocity $-U$ at infinity, as shown in Fig. 7.2.2, with the electric field positive in the direction of the positive x axis. The polar coordinate system used is shown in Fig. 7.2.2.

We begin by writing the solution for the electric field about a nonconducting sphere, remembering that this field will simply be added to that produced by the charge separation in the double layer. With $\mathbf{E} = -\nabla\phi$, the applied electric field satisfies the Laplace equation (7.2.4) subject to the boundary conditions $\phi = -E_x r \cos\theta$ as $r \to \infty$, and that the normal component of the current density vanishes at the surface, whence $\partial\phi/\partial r = 0$ at $r = a$. This well-known electrostatics solution is given by

$$\phi = -E_x\left(r + \frac{1}{2}\frac{a^3}{r^2}\right)\cos\theta \qquad (7.2.7)$$

Now the total electric potential at any point is taken equal to the sum of the potential due to the applied electric field plus the potential in the double layer, and this sum must satisfy Poisson's equation. Denoting the potential in the double layer by ψ, we see that

$$\nabla^2\psi = -\frac{\rho_E}{\epsilon} \qquad (7.2.8)$$

since $\nabla^2\phi = 0$. In accordance with the assumptions of the model set out above, the potential $\psi = \psi(r)$ is evaluated by using the Debye-Hückel approximation, which for spherical symmetry from Eqs. (7.1.7) and (7.1.8) gives

$$\psi = \zeta\left(\frac{a}{r}\right)e^{-(r-a)/\lambda_D} \qquad (7.2.9)$$

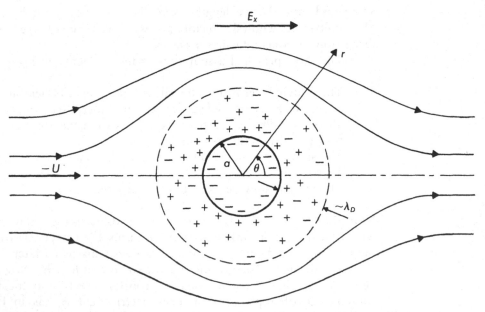

Figure 7.2.2 Electrophoretic motion in reference frame in which particle is stationary. Arrows indicate directions of streamlines and electric field lines.

It remains to evaluate the inertia free, viscous velocity field subject to the combined electric field force. The Navier-Stokes equation in this case is

$$-\mu\nabla^2 u + \nabla p = -\rho_E \nabla(\phi + \psi) \qquad (7.2.10)$$

where ρ_E is expressible in terms of ψ through Eq. (7.2.8). This equation, together with the continuity equation

$$\nabla \cdot u = 0 \qquad (7.2.11)$$

and the boundary conditions

$$u_r = -U\cos\theta \quad u_\theta = U\sin\theta \quad \psi = 0 \qquad \text{as } r \to \infty \qquad (7.2.12a)$$

$$u_r = u_\theta = 0 \quad \psi = \zeta \qquad \text{at } r = a \qquad (7.2.12b)$$

define the velocity distribution about the particle.

At this point the solution procedure becomes somewhat lengthy and detailed, although examination of the linear system shows that a solution can "in principle" be obtained. See Henry (1931) and Russel et al. (1989) for details of two approaches. Employing the solution so obtained for the pressure distribution and the velocity distribution about the sphere, we can calculate the x component of the stress normal to any point of the surface τ_{rx}. The hydrodynamic force on the sphere is then $2\pi a^2 \int_0^\pi \tau_{rx} \sin\theta \, d\theta$. Adding to the hydrodynamic force the force due to the fixed surface charge $-4\pi\epsilon a^2(\partial\psi/\partial r)_{r=a} E_x$ (Eq. 7.1.3) gives the total force on the sphere:

$$-6\pi\mu a U + 6\pi\zeta\epsilon E_x a\left(1 + 5a^5 \int_\infty^a \frac{\psi/\zeta}{r^6} \, dr - 2a^3 \int_\infty^a \frac{\psi/\zeta}{r^4} \, dr\right) = 0 \qquad (7.2.13)$$

This total force must vanish for steady motion, so it has been set equal to zero.

Using the Debye-Hückel solution for the potential distribution in the double layer enables Eq. (7.2.13) to be integrated to give a result that may be written in the form

$$U = \frac{2}{3}\frac{\zeta\epsilon E_x}{\mu} f(\alpha) \qquad (7.2.14a)$$

where

$$f(\alpha) = 1 + \frac{1}{16}\alpha^2 - \frac{5}{48}\alpha^3 - \frac{1}{96}\alpha^4 + \frac{1}{96}\alpha^5$$

$$+ \frac{1}{8}\alpha^4 e^\alpha\left(1 - \frac{\alpha^2}{12}\right)\int_\infty^\alpha \frac{e^{-t}}{t} \, dt \qquad (7.2.14b)$$

and where $\alpha = a/\lambda_D$ is the inverse of the Debye length ratio used earlier. This equation is known as the *Henry equation*. The Henry equation has the property that in the limits of large and small Debye lengths it gives, respectively, the Hückel and Helmholtz-Smoluchowski velocities. The function $f(\alpha)$ increases monotonically with α from 1 to $\frac{3}{2}$, as seen in Fig. 7.2.3.

We return now to consider one of the other two electrophoretic effects mentioned earlier, that of *surface conductance*, which it was noted was not taken into account in Henry's 1931 paper. This phenomenon arises because with a finite-thickness double layer there is a region of the flow near the surface in which charge neutrality is absent and in which there is an excess of counterions compared with the bulk of the electrolyte. The excess counterion concentration gives rise to a region of higher conductivity in which the applied electric field is reduced. In what follows we estimate the effect of this surface conductance in the limit where $\lambda_D/a \ll 1$ but finite. This calculation was published independently by Booth (1948) and Henry (1948).

To determine the distribution of electric potential, as modified by surface conductance, we again take the electric field to be spherically symmetric and to satisfy the Laplace equation. A thin spherical double layer shell is considered to surround the particle, and the conductivity of this shell is taken to have the mean value σ_s'. In reality the conductivity in the thin double layer varies continuously. Outside of the double layer shell the bulk conductivity σ_b is that of the electrolyte. This electrostatics problem is a straightforward one in which, from the Laplace equation, the solution for the potential is

$$\phi_f = -E_x\left(r + \frac{Aa^3}{r^2}\right)\cos\theta \qquad \text{in fluid} \qquad (7.2.15a)$$

and

$$\phi_s = -E_x\left(Br + \frac{Ca^3}{r^2}\right)\cos\theta \qquad \text{in double layer} \qquad (7.2.15b)$$

where A, B, and C are constants to be determined by the boundary conditions. Note that this is the solution form previously given by Eq. (7.2.7) for the applied potential.

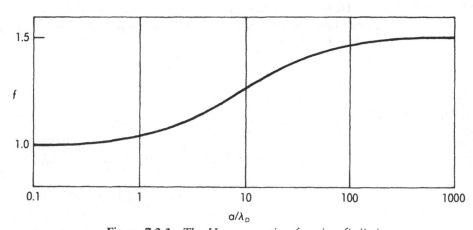

Figure 7.2.3 The Henry equation function $f(a/\lambda_D)$.

The boundary conditions appropriate to the above solution are continuity of ϕ and continuity of current density $\sigma \nabla \phi$ at the nonconducting sphere surface $r = a$, and at the edge of the double layer $r = a + \delta$, where δ is the thickness of the double layer shell. From straightforward calculation

$$\frac{Aa^3}{(a+\delta)^3} = \frac{2(a+\delta)^3(\sigma_b - \sigma_s') + a^3(\sigma_b + 2\sigma_s')}{2(a+\delta)^3(2\sigma_b + \sigma_s') + 2a^3(\sigma_b - \sigma_s')} \tag{7.2.16}$$

We digress for a moment to recall that surface current density, in parallel with the definition of surface charge, is the limit of the current distribution flowing within a surface layer of finite thickness δ when $\delta \to 0$, where the current per unit width of surface in the direction normal to that of the current flow remains finite. The surface current density is expressed through the relation

$$\mathbf{i}_s = -\sigma_s \nabla_s \phi \tag{7.2.17}$$

where σ_s is the surface conductance and ∇_s is the surface gradient.

In Eq. (7.2.16) identifying the product $\sigma_s' \delta$ with the surface conductance σ_s, expanding for δ/a small, and taking the limit as $\delta/a \to 0$ lead to the result

$$A = \frac{\sigma_b - 2\sigma_s/a}{2(\sigma_b + \sigma_s/a)} \tag{7.2.18}$$

In the limit $\delta/a \to 0$, it can be assumed that the exterior applied field extends to the surface $r = a$, and the potential there is given by

$$\phi_f = -(1 + A)E_x a \cos \theta \tag{7.2.19}$$

Without surface conductance, corresponding to $\sigma_s/a = 0$, the constant $A = \frac{1}{2}$, and the potential at the surface is

$$\phi = -\tfrac{3}{2} E_x a \cos \theta \tag{7.2.20}$$

This is the value given by the potential solution used in the electrophoretic retardation analysis (Eq. 7.2.7).

Examination of Eqs. (7.2.19) and (7.2.20) shows that if a new reduced "effective" electric field $(E_x)_{eff} = \frac{2}{3}(1 + A)E_x$ is defined, then the solution for the electrophoretic velocity with surface conductance will be exactly the same as given by Eq. (7.2.14) with E_x replaced by $(E_x)_{eff}$, whence

$$U = \frac{2}{3} \frac{\zeta \epsilon E_x}{\mu(1 + \sigma_s/\sigma_b a)} f\left(\frac{\lambda_D}{a}\right) \tag{7.2.21}$$

Expressed in terms of surface charge by means of Eq. (7.1.10),

$$U = \frac{2}{3} \frac{q_s \lambda_D E_x}{\mu(1 + \sigma_s/\sigma_b a)} \frac{f(\lambda_D/a)}{(1 + \lambda_D/a)} \tag{7.2.22}$$

What can be seen is that the effect of surface conductance is strongest in poorly conducting solutions.

The order of magnitude of the surface conductance may be estimated in the small Debye length limit by noting that approximately

$$i_s = q_s U = \frac{q_s^2 \lambda_D E_x}{\mu} = \sigma_s E_x \tag{7.2.23}$$

from which

$$\frac{\sigma_s}{\sigma_b a} = \frac{q_s^2}{\mu \sigma_b} \frac{\lambda_D}{a} \tag{7.2.24}$$

This shows that the first-order correction for surface conductance is $O(\lambda_D/a)$. For $\lambda_D/a \ll 1$ with $q_s^2/\mu\sigma_b \lesssim O(1)$, Eq. (7.2.22) reduces to the Helmholtz-Smoluchowski result.

The last effect mentioned at the outset in connection with a finite-thickness double layer is that of *relaxation*. Under the action of the applied electric field the ions in the double layer have a net movement opposite to the particle. The ions drag along the liquid, setting up a local electroosmotic flow that opposes the particle motion. It is this effect that is taken into account in the analysis of Henry. However, the motion of the ions also distorts the double layer from sphericity; that is, the electric field tends to strip the double layer from the particle and make it asymmetric so that the center of the double layer lags behind the center of the particle (Fig. 7.2.4). The relaxation effect has its origin in this asymmetry, the name being used because it is a consequence of the finite time required for the adjustment of the double layer to the original symmetry by electromigration and diffusion in the moving system.

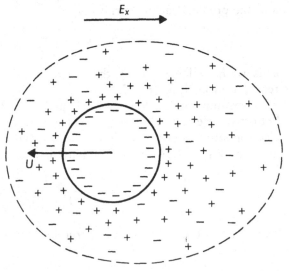

Figure 7.2.4 Distortion of double layer around particle in electrophoretic motion.

For the example of Fig. 7.2.4, as the particle moves to the left the cations in the liquid are left behind; restoring the symmetric double layer from the cations in the new region of liquid entered by the particle requires a finite time. On average then, the double layer is not concentric with the particle, and the displacement of charge sets up a counter emf that reduces the electrophoretic velocity.

A proper calculation of the relaxation effect requires that the velocity distribution in the liquid be explicitly taken into account. Moreover, in general, there will be a mutual interaction between the relaxation effect and the electrophoretic retardation, which must now be determined for an asymmetric atmosphere. If the total electric potential is calculated with the charge density found from solving the Nernst-Planck equation (together with the Navier-Stokes equation), the surface conductance will then also be accounted for. Analytic solution procedures are fairly complex for the geometry at hand. Numerical solutions have, however, been carried out by Wiersema et al. (1966). At low ζ potentials, for which analytic solutions have also been determined, the behavior of the results is much the same as given by the Henry equation (Fig. 7.2.3). At large and small Debye lengths ($a/\lambda_D \lesssim 0.1$ and $a/\lambda_D \gtrsim 300$) the relaxation effect can generally be neglected. For these reasons, we shall not discuss this effect further, other than to refer the interested reader to Wiersema et al., where numerical results are given and where reference to other work on the subject may also be found. Additional references are in Russel et al. (1989).

7.3   Electrophoretic Separations

Electrophoresis is a powerful tool in the separation and analysis of colloids, proteins, and nucleic acids. There are three major electrophoretic techniques as well as variations, each with its own name. Which one is applicable or most appropriate depends on the size and characteristics of the dissolved or suspended material and the type of information desired. The three principal techniques are *microelectrophoresis, moving boundary electrophoresis,* and *zone electrophoresis* (Shaw 1980).

In microelectrophoresis the particle velocity is observed directly with a microscope (or ultramicroscope). It is applicable to reasonably stable colloidal suspensions or emulsions containing microscopically visible particles or droplets in sufficiently dilute concentration that the individual particles can be distinguished. The electrophoretic behavior is then measured directly.

The measurement of particle velocity in microelectrophoresis is carried out with a type of apparatus shown schematically in Fig. 7.3.1 (Shaw 1969). The apparatus consists basically of a microscope with a calibrated reticule for the observation of the individual particles, a thin flat cell of rectangular cross section, although long cylindrical cells are also used, electrodes across which the potential is applied, and a system of tubing and stopcocks for filling and cleaning the cell and electrode compartments.

Because the internal glass surfaces of the cell generally acquire a charge, there is a double layer formed on the cell walls that gives rise to an electro-

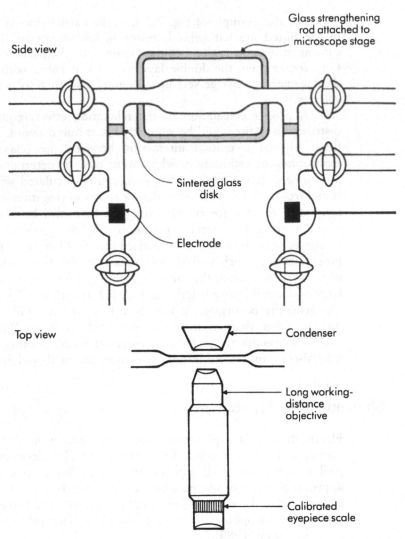

Figure 7.3.1 Vertically mounted flat microelectrophoresis cell. [Shaw, D.J. 1969. *Electrophoresis*. New York: Academic. With permission.]

osmotic flow of the liquid near the walls. If the double layer is sufficiently thin, the electroosmotic flow may be taken to be a constant "slip velocity" at the walls. In fact, because of viscous effects, the velocity must drop to zero right at the walls in a layer on the order of a Debye length in thickness. Since the cell is closed, the electroosmotic flow in turn causes a compensating return flow of liquid with a maximum velocity at the center. The velocity distribution is parabolic because of the Poiseuille character of the flow. However, the liquid moves in one direction at about a Debye length from the wall, more slowly further from the wall, and in the opposite direction at the center of the cell (see Fig. 7.3.2). At some point in the cell there is no net motion of the liquid as the result of the ζ potential of the cell wall–liquid interface. This point where the

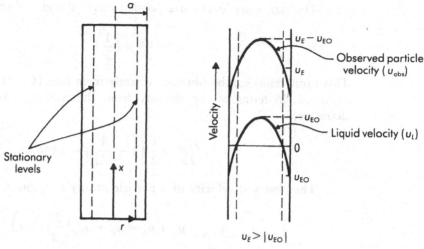

Figure 7.3.2 Stationary levels and velocity distributions in cylindrical electrophoresis cell.

electroosmotic flow and the return flow of the liquid cancel is called the *stationary level*. At the stationary level the observed velocity of the particles is equal to their electrophoretic velocity (Fig. 7.3.2).

We next calculate the stationary level and observed particle velocity in a cylindrical capillary of radius a, assuming $\lambda_D/a \ll 1$. The electroosmotic effect gives rise to a velocity u_{EO} across the cross section of the tube toward the electrode of the same polarity as the charge on the cell wall. The liquid velocity u_L across the tube according to the description above, is the vector sum of the electroosmotic velocity and the reverse Poiseuille flow:

$$u_L = u_{EO} - c(a^2 - r^2) \tag{7.3.1}$$

where c is a constant.

The value of c is determined by the condition that the net liquid flux across the cross section must be zero:

$$\int_0^a 2\pi r u \, dr = 0 \tag{7.3.2}$$

Substituting the velocity distribution of Eq. (7.3.1) into (7.3.2) and integrating, we get

$$c = \frac{2u_{EO}}{a^2} \tag{7.3.3}$$

from which

$$\frac{u_L}{u_{EO}} = \frac{2r^2}{a^2} - 1 \tag{7.3.4}$$

The stationary level corresponds to $u_L = 0$ and is therefore at

$$\frac{r_{\text{stat}}}{a} = \left(\frac{1}{2}\right)^{1/2} \tag{7.3.5}$$

This same result can be obtained directly from Eqs. (6.5.24) and (6.5.23) with $\lambda_D/a \ll 1$. In terms of the distance from the wall y_{stat} as a fraction of the diameter,

$$\frac{y_{\text{stat}}}{2a} = \frac{1}{2}\left(1 - \frac{1}{\sqrt{2}}\right) = 0.146 \tag{7.3.6}$$

The observed velocity of a particle at any position in the cell is

$$u_{\text{obs}} = u_E + u_L = u_E + u_{\text{EO}}\left(\frac{2r^2}{a^2} - 1\right) \tag{7.3.7}$$

where u_E is the true electrophoretic velocity. If u_E is known from observation at the stationary level, the electroosmotic velocity can be calculated from the particle velocities at other positions in the cell. Note that with the small Debye length assumption, if the particles and the cell wall have the same ζ potential then from the Helmholtz-Smoluchowski equation $u_E = -u_{\text{EO}}$, and according to Eq. (7.3.7) the velocity at the center of the cell is $2u_E$. This should not be surprising, since it is equivalent to the maximum velocity in an ordinary Poiseuille flow being equal to twice the mean velocity.

Similarly, for a plane cell of depth $2h$ (in place of $2a$) with y distance from the bottom or top of the cell, we have from Eq. (4.2.14) that

$$u_L = u_{\text{EO}} - c(2yh - y^2) \tag{7.3.8}$$

Again, setting the net flux equal to zero, we find

$$c = \frac{3}{2}\frac{u_{\text{EO}}}{h^2} \tag{7.3.9}$$

from which

$$\frac{u_L}{u_{\text{EO}}} = 1 - \frac{3}{2}\left(\frac{2y}{h} - \frac{y^2}{h^2}\right) \tag{7.3.10}$$

For the plane cell, $u_L = 0$ when

$$\frac{y_{\text{stat}}}{2h} = \frac{1}{2}\left(1 - \frac{1}{\sqrt{3}}\right) = 0.211 \tag{7.3.11}$$

The observed velocity is again $u_E + u_L$, whence

$$u_{\text{obs}} = u_E + u_{\text{EO}}\left[1 - \frac{3}{2}\left(\frac{2y}{h} - \frac{y^2}{h^2}\right)\right] \tag{7.3.12}$$

It follows that at the channel center $y = h$, the observed particle velocity $u_{obs} = u_E - u_{EO}/2$. Therefore, in the special case when the particles and the cell walls have the same ζ potential so that $u_{EO} = -u_E$, the observed velocity at the cell center is $\frac{3}{2}u_E$. Again this is equivalent to the maximum velocity in Poiseuille flow in a two-dimensional channel being equal to three halves the mean velocity.

In microelectrophoresis the measurement is carried out by filling the cell with the suspension and applying a known potential. With an objective lens that gives a small depth of focus, the microscope is focused at the stationary level and the time for a particle to move a known distance is measured. This technique is also applicable with several different types of particles present.

An alternative technique when the particles are too small to be readily seen by an ultramicroscope is that of *moving boundary electrophoresis*, which has its parallel in hindered settling. Here, the motion of the interface formed between a zone of the suspension and the solvent or dispersion medium is measured under the influence of an electric field. This technique is particularly useful for separating and identifying dissolved macromolecules such as proteins.

A schematic illustration of the apparatus used in this technique, in its simplest form, is shown in Fig. 7.3.3. It consists of a U-tube fitted with electrodes at the top of each arm of the U. The suspension is added to the bottom of the U. A voltage is applied; that is, a current is passed through the cell, as a consequence of which the boundary will migrate. This migration is followed by an optical technique, usually a schlieren system as with sedimentation measurements in a centrifugal field.

In the late 1930s A. Tiselius modified the apparatus (Fig. 7.3.4) so that the cell arms are rectangular rather than circular, thus permitting better visibility of

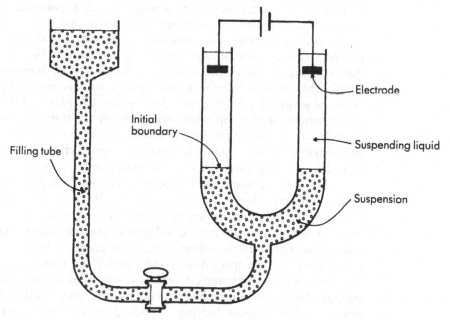

Figure 7.3.3 Moving boundary electrophoresis.

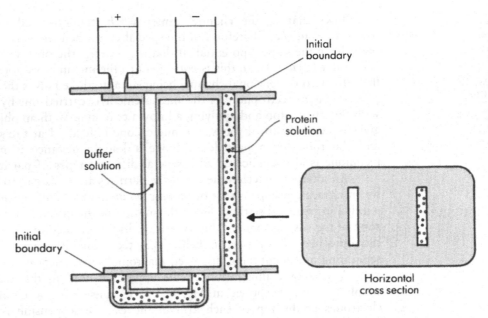

Figure 7.3.4 Tiselius moving boundary electrophoresis apparatus. [After Shaw, D.J. 1980. *Introduction to Colloid and Surface Chemistry*, 3rd edn. London: Butterworths. With permission.]

the interface. In addition, the boundary surface is established by sliding together previously filled upper and lower portions of the cell. The upper portion contains the suspending liquid, and the lower portion contains the suspension. In this way the initial boundary is well defined, although as it moves diffusion will in time broaden the interfaces (ascending and descending boundary), as discussed, for example, in connection with centrifugal sedimentation. In this regard, sedimentation effects should be minimized by keeping the particle density close to that of the suspending liquid and by keeping the particle sizes small. This latter requirement will tend to amplify any diffusional effects. As with the sedimentation processes discussed in Chapter 5, if the solution consists of a number of electrophoretically different fractions, the sharp peak corresponding to the initial boundary, either ascending or descending, will not only broaden but will split up into different broadened characteristic peaks moving at different speeds, giving rise to a chromatographic pattern.

The moving boundary method has the same complication of a return flow as the capillary. However, because the U-tube is relatively wide, the resultant induced flow is essentially uniform over the cross section. The Tiselius method was for some time after its development an important analytical technique where electrophoretic mobilities were not required, as in protein analysis. Today it has been largely superseded by the simpler and more effective zone method.

In *zone electrophoresis* a homogeneous solid or gel framework is used to support the solution and minimize convective effects arising from temperature gradients due to Joule heating. In this procedure the supporting material is saturated with the buffer solution, and a small amount of the macromolecular solution being examined is applied as a narrow band. When a potential difference is applied between the ends of the strip, each component of the solution migrates at a rate determined primarily by its electrophoretic mobility. After an appropriate time the strip is removed, dried, and developed as appropriate to reveal the positions and concentrations of the various components in the initial solution. The procedure is seen to be analogous to zonal sedimentation discussed in Section 5.5 in connection with centrifugal sedimentation.

Some supports interact only weakly with the macromolecular solutes, and among these are filter paper, cellulose, and cellulose acetate membranes. Other supports retard the motion of some molecules with respect to others: for example, polyacrylamide and agarose gels, which discriminate by size and have a molecular sieving effect, and ion exchange papers, which retard charged molecules. Their use here is analogous to size exclusion chromatography discussed in Section 4.7. The use of retarding supports will tend to increase the sharpness of the separations and minimize the diffusion and dispersion effects that tend to broaden the bands. When retarding gels are used, the process is termed *gel electrophoresis*.

Even without molecular sieving or charge retardation associated with the support, observed electromigration velocities will generally be affected by electroosmotic flow and by capillary flow through the porous medium. These flow effects make the process unsuitable for mobility measurements. However, by somewhat empirical means, it is today the principal analytical procedure used for protein and amino acid analysis because it is simple, cheap, enables complete separation of all electrophoretically different components, and because small samples can be studied, which is often important for biochemical analyses.

Another important variant on zone electrophoresis is *isoelectric focusing* (Righetti 1983). Its operation is based on the fact that, as discussed in Section 7.1, the charge on proteins, macromolecules, and many colloidal particles depends on pH. Such particles exhibit an *isoelectric point*, which is the pH at which there is no average net charge on the particle. Below the isoelectric pH the particle will be positively charged; above it, negatively charged. Evidently at the isoelectric point the electrophoretic mobility will be zero, and above this point the particle will move toward the cathode and below this point it will move toward the anode.

In isoelectric focusing, electrophoresis measurements are carried out in a pH gradient so that the particle being examined will migrate until it reaches the pH of its isoelectric point. Generally, gels are used, and the pH gradient is established through the use of a mixture of ampholytes, each of which under the action of an electric field comes to rest near its isoelectric point. *Ampholytes* are molecules that have positive and negative charges, for example, polymers containing significant amino and carboxyl groups. At the isoelectric point, since the net migration of the particle under study is zero, convection of the particle by diffusion must just be balanced by the particle migration in the electric field; thus

$$D \nabla c_i = v_i z_i F c_i \mathbf{E} \qquad (7.3.13)$$

The focusing phenomenon discussed in Section 6.6 in connection with the removal of metal ions from soils under the action of an electric field is directly analogous to isoelectric focusing. There, the dominant transport processes are electromigration and diffusion, and pH gradients are set up between the anode and cathode as a consequence of the electrolysis of water. Precipitated metal was found to accumulate at the pH of minimum solubility which coincided with a point between the electrodes where a sharp jump in the pH occurred. This point of minimum solubility generally occurs at the isoelectric point, that is the point at which the concentration of the negative ions and positive ions are equal. For the zinc removal discussed, these were mainly zinc ions (Zn^{2+}) which migrated from the anode toward the cathode and zincate ions ($HZnO_2^-$) which formed in the high pH cathode region and then migrated back toward the anode. The isoelectric focusing effect reported in Probstein & Hicks (1993) was suggested much earlier by Gray & Schlocker (1969).

Many readers will by now have recognized that the procedure of isoelectric focusing has another analogue in density-gradient centrifugation discussed in Section 5.5.

A last variant we mention is *capillary zone electrophoresis* (Gordon et al. 1988). It employs an electroosmotically driven flow in a capillary, arising from an electric field applied parallel to the capillary, which is charged when in contact with an aqueous solution (Section 6.5). The flow has a nearly flat velocity profile (Fig. 6.5.1), thereby minimizing broadening due to Taylor dispersion of the electrophoretically separated solute bands.

In concluding this discussion of electrophoretic separations, we note a general model of these processes developed by Saville & Palusinski (1986) in which the effect of chemical reactions on the dissolved species is taken into account. The approach is analogous to that described for the effects of chemical reactions on electroosmosis and electromigration in Section 6.6.

7.4   Sedimentation Potential and Streaming Potential

In electrophoresis, where charged particles move relative to a stationary liquid, a potential is developed. In the case of sedimentation under gravity, pictured in Fig. 7.4.1, or sedimentation in a centrifugal field, the potential so developed by the particle motions is termed the *sedimentation potential*. The nature and magnitude of the potential can be understood by reference to the complementary phenomena of electroosmosis for the capillary flow considered in Section 6.5. Following the description of Shaw (1969), we note that the liquid flowing relative to the fixed charged wall carries a net charge, that of the electric double layer. The flow therefore gives rise to what is termed a *streaming current* and consequently a potential difference. The potential opposes the mechanical transfer of charge by causing back conduction by ion diffusion and, to a much lesser extent, by electroosmotic flow. The transfer of charge due to these two effects is called the *leak current*. The potential measured at the equilibrium condition when the streaming current and the leak current cancel is termed the *streaming potential*.

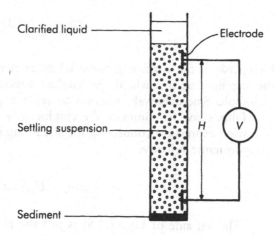

Figure 7.4.1 Sedimentation potential.

We may calculate the streaming potential for a capillary from the axial current flow, which, according to Eq. (3.4.4), for a fully dissociated salt is

$$i = Fz(c_+ - c_-)u - DFz \frac{\partial}{\partial x}(c_+ - c_-) - \frac{z^2F^2D}{RT}(c_+ - c_-)\frac{\partial \phi}{\partial x} \quad (7.4.1)$$

Here, for simplicity we have taken the mobilities to be equal; that is $D_+ = D_- = D$. The total current is then given by integrating across the capillary cross section:

$$I = \int_0^a i2\pi r\, dr \quad (7.4.2)$$

Using the concentration and velocity distributions calculated in Section 6.5, we obtain the total current by quadrature. We do not carry out the integrations here but merely note that within the Debye-Hückel approximation the current corresponding to the constant surface potential solution represented by Eqs. (6.5.23) and (6.5.24) is

$$I = -\pi a^2 \frac{d\Phi}{dx}\left\{\sigma_b + \frac{2}{a}\left[\frac{\epsilon^2\zeta^2}{2\mu\lambda_D} + \frac{D\epsilon}{\lambda_D}\left(\frac{zF\zeta}{RT}\right)^2\right]\right\} + \frac{\pi a^2\epsilon\zeta}{\mu}\frac{dp}{dx} \quad (7.4.3)$$

where, as before, σ_b is the conductivity of the bulk solution. Note further that the coefficient of dp/dx in the expression for I is identical to the coefficient of $d\Phi/dx$ in the expression for the electroosmotic volume flow rate Q with $\lambda_D/a \ll 1$ (Eq. 6.5.24). This is an example of the Onsager reciprocal relation of thermodynamics.

With $\lambda_D/a \ll 1$ and, from Eq. (7.1.12), $\zeta = q_s\lambda_D/\epsilon$, it follows that the terms multiplied by $2/a$ in Eq. (7.4.3) may be neglected. From the definition of the streaming potential as the equilibrium condition corresponding to the net current being zero, we may write

$$\Delta\Phi_{st} = \frac{\epsilon\zeta}{\sigma_b\mu}\,\Delta p \tag{7.4.4}$$

Here, $\Delta\Phi_{st}$ is the streaming potential difference developed between the ends of the capillary across which the applied pressure difference is Δp. Using the Helmholtz-Smoluchowski relation to replace ζ by the electroosmotic velocity and Ohm's law to eliminate the conductivity and electric field we obtain the following expression connecting the streaming current, streaming potential, and electroosmotic velocity:

$$I_{st}\Delta\Phi_{st} = U_{EO}\,\pi a^2 \Delta p \tag{7.4.5}$$

The left side of Eq. (7.4.5) is just the power developed by the streaming current, and the right side is the rate of work done by the shear force in causing the electroosmotic flow. This latter statement follows from the fact that, from an overall balance of forces in a steady, fully developed, viscous capillary flow in a circular capillary of radius a,

$$F \equiv 2\pi a\tau_w L = \pi a^2 \Delta p \tag{7.4.6}$$

where F = total shear force on the fluid
τ_w = wall shear
L = capillary length

We may therefore rewrite Eq. (7.4.5) as

$$P_{st} = FU_{EO} \tag{7.4.7}$$

where P_{st} is the power developed by the streaming current.

Let us return now to the evaluation of the sedimentation potential. Considering the inertia free fall of a charged particle in a liquid, the shear stress near the particle will cause a surface current to flow in a direction defined by the charge of the particle (front to back for a negatively charged particle, and vice versa). This current must then flow through the bulk of the solution in the opposite direction. In complete analogy with the electroosmotic case there is therefore a *sedimentation current* and corresponding sedimentation potential set up by the falling particle.

Since for small Debye length the situation is completely complementary to the electroosmotic case, we may apply the result of Eq. (7.4.5), identifying Δp as the force per unit particle cross-sectional area exerted by the fluid on the particle. Assuming Stokes flow, we use $6\pi\mu aU$ for the force on a single particle; if there are n particles per unit volume, then the total force per unit volume is taken to be nF. With the potential drop measured over the suspension height H, it readily follows that

$$\frac{\Delta \Phi_{sed}}{H} = - \frac{6\pi n U a \epsilon \zeta}{\sigma_b} \tag{7.4.8}$$

where U is the particle fall speed given by Eq. (5.4.6), and a is its radius.

For small Debye length the particle fall speed is not affected by the induced electric field (Newman 1991). Therefore,

$$\frac{\Delta \Phi_{sed}}{H} = \frac{\alpha g (\rho - \rho_{fl}) \epsilon \zeta}{\sigma_b \mu} \tag{7.4.9}$$

where to avoid confusion with potential we have used α to denote the volume fraction of the particles. From Eq. (7.4.9) the potential produced by sedimenting particles of uniform size could in principle be used to evaluate the ζ potential. However, the magnitude of this potential is small, and its measurement is difficult.

References

BOOTH, F. 1948. Surface conductance and cataphoresis. *Trans. Faraday Soc.* **44**, 955–959.

GORDON, M.J., HUANG, X., PENTONEY, S.L., JR. & ZARE, R.N. 1988. Capillary electrophoresis. *Science* **242**, 224–228.

GRAY, D.H. & SCHLOCKER, J. 1969. Electrochemical alteration of clay soils. *Clays Clay Miner.* **17**, 309–322.

HENRY, D.C. 1931. The cataphoresis of suspended particles. Part I–The equation of cataphoresis. *Proc. Roy. Soc.* **A133**, 106–129.

HENRY, D.C. 1948. The electrophoresis of suspended particles. IV. The surface conductivity effect. *Trans. Faraday Soc.* **44**, 1021–1026.

LEVICH, V.G. 1962. *Physicochemical Hydrodynamics.* Englewood Cliffs, N.J.: Prentice-Hall.

MORRISON, F.A., JR. 1970. Electrophoresis of a particle of arbitrary shape. *J. Colloid Interface Sci.* **34**, 210–214.

NEWMAN, J.S. 1991. *Electrochemical Systems*, 2nd edn. Englewood Cliffs, N.J.: Prentice-Hall.

PROBSTEIN, R.F. & HICKS, R.E. 1993. Removal of contaminants from soils by electric fields. *Science* **260**, 498–503.

RIGHETTI, P.G. 1983. *Isoelectric Focusing: Theory, Methodology and Applications.* Amsterdam: Elsevier Biomedical.

RUSSEL, W.B., SAVILLE, D.A. & SCHOWALTER, W.R. 1989. *Colloidal Dispersions.* Cambridge: Cambridge Univ. Press.

SAVILLE, D.A. & PALUSINSKI, O.A. 1986. Theory of electrophoretic separations. Part I: Formulation of mathematical model. *AIChE J.* **32**, 207–214.

SHAW, D.J. 1969. *Electrophoresis.* New York: Academic.

SHAW, D.J. 1980. *Introduction to Colloid and Surface Chemistry*, 3rd edn. London: Butterworths.

WIERSEMA, P.H., LOEB, A.L. & OVERBEEK, J.TH.G. 1966. Calculation of the electrophoretic mobility of a spherical colloid particle. *J. Colloid Interface Sci.* **22**, 78–99.

Problems

7.1   In problem 2.4 the movement of charged colloidal particles in water contained in an annular gap between two oppositely charged infinitely long cylinders was considered. The potential distribution across the gap $d\phi/dr$ was assumed constant, where ϕ is the potential and r is the radial cylindrical coordinate. It is desired here to take the analysis one step further and determine the potential and concentration distribution within the gap, neglecting any electrode effects.

a.   Write the mass flux j $(\text{kg m}^{-2}\,\text{s}^{-1})$ in terms of $d\phi/dr$, the particle concentration ρ, and the known parameters particle charge q, particle radius a, and water viscosity μ. Use the fact derived in Problem 2.4 that *the effect of diffusion can be neglected.*

b.   What is Poisson's equation in cylindrical coordinates for the spatial variation of the potential ϕ, where the charge distribution is expressed in terms of the permittivity ϵ, particle mass m, charge q, and concentration ρ?

c.   Show from conservation of mass that $jr = b$, where b is a constant.

d.   By integration determine the concentration distribution ρ and potential distribution ϕ across the annular gap in terms of r, two integration constants, the parameters $A = -q/6\pi\mu a$ and $B = -q/m\epsilon$, and the constant b.

7.2   It is desired to estimate the electrophoretic velocity U of a long, nonconducting, charged cylindrical particle of length L and radius a and with a low surface potential ζ, as a result of the application of an electric field E_x parallel to the symmetry axis. The Debye length is arbitrary but finite, and the flow is a low Reynolds number, inertia free one.

a.   Verify that the potential distribution about the cylinder takes the form $\phi = AK_0(r/\lambda_D)$, where r is the radial cylindrical coordinate and K_0 is the modified Bessel function of the second kind of order zero. Evaluate the constant A.

b.   Derive a relationship between the surface potential ζ and particle charge q. Note that $\partial/\partial x[K_0(x)] = -K_1(x)$ where K_1 is the modified Bessel function of order one.

c.   Assume that the resistance to the cylinder motion is due to the shear stress associated with the electroosmotic flow that is generated, so that the Navier-Stokes equation reduces to a balance between viscous and electrical forces. Show that the solution for the electrophoretic velocity of the cylinder is the same as that for a sphere of the same zero potential with the Debye length small.

d.   Why is the expression for the electrophoretic velocity, derived in part c, valid for all Debye lengths, if the cylinder length can be considered infinite?

e.   Brownian motion would tend to give the slender cylindrical particle a random orientation. If the applied field is sufficiently strong, will the particle acquire a preferred orientation? Explain.

7.3   A long thin cylindrical tube of radius $a = 5 \times 10^{-3}$ m and length $L = 0.1$ m
is closed at both ends by electrodes, across which a voltage drop $\Delta\phi = 10$ V is applied. The tube contains an ideal dilute aqueous solution of a
fully dissociated doubly charged symmetrical binary salt $(z = 2)$ at a
concentration $c_0 = 1$ mol m^{-3}. The tube wall has a fixed surface potential
$\zeta = 1.43 \times 10^{-3}$ V. The temperature $T = 25°C$, the permittivity $\epsilon = 7 \times 10^{-10}$ C V^{-1} m^{-1}, and the viscosity $\mu = 10^{-3}$ Pa s.
   a.   Determine the Debye length λ_D at the tube wall.
   b.   Using Eqs. (6.5.23) and (6.5.24) find the radial distance at which the
        water velocity is zero. Compare the result with Eq. (7.3.5).
   c.   Using Eqs. (6.5.23) and (6.5.24) find the water velocity at the tube
        center.
   d.   Find the water velocity at a distance λ_D from the wall $(r = a - \lambda_D)$.
        Sketch the velocity profile in the tube *down to the wall* as a function
        of the radial distance r.
   e.   Assume that spherical particles of radius R and charge q are—
        introduced with the number of particles small enough that the suspen-
        sion is dilute. What is the particle velocity at the radius where— the
        water velocity is zero, assuming $R = 10^{-6}$ m and $q = 26.3 \times 10^{-16}$ C.

7.4   A combination ultracentrifuge-electrophoresis apparatus has been proposed for macromolecular separations by placing electrodes at the top and
bottom of the centrifuge cell. Such a device takes advantage not only of
difference in mass but also charge.
   a.   How would Eq. (5.5.1) for the particle drift velocity in an ultracen-
        trifuge be modified if the particle charge is q and the strength of the
        applied electric field is E_r? Assume that the potential distribution is
        constant along the cell.
   b.   What potential gradient is required to keep a sedimenting band of
        spherical particles stationary at 0.1 m from the axis of rotation if the
        particle radius $a = 10^{-7}$ m, charge $q = 10^{-17}$ C, and the sedimentation
        coefficient is 10^{-12} s. The fluid viscosity $\mu = 10^{-3}$ Pa s, and the rotor
        speed is 6000 rad s^{-1}.

7.5   The process of employing electrophoretic migration to reduce buildup of
suspended charged particles on a filter surface through which the suspending liquid flows is termed *electrofiltration*. Consider the separation of a
charged macromolecular solute by ultrafiltration. The particles are negatively charged, and an electric field E_y is applied normal to the membrane
surface and directed away from the surface so as to augment the back
transport of particles away from the membrane.
   a.   Assuming a thin film model with gel formation at the membrane and
        that the potential distribution remains constant in the concentration
        polarization layer, estimate how the limiting flux would be altered.
   b.   Can concentration polarization be prevented altogether? Explain.

7.6   An experiment on sedimentation potential is to be carried out. The medium is water at a temperature $T = 25°C$, with a viscosity $\mu = 10^{-3}$ Pa s and a permittivity $\epsilon = 7 \times 10^{-10}$ C V^{-1} m^{-1}. The particle concentration in the water is $c_0 = 10$ mol m^{-3}, and the particle diffusivity is $D = 10^{-8}$ m^2 s^{-1}. The particles are spherical, have a density twice that of water, a radius $a = 10^{-5}$ m, and a volume concentration in the water of $\alpha = 0.02$. If the particle charge number $z = 1$ and if each particle carries a charge of $q = 2.89 \times 10^{-13}$ C, what potential difference will be detected across the electrodes, assuming the electrodes are separated by a distance $H = 1$ m.

8 Suspension Stability and Particle Capture

8.1 Colloid Stability

Up to this point we have considered distributed dilute dispersions of colloidal size particles and macromolecules in continuous liquid media. Where the particles are uncharged and of finite size, they are always separated by a fluid layer irrespective of the nature of the hydrodynamic interactions that take place. In the absence of external body forces such as gravity or a centrifugal field or some type of pressure filtration process, the uncharged particles therefore remain essentially uniformly distributed throughout the solution sample. We have also considered the repulsive electrostatic forces that act between the dispersed particles in those instances where the particles are charged. These repulsive forces will tend to maintain the particles in a uniform distribution. The extent to which a dispersion remains uniformly distributed in the absence of applied external forces, such as those noted above, is described in colloid science by the term *stability*, whereas colloidal systems in which the dispersed material is virtually insoluble in the solvent are termed *lyophobic colloids*.

In general it is an observed fact that given sufficient time, many two-phase colloidal dispersions will change into a smaller number of larger particles, as, for example, a suspension of oil droplets in water. This aggregation or "coarsening" is a consequence of the fact noted in Section 5.1 that an important characteristic in dealing with the microhydrodynamics of colloidal or macromolecular suspensions is the importance of surface forces, which generally increase in inverse proportion to the characteristic particle size. Specifically, particles may *coalesce*, wherein two small particles fuse together to form a single larger particle, with a reduction in total surface area. The particles may also *flocculate*, wherein they clump together to form a "floc" but do not really fuse into a new particle, although there may be a small area reduction where the particles of the floc touch (Hiemenz 1986). The attractive force between particles that gives rise to aggregation is principally a consequence of van der Waals surface forces, which become stronger as the particles approach each other. If the particles in the solution tend to aggregate, the solution is said to be *unstable*.

Therefore, whether a colloidal or macromolecular solution is stable or unstable in the absence of hydrodynamic or applied external forces will be a consequence of the net interaction between the particles arising from the combined attractive van der Waals force and repulsive contribution arising from electrostatic forces because of any particle charge. Both of these forces are fairly long range.

Yet another repulsion, *steric repulsion*, can be associated with the particle surfaces being covered through adsorption or chemical reaction, usually by long-chain molecules, thus giving the particles a "hairy" surface (Overbeek 1982). If the hairs are soluble in the medium, they repel one another and cause a steep repulsion between the particles, with a range about the size of the randomly coiled hairs. Steric repulsion is important in the stabilization of nonaqueous, nonpolar media and will not be discussed further here.

In order that the particles either repel or attract each other, they must be brought into sufficiently close encounters. This may be a consequence of Brownian motion or hydrodynamic transport. However, the solution stability is determined by the interactions during these encounters. We therefore first examine the nature of the physical-chemical interactions, followed by the transport phenomena that can bring about the encounters. Subsequent to these discussions we shall consider the combined effects of transport and surface interactions.

Whether the colloidal particles encountering each other will flocculate (or coalesce) will generally depend on the net interaction resulting from the combined attractive van der Waals forces and repulsive electrostatic forces resulting from the overlap of the electric double layers. This theory of colloid stability, in considerably more detail than given here, is known as the Derjaguin, Landau, Verwey, Overbeek (DLVO) theory of colloid stability (Hiemenz 1986, Verwey & Overbeek 1948).

The energy of repulsion between two spherical colloid particles may be estimated by using the electric double layer ideas discussed in Section 6.4. However, let us first consider the simple case of the electrostatic repulsive force between parallel plates of like surface potential immersed in an infinite reservoir of electrolyte of concentration c_0 (Fig. 8.1.1). Each plate will develop its own double layer as shown, and a significant interaction will occur if the double layers overlap at a distance approximately equal to the Debye length. Because electroneutrality must be maintained in the region between the plates, the counterion concentration is greater than that in a single free double layer, and the potential does not have a zero value anywhere in the region. By symmetry the potential will have a minimum at the midplane halfway between the plates.

In the plane of symmetry between the parallel double layers, a charged particle experiences no electric force since the electric field is zero. The charged particle (say ion) concentration at this plane, however, is in excess of that in the bulk, so there is an excess pressure (osmotic pressure) at this plane compared with the bulk, which tends to push the surfaces apart.

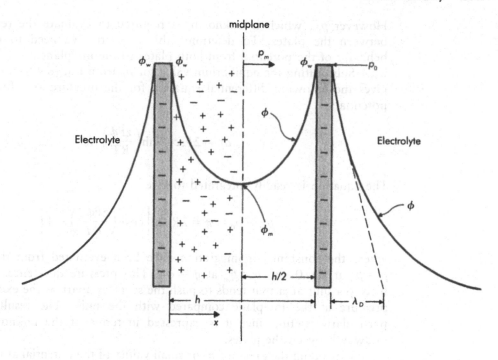

Figure 8.1.1 Schematic of potential distribution resulting from the overlap of the double layers from opposing plates.

At equilibrium in the absence of flow, the only forces acting are the electric force and the pressure gradient, which balance each other to give

$$-\nabla p + \rho_E \mathbf{E} = 0 \qquad (8.1.1)$$

where $\mathbf{E} = -\nabla \phi$. For the one-dimensional problem considered,

$$\frac{dp}{dx} + \rho_E \frac{d\phi}{dx} = 0 \qquad (8.1.2)$$

which from Poisson's equation becomes

$$\frac{dp}{dx} - \epsilon \frac{d^2\phi}{dx^2} \frac{d\phi}{dx} = 0 \qquad (8.1.3)$$

This equation admits the particularly simple first integral

$$p - \frac{\epsilon}{2} \left(\frac{d\phi}{dx} \right)^2 = \text{constant} \qquad (8.1.4)$$

which states that the difference between the hydrostatic pressure and electric pressure is a constant. The constant can be defined by noting that $d\phi/dx = 0$ at $x = h/2$ where $p = p_m$, and where the subscript m denotes the midplane.

However p_m, which is unknown, is required to evaluate the repulsive force between the plates. To determine this pressure, we need to calculate the behavior of the potential from both plates to the midplane.

Substituting the equilibrium value for ρ_E from Eq. (6.4.8) into Eq. (8.1.2) gives the following differential equation for the pressure as a function of the potential:

$$dp = 2Fzc_0 \sinh\left(\frac{zF\phi}{RT}\right) d\phi \qquad (8.1.5)$$

The equation is readily integrated to give

$$p_m - p_0 = \pi = 2RTc_0\left[\cosh\left(\frac{zF\phi_m}{RT}\right) - 1\right] \qquad (8.1.6)$$

where the constants of integration have been evaluated from the conditions $p = p_0$ at $\phi = 0$ and $p = p_m$ at $\phi = \phi_m$. This pressure difference, which is the force per unit area that tends to push the surfaces apart, is the excess (osmotic) pressure at the midplane compared with the bulk. The result is still not particularly useful, since it is expressed in terms of the unknown potential midway between the plates.

Expanding the pressure π for small values of the potential at the midplane, we get

$$\pi \approx RTc_0\left(\frac{zF\phi_m}{RT}\right)^2 \qquad (8.1.7)$$

If, in addition, we assume that the surface potential ϕ_w is also small (we have generally identified this potential with the ζ potential), we may use the Debye-Hückel result for the potential distribution given by $\phi_w \exp(-x/\lambda_D)$. The potential at the midplane is then obtained from the sum of the potentials from each of the two opposing plates to give

$$\phi_m = \phi_1 + \phi_2 = 2\phi_w \exp\left(-\frac{h}{2\lambda_D}\right) \qquad (8.1.8)$$

from which the force per unit area between the plates (excess pressure) is

$$\pi = \frac{2\epsilon\phi_w^2}{\lambda_D^2} \exp\left(-\frac{h}{\lambda_D}\right) \qquad (8.1.9)$$

where we have expressed the coefficient of the exponential in terms of the Debye length (Eq. 6.4.5).

The repulsion or attraction between two particles is most conveniently characterized in terms of potential energy rather than force. The repulsive potential energy (per unit plate area) for the case considered of opposing plates of the same charge is simply the integral of the excess pressure from ∞ to h (the integral of the force over the distance which it acts):

$$V_R^{pl} = -\int_\infty^h \pi \, dh \qquad (8.1.10)$$

where we note that $V_R^{pl} = 0$ when $h \to \infty$. Integrating, we obtain

$$V_R^{pl} = \frac{2\epsilon\phi_w^2}{\lambda_D} \exp\left(-\frac{h}{\lambda_D}\right) \qquad (8.1.11)$$

where we remind the reader that this expression is only valid for small ϕ_m and small ϕ_w (ζ potential). This result shows that the repulsive potential decreases strongly with the reservoir electrolyte concentration c_0, since, from Eq. (8.1.11), $V_R^{pl} \sim c_0^{1/2} \exp(-\text{const } c_0^{1/2})$.

The corresponding result for the repulsive potential valid for small ϕ_m but for large ϕ_w (ζ potential) is given by (Overbeek 1972)

$$V_R^{pl} = \frac{2\epsilon}{\lambda_D} \left(\frac{4RT\gamma}{zF}\right)^2 \exp\left(-\frac{h}{\lambda_D}\right) \qquad (8.1.12a)$$

where

$$\gamma = \tanh\left(\frac{zF\phi_w}{4RT}\right) \qquad (8.1.12b)$$

Both Eqs. (8.1.11) and (8.1.12) show that the repulsive potential is inversely proportional to the Debye length and to $\exp(h/\lambda_D)$, which characterizes the interaction of the "tails" of each plate's surface double layer. A principal difference is that at low surface potential the repulsion is proportional to ϕ_w^2, while at high surface potential the repulsion is independent of the surface potential ($\tanh \sim 1$) and is inversely proportional to z^2.

For interacting identical spherical particles with spherical double layers, a similar calculation of the repulsive potential can be carried out (Overbeek 1972). Provided the thickness of the double layers is small compared with the particle size, the interaction between the double layers on the spherical particles can be assumed to be made up of contributions from infinitesimally small parallel rings, each of which can be considered as a flat plate (see Fig. 8.1.2). The energy of repulsion between the spherical double layers is then

$$V_R^{sph} = -\int_\infty^0 V_R^{pl} 2\pi H \, dH \qquad (8.1.13)$$

From geometrical considerations for $(h - h_0)/2 \ll a$, it follows that $2H \, dH \approx a \, dh$, whence

$$V_R^{sph} = -\pi a \int_\infty^{h_0} V_R^{pl} \, dh \qquad (8.1.14)$$

For small ϕ_m and small ϕ_w, we find, upon substituting Eq. (8.1.11) into Eq. (8.1.14) and integrating,

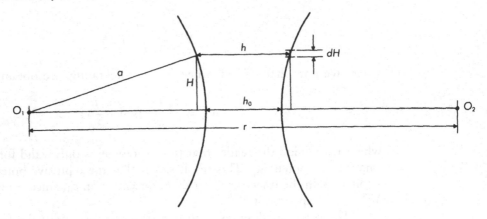

Figure 8.1.2   Geometry for calculation of repulsion between identical spheres.

$$V_R^{sph} = 2\pi a\epsilon\phi_w^2 \exp\left(-\frac{h_0}{\lambda_D}\right) \qquad (8.1.15)$$

For small ϕ_m and large ϕ_w, we obtain, on substituting Eq. (8.1.12) into Eq. (8.1.14) and integrating, the repulsion potential

$$V_R^{sph} = 2\pi a\epsilon\left(\frac{4RT\gamma}{zF}\right)^2 \exp\left(-\frac{h_0}{\lambda_D}\right) \qquad (8.1.16)$$

As van der Waals postulated, the attractive forces between neutral molecules also originate from electrical interactions (Hiemenz 1986). Although there are several types of van der Waals attractive forces that originate from electrical interactions, the most important for colloids is that operating between nonpolar molecules. These forces are due to the polarization of one molecule by quantum fluctuations in the charge distribution in the second molecule, and vice versa. They are known as the London dispersion forces, their origin having first been explained by F. London in 1930.

The London attractive energy between two molecules is long-range and may be written in the form

$$V_A = \frac{b_{12}}{r^6} \qquad (8.1.17)$$

The distance r is the separation distance between the molecular dipole moments, and the quantity b_{12} is defined in terms of the material properties, especially the optical properties. Note that the attractive energies corresponding to all of the van der Waals forces are inversely proportional to r^6, but, as noted above, the London force generally dominates.

The total attractive energy for any geometry is obtained by integrating over pairs of volume elements $d\mathcal{V}_1$ and $d\mathcal{V}_2$ and is given by

$$V_A = -b_{12}n_1n_2 \int \frac{d\mathcal{V}_1\,d\mathcal{V}_2}{r^6} = -\frac{A}{\pi^2}\int \frac{d\mathcal{V}_1\,d\mathcal{V}_2}{r^6} \qquad (8.1.18)$$

where n_1 and n_2 are the number of molecules per unit volume, and A is the *Hamaker constant*. The Hamaker constant has a typical value between 10^{-20} and 10^{-19} J with a range of an additional order of magnitude at the low and high ends.

J.H. De Boer and H.C. Hamaker computed the London interaction between a pair of semi-infinite plates and between spherical particles. They showed that, despite the relatively long-range attractive nature of the force between molecular pairs, the total attractive force between the bodies decays much less rapidly.

For two parallel semi-infinite plates the attractive force between the plates was found, after integrating Eq. (8.1.18), to be

$$V_A^{pl} = -\frac{A}{12\pi h^2} \tag{8.1.19}$$

Here, the attractive force is seen to fall off as the square of the separation distance, a consequence of the fourfold integration. For two equal spheres of radius a, where $h_0 \ll a$, it was shown that

$$V_A^{sph} \approx -\frac{aA}{12h_0} \tag{8.1.20}$$

This result can be easily obtained by substituting Eq. (8.1.19) for the attractive force between parallel plates into Eq. (8.1.14) and integrating, rather than integrating Eq. (8.1.18) directly and simplifying the result for small particle spacing. Clearly, as noted above, the attractive force in both cases falls off much more slowly than the r^{-6} behavior for a molecular pair.

The total potential energy of interaction, say between two spherical particles, is obtained by summing the attractive and repulsive energies. This is illustrated schematically in Fig. 8.1.3, where three different total interaction energy curves are shown, each having been obtained by summing an attraction curve V_A with three different electrostatic repulsion curves.

The repulsive energy is an exponential function of the interparticle distance with a range of the order of λ_D, and the attractive energy decreases as an inverse power of the interparticle distance. Therefore, the London force will predominate at small and large interparticle distances, whereas at intermediate distances double layer repulsion dominates.

In Fig. 8.1.3 the curve $V(1)$ represents a well-stabilized solution with a repulsive energy maximum. The curve $V(3)$ represents a situation where the colloidal solution is unstable and rapid flocculation will occur, double layer repulsion not dominating at any interparticle distance. The curve $V(2)$ represents the transition between stability and flocculation at the primary maximum. If the potential energy maximum is large compared with the thermal energy kT of the particles, the system should be stable; otherwise it should flocculate. The height of the energy barrier depends on the ζ potential and the range of forces on the Debye length. For small ζ potentials the repulsion decreases as the ζ potential decreases and as the Debye length decreases. Another feature of the potential energy curves is the presence of a secondary minimum at relatively large interparticle distances. If this minimum is relatively deep compared with

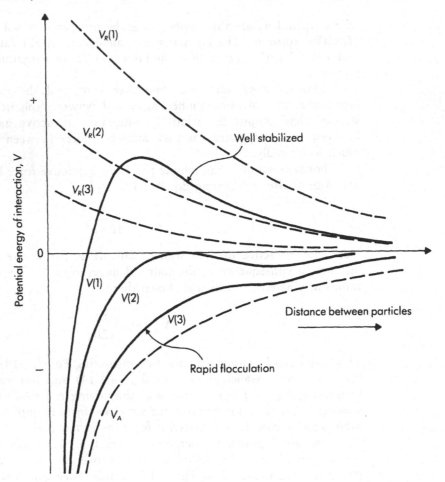

Figure 8.1.3 Total interaction energy curves obtained by the summation of an attraction curve with different repulsion curves (after Shaw 1980).

kT, it should give rise to a loose, easily reversible form of flocculation (Shaw 1980).

Another feature of two-phase colloidal systems (lyophobic colloids) is their sensitivity to flocculation by small amounts of added electrolyte. The electrolyte causes the diffuse part of the double layer to compress. When the double layer is reduced in thickness, the colloid flocs because the particles approach close enough for London forces to take over.

The concentration of a nonadsorbing indifferent electrolyte just sufficient to rapidly flocculate a lyophobic colloid is known to be strongly dependent on the charge number of the counterions (here, the ions of charge opposite to that of the colloid). However, the colloid stability is largely independent of the charge number of the coions and of the concentration of the colloid. These observations are embodied in what is known as the *Schulze-Hardy rule*, which states that the valence of the counterions has the principal effect on the stability of a lyophobic colloid.

The essentials of the Schulze-Hardy rule are readily derived from the DLVO theory presented in this section. With reference to Fig. 8.1.3 it can be seen that the primary maximum of the curve V(2) is the demarcation point between stability and instability. At that point

$$V = V_R + V_A = 0 \tag{8.1.21}$$

and

$$\frac{dV}{dh} = \frac{dV_R}{dh} + \frac{dV_A}{dh} = 0 \tag{8.1.22}$$

Using the repulsive and attractive potentials derived for identical spherical particles with large surface potentials, as given by Eqs. (8.1.16) and (8.1.20), respectively, we find from the condition of the total potential being equal to zero that

$$2\pi a \epsilon \left(\frac{4RT\gamma}{zF}\right)^2 \exp\left(-\frac{h_0}{\lambda_D}\right) - \frac{aA}{12h_0} = 0 \tag{8.1.23}$$

The condition of zero slope (Eq. 8.1.22) gives

$$-\frac{V_R}{\lambda_D} - \frac{V_A}{h_0} = 0 \tag{8.1.24}$$

From the last equation the location of the primary maximum is seen to be at an interparticle distance $h_0 = \lambda_D$. Inserting this result into Eq. (8.1.23) gives the critical Debye length by the proportionality

$$(\lambda_D)_{crit} \sim \frac{Az^2}{\gamma^2} \tag{8.1.25}$$

But from the definition of the Debye length, $\lambda_D \sim (cz^2)^{-1/2}$, from which the criterion for the critical flocculating electrolyte concentration becomes

$$c_{crit} \sim \frac{\gamma^4}{A^2 z^6} \tag{8.1.26}$$

For dilute aqueous solutions at 25°C the coefficient in the proportionality is 3.8×10^{-36} J^2 mol m^{-3}, with the Hamaker constant in joules and the concentration in mol m^{-3}.

At high potentials γ approaches unity, so the theory predicts that the critical flocculating concentration of indifferent electrolytes containing counterions with charge numbers 1, 2, and 3 will be in the ratio $1:2^{-6}:3^{-6}$ or $100:1.56:0.137$. In other words, successively lower concentrations of salts of Na$^+$, Ca^{2+}, Al^{3+} are needed to cause spontaneous flocculation. Experiment closely bears out the theoretical results given (Shaw 1980, Hiemenz 1986). We also note from these results that the flocculating concentration strongly depends on the ζ potential at low potentials but is practically independent of it at high potentials. Moreover, the flocculating concentration is independent of the particle size for a given potential.

8.2   Brownian and Velocity Gradient Flocculation

In the last section we wanted to know whether two particles will flocculate on the basis of physical-chemical considerations when they are brought together. However, if the suspended particles in solution are not completely stabilized, then the rate at which they will flocculate must depend on the frequency with which they encounter one another as a consequence of the fluid and particle motions.

Among the primary collision mechanisms is Brownian flocculation, also termed *perikinetic flocculation*, which dominates for submicrometer particles at relatively high number densities. The second principal collision mechanism is that of velocity gradient flocculation, also termed *orthokinetic flocculation*, which dominates for particles of micrometer size and larger. Evidently, the presence of any stabilizer in the solution will reduce the number of particle encounters and subsequent floccing, as discussed in the last section, resulting in *slow flocculation*. In our discussion we shall separate the transport and stability problems by assuming that the suspension is completely destabilized, so flocculation occurs on encounter (*rapid flocculation*). Our concern here is with the effect of the particle motion alone on the number of encounters between the suspended particles.

Brownian Flocculation

We begin by examining the rate of collision of suspended spherical particles in a static fluid due to Brownian motion. This theory was first put forward by the great Polish physicist M. Von Smoluchowski, to whom we have often referred. In consequence of the equivalence between diffusion and Brownian motion, we consider the relative motion between the particles as a diffusion process. The particles are assumed to be in sufficiently dilute concentration that only binary encounters need be considered. To further simplify the calculation, we consider the suspension to be made up of only two different-sized spherical particles, one of radius a_1 and the other of radius a_2.

Let us consider the diffusional flux of particles a_2 diffusing toward a particle of radius a_1, which we take to be our "test" particle (see Fig. 8.2.1). That is, we consider the sphere of radius a_1 to be fixed at the origin of the coordinate system in an infinite medium containing suspended spheres each of radius a_2. The particles of radius a_2 are in Brownian motion and diffuse to the surface of a_1, which we assume to be a perfect sink. Clearly, any particle a_2 will suffer a collision with the test particle a_1 whenever the center of a_2 approaches a distance a_2 from the surface of the a_1 particle. Therefore the test molecule a_1 carries a sphere of influence of radius $a_1 + a_2$. By this argument the con-

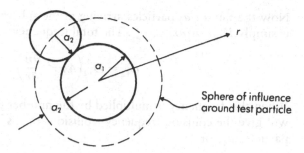

Figure 8.2.1 Sphere of influence of fixed test particle of radius a_1 among moving particles of radii a_2.

centration of a_2 particles vanishes at the radius of the influence of the test sphere.

For particles of uniform radius a the characteristic time for Brownian diffusion is

$$\tau_{Br} \sim \frac{a^2}{D_{Br}} = \frac{6\pi\mu a^3}{kT} \tag{8.2.1}$$

With the particle radius $a \sim 0.1\ \mu m$ and an aqueous solution, this time is approximately 5×10^{-3} s at standard conditions. Therefore, if we examine the flocculation process for times larger than this, we may consider the diffusion process as a steady-state one for which we may write

$$D_{12}\frac{1}{r^2}\frac{d}{dr}\left(r^2\frac{dn}{dr}\right) = 0 \tag{8.2.2}$$

Here, n represents the local number density of the particles of radius a_2, and D_{12} is the Brownian diffusion coefficient describing the relative motion of the two particles. However, since the motion of the particles is assumed to be independent, it is readily shown that $D_{12} = D_1 + D_2$ (Friedlander 1977).

The boundary conditions are that at large distances n approaches the bulk concentration of the particles of radius a_2 (denoted by n_2), and, as noted above, the particles vanish at the radius of the sphere of influence of the test particle; that is,

$$n = n_2 \quad \text{as } r \to \infty \tag{8.2.3a}$$

$$n = 0 \quad \text{at } r = a_1 + a_2 \tag{8.2.3b}$$

Integrating Eq. (8.2.2) subject to the boundary conditions Eq. (8.2.3), we obtain the relatively simple solution for the number density

$$n = n_2\left(1 - \frac{a_1 + a_2}{r}\right) \tag{8.2.4}$$

Now the flux of a_2 particles arriving at the sphere of influence of the a_1 particle is simply $D_{12}(\partial n/\partial r)_{r=a_1+a_2}$. The total frequency of arriving particles is therefore

$$\omega_{12} = D_{12}\left(4\pi r^2\,\frac{\partial n}{\partial r}\right)_{r=a_1+a_2} \tag{8.2.5}$$

When this frequency is multiplied by the number density of a_1 test particles, this will give the collision frequency (collisions $m^{-3}\,s^{-1}$) of a_2 particles with the test particles a_1, or

$$\beta_{12} = 4\pi n_1 n_2 D_{12}(a_1 + a_2) \qquad a_1 \neq a_2 \tag{8.2.6}$$

For a monodisperse system of equal-sized particles, the self-collision frequency requires dividing by 2 to avoid double counting, once as a "diffusing" particle and once as a "test" particle. In this special case Eq. (8.2.6) reduces to

$$\beta_{ii} = 8\pi D_i a_i n_i^2 = \frac{4}{3}\,\frac{kTn_i^2}{\mu} \tag{8.2.7}$$

where $i = 1$ or 2 and where we have substituted for D_i the Brownian diffusion coefficient given by the Stokes-Einstein equation.

For a simple calculation of the number concentration evolution in a batch experiment due to Brownian flocculation of a destabilized suspension, we may suppose that the particles grow with uniformly equal size, which is determined by constant total volume distributed over all the particles. Dropping the subscript on n, we then have from Eq. (8.2.7) that the rate of decrease of the initial number density distribution is described by second-order kinetics; that is,

$$-\frac{dn}{dt} = \frac{4}{3}\,\frac{kT}{\mu}\,n^2 \tag{8.2.8}$$

With the initial condition $n = n_0$ at $t = 0$ the equation is readily integrated to give

$$n = n_0\left(1 + \frac{t}{\tau}\right)^{-1} \tag{8.2.9a}$$

where

$$\tau = \left(\frac{4}{3}\,\frac{kT}{\mu}\,n_0\right)^{-1} \tag{8.2.9b}$$

The characteristic time τ is known as the *flocculation time* and is seen to be the time for the concentration to halve itself.

Gradient Flocculation

In the Brownian flocculation example the effect of particle collisions due to fluid mixing was not considered. The simplest example one might consider is that of constant laminar shear with a linear velocity profile (Levich 1962). Such a flow may be associated with a Couette motion or the region near the wall in a steady laminar flow. Neglecting inertia, the particles follow the straight streamline paths; then if a particle is moving in a region of higher velocity and another in a region of lower' velocity and if the distance between the particles does not exceed the sum of their radii, the particles can collide. The collision takes place because the particle in the higher-velocity layer overtakes the particle in the lower-velocity layer below it. This case was also first analyzed by Von Smoluchowski. In calculating the collision frequency, we neglect the effect of Brownian motion and the hydrodynamic interaction between the particles.

Again consider a test particle of radius a_1, which we suppose to be stationary, with the particles of radii a_2 moving relative to it. The geometry is shown in Fig. 8.2.2, with the test particle of radius a_1 taken to have its origin at the center of the coordinate system. The sphere of influence around the test particle has a radius $a_1 + a_2$. A collision is possible if the distance along the y axis between an a_2 particle and the test particle is $\leq (a_1 + a_2)\sin \theta$.

The velocity of the particles a_2 at any point in the shear layer relative to the test particle may be written

$$u = \frac{du}{dy} \, y = \dot{\gamma} y \tag{8.2.10}$$

Now the number of a_2 particles entering the strip dy per unit time is given by

$$d\omega_{12} = n_2 u 2(a_1 + a_2) \cos \theta \, dy \tag{8.2.11}$$

The total collision frequency with the test particle (collisions s^{-1}) is therefore

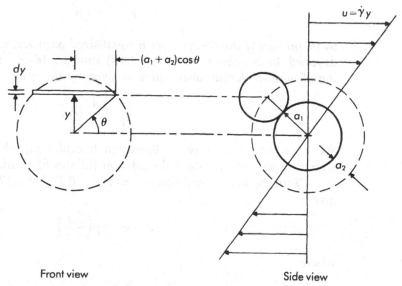

Front view   Side view

Figure 8.2.2 Geometry for flocculation in a uniform shear flow. Test particle of radius a_1 is stationary at origin and sphere of influence is of radius $a_1 + a_2$.

$$\omega_{12} = 2 \int_0^{a_1 + a_2} n_2 u 2(a_1 + a_2) \cos \theta \, dy \qquad (8.2.12)$$

where the factor of 2 multiplying the integral takes into account the flow into the upper hemisphere from left to right plus the flow into the bottom hemisphere from right to left.

From Eq. (8.2.10) $u = \dot{\gamma} y$, and from the geometry $y = (a_1 + a_2) \sin \theta$, from which the total frequency of a_2 particles entering the sphere of influence may be written,

$$\omega_{12} = 4 n_2 \dot{\gamma} (a_1 + a_2)^3 \int_0^{\pi/2} \cos^2 \theta \sin \theta \, d\theta \qquad (8.2.13)$$

Multiplying this frequency by the number density of test particles a_1 and integrating results in the following expression for the collision frequency per unit volume of a_2 particles with all the test particles a_1:

$$\beta_{12} = \tfrac{4}{3} n_1 n_2 \dot{\gamma} (a_1 + a_2)^3 \qquad (8.2.14)$$

As in the Brownian motion case, and for the same reason, to obtain the self-collision frequency for a monodisperse system requires dividing the above result by 2 after setting $1 = 2 = i$:

$$\beta_{ii} = \tfrac{16}{3} \dot{\gamma} a_i^3 n_i^2 \qquad (8.2.15)$$

A simplified calculation of the number concentration evolution with time due to flocculation by laminar shear may be carried out by assuming a total constant solids volume fraction

$$\phi = \tfrac{4}{3} \pi a^3 n \qquad (8.2.16)$$

to be uniformly distributed over n equal-sized particles, where the subscript i is dropped. In this case from Eqs. (8.2.15) and (8.2.16) the rate of decrease of the initial number density distribution is given by the equation

$$-\frac{dn}{dt} = \frac{4}{\pi} \dot{\gamma} \phi n \qquad (8.2.17)$$

We note that in contrast to Brownian flocculation, which was described by second-order kinetics, shear flocculation follows first-order kinetics.

With the initial condition $n = n_0$ at $t = 0$ Eq. (8.2.17) can be integrated to give

$$n = n_0 \exp\left(-\frac{t}{\tau}\right) \qquad (8.2.18a)$$

where

$$\tau = \frac{\pi}{4 \dot{\gamma} \phi} \qquad (8.2.18b)$$

Here, the characteristic flocculation time is seen to be inversely proportional to the velocity gradient and the solids volume fraction.

We may compare the flocculation rates due to shear with those due to Brownian motion by ratioing the right side of Eq. (8.2.17) to the right side of Eq. (8.2.8). The result is

$$\frac{\omega_{\text{shear}}}{\omega_{\text{Brown}}} = \frac{3}{\pi} \frac{\dot{\gamma}\phi}{(kT/\mu)n} = \frac{4a^3\dot{\gamma}}{kT/\mu} \qquad (8.2.19)$$

where we have replaced ϕ by $\frac{4}{3}\pi a^3 n$. The ratio is seen to increase linearly with the shear rate but as the cube of the particle size. Friedlander (1977) has considered simultaneous shear and Brownian motion by the flocculation rates ω_{shear} and ω_{Brown} to be additive and shown good agreement with experiment, although such a linearly independent behavior is open to question (Schowalter 1984). It is also to be observed that Eq. (8.2.19) is just a Peclet number defined as the ratio of the Brownian time scale to shear rate time scale (Eq. 5.3.25).

8.3   Particle Capture by Brownian Diffusion and Interception

An important means by which small particles in suspension are separated from solutions is through capture by collectors, which may be larger particles, or granular, porous, or fibrous media. An example of such collection is filtration. The separated solids may be collected as a cake on the surface of the filter medium (much like ultrafiltration), and this is termed *cake filtration*. Alternatively, the solids may be retained within the pores of the medium, and this is termed *depth filtration*. It is important to recognize that particle collection in a porous medium is not simply a matter of "straining"; that is, the capture is not purely steric, since, in filtration, particles are captured that are much smaller than pores of the medium. The capture of small suspended particles from fluids in laminar flow by a collector is a consequence of the simultaneous action of fluid mechanical forces and forces between the particle and collector, such as van der Waals or electrostatic forces. It is the combined forces, at least close to the collector, that govern the particle trajectories and determine whether a particle will be transported to and retained at the surface of a collector that is fixed in the flow (Spielman 1977).

Following Spielman and the aims of this book, our discussion is confined to the capture of particles in liquid suspension from low-speed laminar flows, where the particles are generally small compared with the collector. The two principal transport mechanisms are (a) Brownian diffusion for submicrometer-size particles, and (b) interception of micrometer-size, nondiffusing, inertia free particles with the collector as a consequence of geometrical collision due to particles following fluid streamlines. Inertial impaction, which can be important for gas-borne particles, is usually unimportant for particles in liquids, because the particle–fluid density difference is smaller and the higher viscosity of liquids resists movement relative to the fluid (Spielman 1977). In this section we shall consider only the hydrodynamic transport problems of Brownian diffusion and geometrical interception, without regard to the surface forces that enter when a particle comes in close enough proximity to the collector that surface forces become important. This latter problem is considered in the following section.

Brownian Diffusion

We treat first the capture by a collector of submicrometer-size particles undergoing Brownian motion in a low-speed flow of velocity U. The collector is taken to be a sphere of radius a and is assumed to be ideal in that all of the particles that impinge on its surface stick to it (Fig. 8.3.1). Because the Brownian particle diffusivities $D = kT/6\pi\mu a_p$, where a_p is the particle radius, are typically about a thousand times smaller than the molecular diffusivities, the diffusion Peclet number (Ua/D) is generally very large compared with unity. The diffusive flux of the particles to the surface is therefore governed by the steady, convective, diffusion boundary layer equation, with the particles treated as diffusing "points."

We may write the convective diffusion boundary layer equation in spherical coordinates (r, θ) in the form

$$u_r \frac{\partial \rho}{\partial r} + \frac{u_\theta}{r} \frac{\partial \rho}{\partial \theta} = D \frac{\partial^2 \rho}{\partial r^2} \tag{8.3.1}$$

where the boundary conditions for an ideal collector are

$$\rho = 0 \qquad \text{at } r = a \tag{8.3.2a}$$

$$\rho \rightarrow \rho_0 \qquad \text{as } r \rightarrow \infty \tag{8.3.2b}$$

Here, we impose the conditions of a perfect sink at the surface, while at large

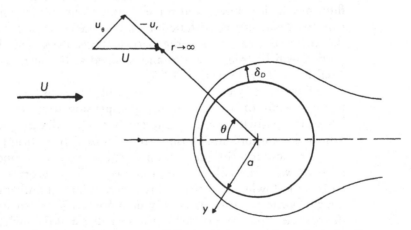

Figure 8.3.1 Brownian diffusion layer around a spherical collector.

distances from the surface the concentration is taken to approach the bulk suspension concentration ρ_0.

In writing Eq. (8.3.1) as above, we have made the usual boundary layer approximations in neglecting derivatives on the right side with respect to θ compared with the radial derivatives. The transverse curvature term $(2/r)\partial\rho/\partial r$ has also been neglected compared with $\partial^2\rho/\partial r^2$, based on the assumption that the concentration boundary layer thickness is small compared with the particle collector radius. In defining the flow about the collector, that is, u_r and u_θ, we will follow Levich (1962) and assume it is a low Reynolds number, inertia free, Stokes flow. This will generally be true because of the large Schmidt numbers. As a consequence of the assumptions made, it is shown below that the mathematical characteristics and solution are quite similar to those for the developing diffusion layer in a channel flow with rapidly reacting walls (Section 4.3).

The well-known solution for Stokes flow about a sphere is (see, e.g., Landau & Lifshitz 1987)

$$u_r = -U\cos\theta\left(1 - \frac{3}{2}\frac{a}{r} + \frac{1}{2}\frac{a^3}{r^3}\right) \tag{8.3.3a}$$

$$u_\theta = +U\sin\theta\left(1 - \frac{3}{4}\frac{a}{r} - \frac{1}{4}\frac{a^3}{r^3}\right) \tag{8.3.3b}$$

where for convenience θ is chosen to be measured positive clockwise so that the forward stagnation point is at $\theta = 0$. Employing the solution in this form is, however, both complicated and unnecessary since, at least over the forward portion of the sphere, the diffusion layer thickness is small compared with the characteristic distance over which the sphere influences the flow (the order of several body radii). This is a statement of the fact that the velocity about the sphere approaches the free stream value U slowly. Our interest therefore centers on the velocity field near the surface, where the radial distance r is close to the sphere radius a. In fact, it is precisely in this region where the Stokes solution is most accurate for finite but low Reynolds numbers.

By setting in Eqs. (8.3.3)

$$\frac{r}{a} = 1 + \frac{y}{a} \tag{8.3.4}$$

and expanding for $y/a \ll 1$, it can be shown that, to first order,

$$u_r = -\frac{3}{2}\left(\frac{y}{a}\right)^2 U\cos\theta \tag{8.3.5a}$$

$$u_\theta = +\frac{3}{2}\left(\frac{y}{a}\right)U\sin\theta \tag{8.3.5b}$$

With the velocity field so defined, Eq. (8.3.1) can in principle be solved subject to the boundary conditions of Eqs. (8.3.2).

Levich (1962) showed that a similarity solution exists to the problem as posed. We present his solution in a somewhat different form with the aim of paralleling our earlier treatments of convective diffusion layers and, in particular, the developing diffusion layer in a channel flow with a rapidly reacting wall. In direct analogy with Eq. (4.3.4) we introduce the new dependent variable

$$\rho^*(\eta) = \frac{\rho}{\rho_0} \tag{8.3.6}$$

where the independent similarity variable η, in partial analogy with Eq. (4.3.5), is

$$\eta = y \sin\theta \left(\frac{3}{2} \frac{U}{Da^2 \int_0^\theta \sin^2\theta \, d\theta} \right)^{1/3} \tag{8.3.7}$$

The functional dependence upon $\sin\theta$ is not self-evident but will not be discussed further here. Instead, the reader is referred to Levich's treatment in Section 14 of his book, where, with some algebra, the form of η as given by Eq. (8.3.7) can be shown to be appropriate from his formulation.

With manipulation, Eqs. (8.3.1) and (8.3.5) and the boundary conditions Eqs. (8.3.2) are reducible to the form

$$\frac{d^2\rho^*}{d\eta^2} + \frac{\eta^2}{3} \frac{d\rho^*}{d\eta} = 0 \tag{8.3.8}$$

$$\rho^* = 0 \qquad \text{at } \eta = 0 \tag{8.3.9a}$$

$$\rho^* \rightarrow 1 \qquad \text{as } \eta \rightarrow \infty \tag{8.3.9b}$$

The differential equation may be seen to be exactly the same as Eq. (4.3.7) governing the developing diffusion layer in channel flow, with the boundary conditions the same as those appropriate to the case of a rapidly reacting wall, for which the solution is given by Eq. (4.3.19).

The diffusional mass flux to the collector from the solution of Eq. (4.3.19) is

$$j = D\left(\frac{\partial\rho}{\partial y}\right)_w = 0.776\rho_0 D\left(\frac{U}{a^2 D}\right)^{1/3} \frac{\sin\theta}{[\theta - (\sin 2\theta)/2]^{1/3}} \tag{8.3.10}$$

(cf. Eq. 4.3.17), where the integral over θ appearing in the definition of η has been evaluated. The above result differs slightly in the value of the numerical coefficient from that given by Levich, a consequence of a slight difference in the evaluation of a Gamma function for which we obtain 1.17 in place of Levich's 1.15. From the result for the flux and the Nernst relation $j = D\rho_0/\delta_D$ the diffusion layer thickness is estimated to be (cf. Eq. 4.3.18)

$$\frac{\delta_D}{a} = \frac{1.29[\theta - (\sin 2\theta)/2]^{1/3}}{\sin \theta} \left(\frac{D}{Ua}\right)^{1/3} \tag{8.3.11}$$

where $Pe = Ua/D$.

The functional dependence on θ in Eq. (8.3.10) is $(3/2)^{1/3}$ at the stagnation point $\theta = 0$, showing the mass flux to be highest there, drops to $(2/\pi)^{1/3}$ at $\theta = \pi/2$, and to 0 at $\theta = \pi$. Correspondingly, from Eq. (8.3.11), this shows that the diffusion layer thickness increases with θ from a finite value at the stagnation point and becomes infinite at $\theta = \pi$. Actually neither δ_D becomes infinite nor does j go to zero at $\theta = \pi$. The anomaly arises from the assumption made in the analysis that $\delta_D \ll a$, whereas the solution shows that for θ close to π the diffusion layer thickness becomes comparable with the collector radius and the approximation breaks down. However, for $\theta \sim \pi$ it is evident from the physics that there is little contribution to the total mass flux to the collector.

The particle mass flow rate to the spherical collector ($kg\ s^{-1}$) can therefore be expressed as

$$I_{sph} = \int j\, dA = 2\pi a^2 \int_0^\pi j \sin \theta\, d\theta \tag{8.3.12}$$

Inserting the flux value given by Eq. (8.3.10) and evaluating the definite integral over θ give the following result for total mass collected by the sphere per unit time:

$$I_{sph} = 7.84 \rho_0 a D \left(\frac{Ua}{D}\right)^{1/3} \tag{8.3.13}$$

The slight difference in the numerical coefficient from that given by Levich is for the reason described above in connection with the result of Eq. (8.3.10).

It is conventional to define a dimensionless *collection efficiency* by comparing the actual diffusional mass flow rate to the mass flow rate of particles to the collector for straight particle trajectories, that is, the mass "swept out" by the projected cross-sectional area of the collector. From this definition the spherical collector efficiency E_{sph} is

$$E_{sph} = \frac{I_{sph}}{\pi a^2 U \rho_0} \tag{8.3.14}$$

Inserting the result of Eq. (8.3.13) with $Pe = Ua/D$ gives

$$E_{sph} = \frac{2.50}{Pe^{2/3}} \tag{8.3.15}$$

which shows that the collection efficiency decreases with increasing particle radius as $a_p^{-2/3}$ (since $D = kT/6\pi\mu a_p$).

The diffusional flow rate to a cylindrical collector, whose axis is normal to the flow direction, has also been determined by a procedure similar to that outlined for the spherical collector. Although no steady, uniformly valid, inertia free, Stokes solution exists for an unbounded medium, a solution valid near the surface can be obtained by using coaxial cell models in which either the shear or vorticity vanish at the finite outer cell boundary. This models, for example, the interference between cylindrical arrays, and we shall discuss this approach further in Section 8.6 when we consider filtration. The result for the diffusional mass flow rate per unit length of cylinder is (Spielman 1977, Levich 1962, Eq. (4.3.12) with exponent 1/3).

$$I_{cyl} = 4.63\rho_0 (UaD^2)^{1/3} \tag{8.3.16}$$

From the above the corresponding collection efficiency for a cylinder normal to the flow, which is conventionally defined for the two-dimensional case considered by

$$E_{cyl} = \frac{I_{cyl}}{2aU\rho_0} \tag{8.3.17}$$

is

$$E_{cyl} = \frac{2.32}{Pe^{2/3}} \tag{8.3.18}$$

The similar behavior of the cylindrical and spherical collection efficiencies is evident. It is a consequence of the similar behavior of the velocity field near the surface.

Interception

Capture by interception assumes that the center of a small nondiffusing spherical particle follows an undisturbed fluid streamline near a larger collector until the particle and collector touch, whereupon the particle is retained by adhesion. We have illustrated this for the case of a cylinder in Fig. 8.3.2, taken from Spielman (1977). This simple model neglects any lubrication effects between the particle and the collector, as well as surface attraction. Electrokinetic phenomena would also need to be considered if the particles and collector were charged.

Let us again first consider collection by a spherical collector assuming the flow to be an inertia free, Stokes flow. The stream function corresponding to the velocity field, defined by Eqs. (8.3.3), is

$$\Psi = \frac{1}{2} Ur^2 \sin^2 \theta \left[1 - \frac{3}{2} \frac{a}{r} + \frac{1}{2} \left(\frac{a}{r} \right)^3 \right] \tag{8.3.19}$$

Expanding the solution for the region near the surface, using $r/a = 1 + y/a$ with $y/a \ll 1$, we obtain

$$\Psi = \tfrac{3}{4} Uy^2 \sin^2 \theta = \tfrac{3}{4} U(r - a)^2 \sin^2 \theta \tag{8.3.20}$$

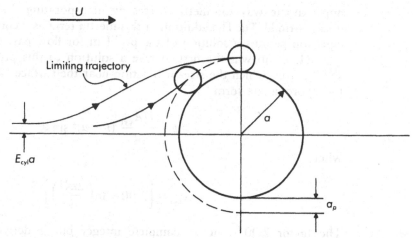

Figure 8.3.2 Trajectories for interception by a cylinder. [After Spielman, L.A. 1977. Particle capture from low-speed laminar flows. *Ann. Rev. Fluid Mech.* 9, 297–319. Copyright © 1977 by Annual Reviews Inc. With permission.]

The mass flow rate to the sphere is found by calculating the rate at which mass impinges on the sphere between the stream surface $\Psi = 0$ and the stream surface defined by the loci of limiting trajectories, illustrated in Fig. 8.3.2 for the cylindrical case. The loci of limiting trajectories are determined by setting $r = a + a_p$ and $\theta = \pi/2$, whence

$$\Psi_{\text{lim}} = \tfrac{3}{4} U a_p^2 \tag{8.3.21}$$

From the definition of the axisymmetric stream function and elementary continuity considerations, the volume flow rate between any two stream surfaces is simply $2\pi\Delta\Psi$. It follows that the mass flux of particles intercepted by the spherical collector is

$$I_{\text{sph}} = 2\pi\rho_0\Delta\Psi = \tfrac{3}{2}\pi\rho_0 U a_p^2 \tag{8.3.22}$$

From the definition $E_{\text{sph}} = I_{\text{sph}}/\pi a^2 U \rho_0$, the efficiency of capture by the spherical collector is (Yao et al. 1971)

$$E_{\text{sph}} = \frac{3}{2}\left(\frac{a_p}{a}\right)^2 \tag{8.3.23}$$

A similar calculation can in principle be carried out for a cylindrical collector. However, a difficulty arises in that there is no solution of the inertia free, Stokes equation for an infinite cylinder in an otherwise unbounded flow. This was already observed in Section 5.1. Nevertheless we can illustrate the low Reynolds number behavior by using the so-called *Stokes-Oseen solution* for uniform flow of velocity U past an infinite circular cylinder of radius a whose symmetry axis is perpendicular to the flow. Oseen's method accounts in an approximate way for inertia forces by incorporating the inertia term in the linear form $U \cdot \nabla u$. The addition of this inertia term as a correction to the Stokes equation permits a solution of the problem for flow past a cylinder.

H. Lamb was the first to give a solution to this problem (see Batchelor 1967), from which the stream function near the surface can be shown to have the approximate form

$$\psi = \frac{U A_{cyl}}{a} (r - a)^2 \sin \theta \qquad (8.3.24a)$$

where

$$A_{cyl} = \left[2.00 - \ln\left(\frac{2aU}{\nu}\right) \right]^{-1} \qquad (8.3.24b)$$

The factor 2.00 is not a complete integer but is derived from the natural logarithm of 7.4. Note that A_{cyl} is a parameter that characterizes the flow model and is a function of the Reynolds number $Re = 2aU/\nu$. As we discuss in Section 8.5, for an assemblage of cylinders where inertia free solutions can be obtained, A_{cyl} can be shown to be a function of the volume fraction of the cylinder assemblage. The solution given by Eq. (8.3.24) is appropriate for Reynolds numbers based on the cylinder diameter of around 1.

Our result for the collector efficiency, using the stream function from above and the definition $E_{cyl} = I_{cyl}/2aU\rho_0$, is

$$E_{cyl} = A_{cyl} \left(\frac{a_p}{a}\right)^2 \qquad (8.3.25)$$

The distance $E_{cyl}a$ of the limiting trajectory from the axis $\theta = 0$ as $r \to \infty$, as shown in Fig. 8.3.2, follows directly from the definition of the stream function and the fact that $\psi_{lim} = E_{cyl}aU$. Note that the corresponding limiting trajectory distance for the spherical collector is $E_{sph}^{1/2}a$.

As with diffusion, the same behavior with particle radius is found for the spherical and cylindrical collection efficiencies. This is again a consequence of the similarity of the velocity fields near the collector surfaces. However, in contrast to the diffusion collection efficiency, which decreases with increasing particle radius as $a_p^{-2/3}$, the interception collection efficiency increases with increasing particle radius as a_p^2. In Fig. 8.3.3 we have sketched a "typical" behavior of the two collection efficiencies with particle radius for a small spherical collecting particle in water at 20°C. For the case of simultaneous diffusion and interception the sum of the two efficiencies is also drawn in.

It may be observed that the spherical collection efficiency for both diffusion and interception satisfies the general functional form

$$E_{sph} \left(\frac{a_p}{a}\right) Pe = fcn\left[\left(\frac{a_p}{a}\right)^3 Pe \right] \qquad (8.3.26)$$

A similar functional relation holds for cylinders with $(a_p/a)^3$ multiplied by A_{cyl}. This general relation can be derived from the appropriate boundary layer

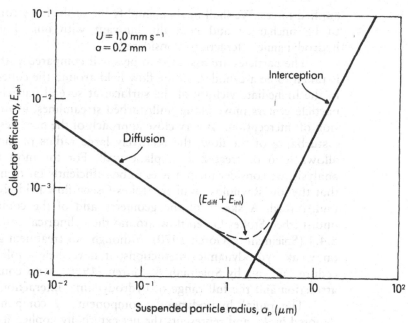

Figure 8.3.3 Behavior of Brownian diffusion and interception collection efficiencies as a function of particle radius for a spherical collecting particle in water at 20°C.

diffusion equation by nondimensionalization and order-of-magnitude considerations. The important key is the observation first made by S.K. Friedlander that Brownian diffusion of "point" particles and interception are limiting cases of the diffusion equation. The principal difference is that in the interception case, $\rho = 0$ at the collision envelope $r = a + a_p$ rather than at the collector surface itself (see Spielman 1977).

8.4   Particle Capture with Surface Forces

In the preceding section our analysis for Brownian diffusion assumed the particles were diffusing points, whereas for interception the center of a particle of finite size was assumed to follow the undisturbed streamline near a large collector. In both cases, no other forces were considered to act on the particles, and when they struck the collector it was assumed that they adhered. In reality, however, even in the absence of inertia there may be other external forces acting on the particles, including London forces of attraction, gravitational hydrodynamic interactions between the particle and collector, and double layer repulsive forces.

Inclusion of the forces mentioned can considerably complicate any detailed mathematical solutions. For that reason, in what follows we shall not consider Brownian motion but discuss only the case of interception, so the particles considered are typically greater than about 1 μm. For interception alone it is somewhat easier to visualize the effect on the collection efficiency of externally applied forces. We shall further simplify the problem by considering the particles to be uncharged and neutrally buoyant with only London attraction and hydrodynamic interactions considered.

The particles are assumed to be small compared with the collector, which is taken to be a cylinder, so the flow field around the collector is disturbed only in the immediate vicinity of the surface; at several particle diameters away the particle centers move along undisturbed streamlines, as in our previous discussion of interception. At very close approach of the particles, where there is some disturbance of the flow, the relatively large radius of the cylindrical collector allows it to be treated as a plane wall. For the most part, to simplify the analysis, we consider the particles to be sufficiently far from the collector surface that the low Reynolds number, Stokes-Oseen flow field given by Eq. (8.3.24) is undisturbed. A sketch of the geometry and of the component flows for the undisturbed Stokes-Oseen flow around the cylindrical collector is shown in Fig. 8.4.1 (Spielman & Goren 1970). Although our treatment generally concentrates on weak hydrodynamic interactions, it nevertheless follows the general procedure laid out by Spielman & Goren (1970), who considered both London attraction and the full range of hydrodynamic interactions.

The radial hydrodynamic component (y component) of the force is denoted by F_{St} and represents the net externally applied hydrodynamic force on the particle resulting from the particle being driven toward (or away from) the collector by the external flow (undisturbed or disturbed) plus any negative resistive "lubrication" force arising from a close approach of the particle to the collector. The attractive molecular London force acting along the line of centers is denoted by F_{Ad} (Ad denotes adhesion). Because of the linearity of the Stokes-Oseen equation, the velocity fields and associated forces may be superposed.

From Eq. (8.3.24) for the Stokes stream function near a cylinder, the undisturbed velocity field is easily shown to be resolvable into two flows. One is a planar stagnation-type flow shown in Fig. 8.4.1B that is associated with the velocity component at infinity along the line of centers of the cylinder and particle; the other is a shear flow normal to the line of centers shown in Fig. 8.4.1C. The respective expressions valid for the cylinder radius $a \gg (x^2 + y^2)^{1/2} \gg a_p$, where a_p is the particle radius, are

$$\mathbf{u}_{St} = \frac{U}{a^2} A_{cyl} \cos \theta_p (2xy\mathbf{i}_\theta - y^2\mathbf{i}_r) \qquad (8.4.1)$$

$$\mathbf{u}_{Sh} = \frac{2U}{a} A_{cyl} \sin \theta_p \, y\mathbf{i}_\theta \qquad (8.4.2)$$

Here, $y = r - a$ and $x = a(\theta - \theta_p)$, where θ_p is the angle of the particle from the forward stagnation point. As shown in Fig. 8.4.1A, the particle center at $x = 0$ is located at $y = y_p = a_p + h$, where h is the gap distance along the line of centers.

Both the hydrodynamic and London forces are additive so that the net externally applied radial force F_n, shown in Fig. 8.4.1A, is given by

$$F_n = F_{St} + F_{Ad} \qquad (8.4.3)$$

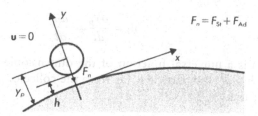

(A) Particle moving under applied force

$$F_n = F_{St} + F_{Ad}$$

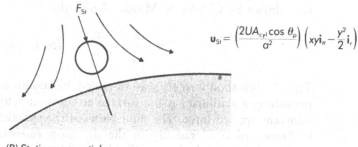

(B) Stationary particle

$$\mathbf{u}_{St} = \left(\frac{2UA_{cyl}\cos\theta_p}{a^2}\right)\left(xy\mathbf{i}_\theta - \frac{y^2}{2}\mathbf{i}_r\right)$$

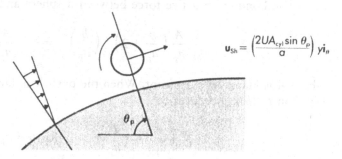

(C) Freely moving particle

$$\mathbf{u}_{Sh} = \left(\frac{2UA_{cyl}\sin\theta_p}{a}\right)y\mathbf{i}_\theta$$

Figure 8.4.1   Particle motion and resolution of undisturbed Stokes–Oseen flow about a larger cylindrical collector. [After Spielman, L.A. & Goren, S.L. 1970. Capture of small particles by London forces from low-speed liquid flows. *Environ. Sci. Technol.* **4,** 135–140. Copyright 1970 American Chemical Society. With permission.]

From the inertia free equations of motion and the boundary conditions that the fluid velocity vanishes at large distances from the particle and on the collector surface, it can be shown from dimensional considerations that the force F_n is expressible in the general form

$$F_n = 6\pi\mu a_p v_r f_1\left(\frac{h}{a_p}\right) \tag{8.4.4}$$

Here, v_r is the radial particle velocity with

$$v_r = \frac{dy_p}{dt} = \frac{dh}{dt} \tag{8.4.5}$$

and $f_1(h/a_p)$ is a universal function of the dimensionless gap width.

When $h \gg a_p$, the particle is far from the collector and must therefore obey Stokes' law; thus

$$f_1\left(\frac{h}{a_p}\right) \approx 1 \quad \text{for } h \gg a_p \tag{8.4.6}$$

On the other hand, when $h \ll a_p$, the particle is very close to the collector, and it was shown by Charles & Mason (1960) that

$$f_1\left(\frac{h}{a_p}\right) \approx \frac{a_p}{h} \quad \text{for } h \ll a_p \tag{8.4.7}$$

This "lubrication" result was obtained by assuming a spherical particle approaching a stationary plane surface at constant velocity under the action of a constant applied force. The fluid between the spherical particle and the surface is "squeezed out" radially in the direction essentially parallel to the plane surface. In their calculation, Charles & Mason approximated the sphere by a parabola of the same radius as the particle at the apex.

The London attractive force between a sphere and a plane is given by

$$F_{Ad} = -\frac{2}{3} \frac{A}{a_p} \left(\frac{h}{a_p} + 2\right)^{-2} \left(\frac{h}{a_p}\right)^{-2} = -\frac{2}{3} \frac{A}{a_p} f_{Ad}\left(\frac{h}{a_p}\right) \tag{8.4.8}$$

where A is Hamaker's constant. When the particle is far from the collector, the function $f_{Ad}(h/a_p)$ reduces to

$$f_{Ad}\left(\frac{h}{a_p}\right) \approx \left(\frac{h}{a_p}\right)^{-4} \quad \text{for } h \gg a_p \tag{8.4.9}$$

When the particle is close to the collector,

$$f_{Ad}\left(\frac{h}{a_p}\right) \approx \left(\frac{2h}{a_p}\right)^{-2} \quad \text{for } h \ll a_p \tag{8.4.10}$$

The hydrodynamic component of the net force F_{St} is obtained, as shown in Fig. 8.4.1B, by considering a second flow in which the particle is held stationary. In this "local" flow the fluid velocity vanishes on the collector surface and on the particle, while far from the particle the velocity approaches the stagnation flow behavior characterized by Eq. (8.4.1). From dimensional considerations

$$F_{St} = -\frac{6\pi\mu a_p^3}{a^2} U A_{cyl} \cos\theta_p \, f_2\left(\frac{h}{a_p}\right) \tag{8.4.11}$$

where $f_2(h/a_p)$ is also a universal function of the dimensionless gap width.

In the case where the particle is far from the collector ($h \gg a_p$), without London attractive forces, Stokes' law is applicable, and from Eqs. (8.4.4) and (8.4.6) we have

$$F_n = F_{St} = 6\pi\mu a_p v_r \qquad (8.4.12)$$

The radial velocity component from Eq. (8.4.1) is

$$v_r = \frac{UA_{cyl}}{a^2}\cos\theta_p\,(h + a_p)^2 \approx UA_{cyl}\cos\theta_p\left(\frac{h}{a}\right)^2 \qquad (8.4.13)$$

Note that this result is also obtainable directly from the solution for the stream function near the surface $\psi = (UA_{cyl}/a)(r - a)^2\sin\theta$ (Eq. 8.3.24) with $v_r = -(1/r)(\partial\psi/\partial\theta)$. Combining Eqs. (8.4.11) to (8.4.13) gives

$$f_2\left(\frac{h}{a_p}\right) \approx \left(\frac{h}{a_p}\right)^2 \qquad \text{for } h \gg a_p \qquad (8.4.14)$$

The analysis of the case where the particle is close to the collector is somewhat more complicated, with f_2 approaching a constant. Goren (1970) has solved this problem and found

$$f_2\left(\frac{h}{a_p}\right) \approx 3.2 \qquad \text{for } h \gg a_p \qquad (8.4.15)$$

From the above results the hydrodynamic force F_{St} is seen to be directed along the inward normal to the cylinder on the upstream side and along the outward normal on the downstream side.

The parameter governing the relative importance of the London attractive force to the hydrodynamic interaction force may be obtained simply by taking the ratio of Eq. (8.4.8) to Eq. (8.4.11) to give

$$\frac{F_{Ad}}{F_{St}} = N_{Ad}^{cyl}\left[\frac{\text{fcn}(h/a_p)}{\cos\theta_p}\right] \qquad (8.4.16)$$

where

$$N_{Ad}^{cyl} = \frac{Aa^2}{9\pi\mu a_p^4 UA_{cyl}} \qquad (8.4.17)$$

Here, N_{Ad}^{cyl} is a dimensionless number termed the *adhesion group*. Clearly the larger the value of the adhesion group the more dominant is the London attractive force for a fixed gap width and particle location. On the other hand, for smaller values of N_{Ad}^{cyl} hydrodynamic interactions dominate.

In the limit of large values of the adhesion group, N_{Ad}^{cyl}, and $h \gg a_p$, the effect of London attraction on the "classical" interception efficiency $A_{cyl}(a_p/a)^2$ (Eq. 8.3.25) can be calculated relatively easily. The particle capture in this case is pictured schematically in Fig. 8.4.2, where it can be seen that the limiting

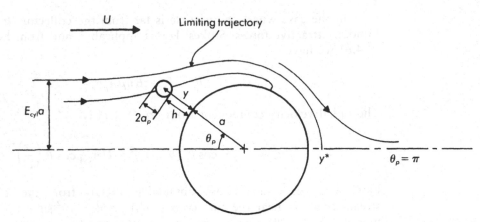

Figure 8.4.2 Trajectories for particle interception by a cylinder with London forces of attraction incorporated. [After Spielman, L.A. & Goren, S.L. 1970. Capture of small particles by London forces from low-speed liquid flows. *Environ. Sci. Technol.* **4**, 135–140. Copyright 1970 American Chemical Society. With permission.]

trajectory, which divides capture trajectories from escape trajectories, is no longer a grazing trajectory as in the classical interception case of Section 8.3, but has a stagnation point at the rear of the collector. The collection efficiency E_{cyl} can be determined by tracing a particle back from $\theta = \pi$, along the limiting trajectory to a point where the gap width is sufficiently large that the trajectory thereafter effectively coincides with a fluid streamline. This displacement of the limiting trajectory, because of N_{Ad}^{cyl} being relatively large, is tantamount to the neglect of the hydrodynamic interactions. Moreover, because $a_p \ll a$, the inertia free stream function near the surface (Eq. 8.3.24) still accurately describes the flow field.

With N_{Ad}^{cyl} large the limiting trajectory falls in the range where $h \gg a_p$, a result that can be confirmed a posteriori. The particle trajectories are found from relations for dh/dt and $d\theta_p/dt$ on eliminating the time t. To obtain dh/dt for the limiting case considered, we use Eqs. (8.4.3) to (8.4.6), (8.4.8), (8.4.9), (8.4.11), and (8.4.14) to derive

$$6\pi\mu a_p \frac{dh}{dt} = -\frac{2}{3}\frac{A}{a_p}\left(\frac{a_p}{h}\right)^4 - \frac{6\pi\mu a_p^3}{a^2} UA_{cyl}\cos\theta_p\left(\frac{h}{a_p}\right)^2 \qquad (8.4.18)$$

Introducing the adhesion group (Eq. 8.4.17), we can simplify Eq. (8.4.18) to

$$\frac{dh}{dt} = -\frac{UA_{cyl}a_p^2}{a^2}\left[N_{Ad}^{cyl}\left(\frac{a_p}{h}\right)^4 + \left(\frac{h}{a_p}\right)^2\cos\theta_p\right] \qquad (8.4.19)$$

For $h \gg a_p$, the tangential velocity $v_\theta = a(d\theta_p/dt)$ is given by Eq. (8.4.2) (see Fig. 8.4.1C), from which

$$\frac{d\theta_p}{dt} = \frac{2UA_{cyl}}{a^2}a_p\left(\frac{h}{a_p}\right)\sin\theta_p \qquad (8.4.20)$$

The limiting trajectory has, as previously noted, a stagnation point ($v_r = 0$, $v_\theta = 0$) at the rear of the collector where $\theta_p = \pi$. At this point, denoted by an asterisk, from Eq. (8.4.19) the gap width y^* (see Fig. 8.4.2) is

$$\frac{y^*}{a_p} \approx \frac{h^*}{a_p} = (N_{Ad}^{cyl})^{1/6} \gg 1 \tag{8.4.21}$$

The limiting trajectory ($\theta_p = \pi$) is seen to lie in the range of large h/a_p for large values of the adhesion group N_{Ad}^{cyl}.

Eliminating time between Eqs. (8.4.19) and (8.4.20) yields, with some algebra, the following first-order linear equation in $(h/a_p)^6$, which describes the particle trajectories under the criteria noted above:

$$\frac{d(h/a_p)^6}{d\theta_p} + 3 \cot \theta_p \left(\frac{h}{a_p}\right)^6 = -3 \frac{N_{Ad}^{cyl}}{\sin \theta_p} \tag{8.4.22}$$

Integrating this equation along the limiting trajectory and taking the cube root of the solution give

$$\left(\frac{h}{a_p}\right)^2 \sin \theta_p = \left(3 N_{Ad}^{cyl} \int_{\theta_p}^{\pi} \sin^2 \theta \, d\theta\right)^{1/3} \tag{8.4.23}$$

With the above result we may now calculate the cylinder capture efficiency, which, from the discussion following Eq. (8.3.25), is

$$E_{cyl} = \frac{\psi_{lim}}{aU} = \lim_{\theta_p \to 0} \left[A_{cyl} \left(\frac{a_p}{a}\right)^2 \left(\frac{h}{a_p}\right)^2 \sin \theta_p \right] \tag{8.4.24}$$

Here, we have again used the solution for the stream function near a cylinder Eq. (8.3.24) with $h \gg a_p$. Substitution of Eq. (8.4.23) with $\theta_p = 0$, on evaluating the definite integral, yields

$$E_{cyl} = A_{cyl} \left(\frac{a_p}{a}\right)^2 \left(\frac{3\pi}{2} N_{Ad}^{cyl}\right)^{1/3} \tag{8.4.25}$$

This result was first derived by G.L. Natanson (see Spielman 1977). We emphasize again that it is valid only for $a_p \ll a$ and for large values of the adhesion group, but not so large that particles are attracted from the area of uniform velocity. It is seen to predict a weaker dependence on particle size than the "classical" solution $A_{cyl}(a_p/a)^2$, in which London attraction and hydrodynamic interactions are not considered. Note that the spherical collection efficiency has a similar behavior, given by (Spielman & Goren 1970)

$$E_{sph} = \frac{3}{2} \left(\frac{a_p}{a}\right)^2 \left[\frac{4}{3} \left(\frac{9}{5} N_{Ad}^{sph}\right)^{1/3}\right] \tag{8.4.26}$$

where N_{Ad}^{sph} is defined by Eq. (8.4.17) with $A_{cyl} = 1$. This approximate solution may be compared with the classical result $\frac{3}{2}(a_p/a)^2$ (Eq. 8.3.23).

In Fig. 8.4.3, taken from Spielman & FitzPatrick (1973), is shown the approximate analytic expression of Eq. (8.4.25) for the cylindrical collector compared with an exact numerical solution in which London attraction and the complete range of hydrodynamic interactions are included. The agreement of the exact solution with the approximate one is seen to be surprisingly good at relatively low values of the adhesion group, indicating that h does not have to be very large compared with a_p before the hydrodynamic interactions become weak.

In concluding this section, we would emphasize again that the problem treated is only illustrative of the approach and results that may be expected with external forces included. For example, if gravity forces are important in interception, as with dense particles, then from Eq. (5.4.5) the additional body force

$$F_{Gr} = -\tfrac{4}{3}\pi a_p^3(\rho - \rho_{fl})g\cos\theta_p \qquad (8.4.27)$$

would have to be added to determine F_n. We would then expect any general solution to also depend on a dimensionless *gravity group*

$$N_{Gr}^{cyl} = \frac{F_{Gr}}{F_{St}} = \frac{2}{9}\frac{(\rho - \rho_{fl})ga^2}{\mu U A_{cyl}} \qquad (8.4.28)$$

where, as before, F_{St} is defined by Eq. (8.4.11). The reader is referred to Spielman (1977) for additional references on both Brownian and interception collection behavior with different external forces included.

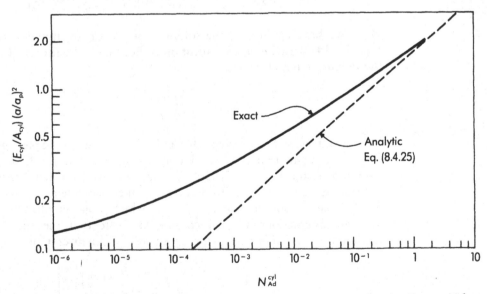

Figure 8.4.3   Collection efficiency versus adhesion group for cylindrical collector. [After Spielman, L.A. & FitzPatrick, J.A. 1973. Theory for particle collection under London and gravity forces. *J. Colloid Interface Sci.* **42**, 607–623. With permission.]

8.5   Filtration and Drag Models of Porous Media

Filtration is the process of separating particles in suspension from a carrier fluid, which we here generally take to be a liquid, by passing the fluid through a permeable material called the filter medium. As noted in the introduction to Section 8.3, the filter may be a porous granular or fibrous medium, and the separated solids may be collected as a cake on the surface or be retained within the pores of the medium.

From the discussion of the preceding two sections, we have seen that particle collection may not simply be a steric action of capturing suspended particles by virtue of the porous medium acting as a constricting "strainer." Particles much smaller than the pores of the medium can be captured and retained. This may be a consequence of hydrodynamic interactions or molecular, electrical, or gravitational forces, acting alone or in combination. Indeed, depending upon the particles and the medium, chemical forces could play a role as we have seen earlier, as could magnetic forces if the particles and medium are magnetic and an appropriate magnetic field gradient is applied.

In Section 4.7 it was discussed how, for a low Reynolds number flow through a porous media or, equivalently, a filter media, the pressure drop follows Darcy's law (Eq. 4.7.7), here rewritten in one dimension:

$$\frac{dp}{dx} = -\frac{\mu}{k}\,U \tag{8.5.1}$$

where we note again that U is the superficial velocity. In Section 4.7, using a capillary model, we derived the Kozeny-Carman relationship between the permeability k and the properties of the medium. The three other principal approaches to determine permeability involve the use of drag models, orifice models, and stochastic models (Scheidegger 1960, Philip 1970). We consider below only the drag model because it, together with the capillary model treated previously, illustrates many of the structural and geometrical effects of flow through porous media.

It has been found as the porosity increases above about 0.8 that the Kozeny constant from the capillary model (Eq. 4.7.15) increases rapidly, becoming indeterminate as the porosity $\varepsilon \to 1$. This is not surprising, since the flow mechanism can be interpreted as changing from one of flow through capillaries to flow around discrete particles. Such a concept leads to the evaluation of the permeability by the so-called drag model. In this model the drag on a single particle, taking into account that it is influenced by the presence of neighboring particles, is summed over the total number of particles in the medium. We shall restrict our discussion of the drag model to Happel's free surface model (Happel 1958, Happel & Brenner 1983).

In Happel's model the porous medium is taken to be a random assemblage that is assumed to consist of a number of "cells," each of which contains a particle surrounded by a fluid envelope. The fluid envelope is assumed to contain the same volumetric proportion of fluid to solid as exists in the entire assemblage. This determines the envelope radius of each cell. For illustration we consider, as did Happel (1958), that the particles of the medium are all spheres of identical size. If a is the radius of the spherical particle in the porous medium, then the envelope radius b of each cell (Fig. 8.5.1) is given by

$$\frac{a}{b} = (1 - \varepsilon)^{1/3} = \phi^{1/3} \tag{8.5.2}$$

where $\phi = 1 - \varepsilon$ is the volume fraction of grains in the porous medium.

Happel further assumed that each cell remains spherical and that the outside surface of the cell is frictionless; that is, the shear stress vanishes at the outer boundary of the cell. The disturbance due to any particle is therefore confined to the fluid cell.

The flow is considered to be inertia free and to obey the Stokes equation, with the solid spherical particle taken to move within the cell with superficial velocity U. This is equivalent to using a coordinate system moving with velocity U. The appropriate boundary conditions are thus

$$u_r = 0 \quad u_\theta = 0 \quad \text{at } r = a \tag{8.5.3a}$$

and

$$u_r = -U \cos \theta \quad \tau_{r\theta} = 0 \quad \text{at } r = b \tag{8.5.3b}$$

Here, $\tau_{r\theta}$ is the shear stress tangential to the cell boundary; that is, the outer cell boundary is assumed to be a free surface. By this artifice the cell model accounts in an approximate way for the interference effects of the neighboring particles.

A closed-form solution can be obtained near the particle surface. The stream function Ψ has the form of Eq. (8.3.20) except that the velocity U, which is here the superficial velocity, is multiplied by a function A_{sph} of the solid volume fraction. The solution may be written

$$\Psi = \tfrac{3}{4} U A_{\text{sph}} (r - a)^2 \sin^2 \theta \tag{8.5.4a}$$

with

$$A_{\text{sph}} = \frac{2(1 - \phi^{5/3})}{2 - 3\phi^{1/3} + 3\phi^{5/3} - 2\phi^2} \tag{8.5.4b}$$

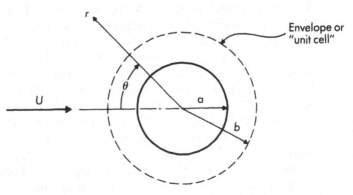

Figure 8.5.1   Happel's cell model for a spherical particle.

Happel (1959) obtained corresponding closed-form inertia free solutions within his model for assemblages of cylinders in those cases where the flow is parallel to the axis of the cylinder and where the flow is at right angles to the cylinder axis. The stream function has the same form as in Eq. (8.3.24), except that A_{cyl} is a logarithmic function of the solid volume fraction, independent of the Reynolds number since the flow is inertia free (Spielman 1977).

The permeability can now be calculated from the drag force per particle, which Happel showed to be given by

$$F = 6\pi\mu a U \frac{3 + 2\phi^{5/3}}{3 - \frac{9}{2}\phi^{1/3} + \frac{9}{2}\phi^{5/3} - 3\phi^2} \qquad (8.5.5)$$

For the drag model the drag force per particle divided by the cell volume $\frac{4}{3}\pi b^3$ must equal the pressure drop per unit length of bed, $-dp/dx$. Solving for U and substituting in Eq. (8.5.1), we obtain

$$k = \frac{3 - \frac{9}{2}\phi^{1/3} + \frac{9}{2}\phi^{5/3} - 3\phi^2}{18\phi(3 + 2\phi^{5/3})} d^2 \qquad (8.5.6)$$

where $d = 2a$ is the particle diameter.

With $\phi = 1 - \varepsilon$, we may expand the above permeability expression for small ε (that is, a tightly packed bed), with the result

$$k = \frac{\varepsilon^3 d^2}{162} \qquad \varepsilon \to 0 \qquad (8.5.7)$$

All terms of $O(1)$, $O(\varepsilon)$, and $O(\varepsilon^2)$ cancel identically to yield the above relation. The Kozeny-Carman formula, Eq. (4.7.16), can be similarly expanded to give

$$k_{K-C} = \frac{\varepsilon^3 d^2}{180} = \frac{\varepsilon^3 d^2}{36K} \qquad \varepsilon \to 0 \qquad (8.5.8)$$

with K the Kozeny constant. The validity of the Happel or Kozeny-Carman models in the limit of very small ε may be questioned. However, comparison of Eqs. (8.5.7) and (8.5.8) gives from Happel's model that $K = 4.5$. Happel pointed out that for packed beds of uniform spheres for $\varepsilon = 0.26$ to $\varepsilon = 0.48$ the best correlation corresponds to $K = 4.80$, with a probable range of variation from 4.8 to 5.1. However, he noted that, in general, $K \approx 5.0$ (the Carman value), independent of particle shape and porosity from $\varepsilon = 0.26$ to $\varepsilon = 0.8$.

We can determine the value of the Kozeny constant as given by Happel's permeability (Eq. 8.5.6) by equating it to the Kozeny-Carman permeability, which from Eq. (4.7.16) may be written

$$k_{K-C} = \frac{(1 - \phi)^3 d^2}{36K\phi^2} \qquad (8.5.9)$$

from which the Kozeny constant, according to Happel's formula, is

$$K_{\text{Hap}} = \frac{(3 + 2\phi^{5/3})(1 - \phi)^3}{2\phi(3 - 4.5\phi^{1/3} + 4.5\phi^{5/3} - 3\phi^2)} \qquad (8.5.10)$$

Table 8.5.1 tabulates the Happel values of the Kozeny constant, and the results are remarkable for the fact that, for a porosity between 0 and 0.6, the constant has a range only between about 4.4 and 5.1. Above a porosity of about 0.7 the Kozeny constant increases rapidly and becomes indeterminate. This is not surprising, since a condition of isolated particles is approached rather than a packed bed. What is, however, most encouraging is the agreement with the early Carman value in the range of porosity for which the model would most likely be expected to hold.

Let us now turn our attention to the behavior of filter media, employing both the characteristics of capture by a single collector, discussed in earlier sections of this chapter, and the understanding of flow characteristics through porous media, which we have just outlined. The suspended particle transport in a filter medium is analogous to transport in flocculation processes with the smaller suspended particles transported by diffusion and the larger ones by interception or, perhaps, by settling; any surface forces present modify the trajectories for particle-collector contact. In what follows we relate the capture of the suspended matter by the porous medium, consisting of an assembly of collectors, to capture by a representative collector.

For a bed of uniform spherical collectors oriented perpendicular to the uniform flow velocity upstream of the medium U, we can write an expression for the average suspended particle balance over a differential slice of bed of depth dx (Spielman & FitzPatrick 1973). The equation for the differential rate

Table 8.5.1
Kozeny Constant According to Happel Model for
Assemblage of Spherical Particles of Uniform Size

Sphere-Cell Radius Ratio $a/b = \phi^{1/3}$	Porosity $\varepsilon = 1 - \phi$	Kozeny Constant K_{Hap}
1.0	0	4.50
0.983	0.05	4.45
0.965	0.10	4.44
0.928	0.20	4.42
0.888	0.30	4.44
0.843	0.40	4.54
0.794	0.50	4.74
0.737	0.60	5.11
0.669	0.70	5.79
0.585	0.80	7.22
0.464	0.90	11.3
0.368	0.95	18.9
0.215	0.99	71.6

of change in average suspended particle number density n (number of particles per unit volume) is

$$\frac{dn}{dt} = -U \frac{dn}{dx} = n_c \frac{nI_{sph}}{\rho_0} \qquad (8.5.11)$$

where n_c = number of collectors per unit volume

I_{sph} = average mass flow rate of particles captured by all spherical collectors in the differential slice (kg s^{-1} (number of collectors)$^{-1}$)

ρ_0 = density of suspended matter equal to n times the mass of a particle

The particles are assumed to be uniform and, for specificity, spherical. The collector volume fraction $\phi = \frac{4}{3} n_c \pi a^3$, where a is the collector radius, whence

$$-U \frac{dn}{dx} = \frac{\phi}{\frac{4}{3}\pi a^3} \frac{nI_{sph}}{\rho_0} \qquad (8.5.12)$$

From the definition of the single particle collection efficiency $E_{sph} = I_{sph}/\pi a^2 U\rho_0$, the above equation can be written as

$$-\frac{dn}{dx} = \left(\frac{3}{4} \frac{\phi E_{sph}}{a}\right) n = \lambda n \qquad (8.5.13)$$

where λ is the *filter coefficient*, with dimensions of inverse length.

For all capture mechanisms where the suspended particles are independent of one another and capture results from individual encounters with the collectors making up the porous medium, the filter coefficient λ is independent of the suspended particle number density n. Examples of this behavior are seen in the single spherical collector efficiencies given by Eqs. (8.3.15), (8.3.23), and (8.4.26). If the incoming suspended particle number density is n_0 and the filter medium is uniform and unclogged, Eq. (8.5.13) integrates to

$$n = n_0 \exp(-\lambda x) \qquad (8.5.14)$$

The filter coefficient λ is thus seen to be a characteristic penetration depth of the suspended matter in the porous medium.

In Sections 8.3 and 8.4 we derived the single spherical collector efficiency under different assumptions as to particle size and surface forces. Three of these relations are given by Eqs. (8.3.15), (8.3.23), and (8.4.26). To apply these relations to porous media, they must be modified to take into account the interactions between the particles making up the porous medium itself. The method for doing this is embodied in the drag model. To illustrate the procedure, we assume that the interaction effect associated with an assemblage of collectors making up the porous medium can be accounted for by using, say, Happel's drag model results. In this case all the single collector results can be translated to porous media behavior simply by multiplying the velocity U appearing in the single collector interception results by the solid volume fraction function A_{sph} appearing in the stream function result for the cell model. This follows from the fact that the stream function characterizes the suspended particle mass intercepted by the medium. For the spherical collector the solid volume fraction function is given by Eq. (8.5.4b). We have already noted that corresponding functions have been derived for cylindrical collectors oriented parallel and normal to the mean flow direction.

As an illustration of the modification, the Brownian collection efficiency for a porous medium made up of an assemblage of spherical collectors would become, from Eqs. (8.3.13) to (8.3.15),

$$E_{sph} = \frac{2.50 A_{sph}}{Pe^{2/3}} \tag{8.5.15}$$

Similarly the interception efficiency of Eq. (8.3.23) would become, from Eq. (8.3.22),

$$E_{sph} = \frac{3}{2} A_{sph} \left(\frac{a_p}{a}\right)^2 \tag{8.5.16}$$

The modification is seen to parallel the incorporation of the Stokes-Oseen function A_{cyl} (Eq. 8.3.24b) into the solution for the collector efficiency of a single cylindrical collector. A similar change would be made for an assemblage of cylindrical collectors, though the volume fraction function that would replace the Stokes-Oseen function A_{cyl} would differ from the one for spherical collectors.

With surface forces present the velocity U appearing in the adhesion group, which in the porous media case represents the superficial velocity, must also be multiplied by A_{sph}. The adhesion group would thus be modified to

$$N_{Ad}^{sph} = \frac{Aa^2}{9 \pi \mu a_p^4 U A_{sph}} \tag{8.5.17}$$

which parallels the single collector definition given for a cylinder (Eq. 8.4.17). This change would be required in addition to multiplying the right side of Eq. (8.4.26) for E_{sph} by A_{sph}. For Eq. (8.4.26) it follows that E_{sph} would be proportional to $A_{sph}^{2/3}$. Similarly, were settling important and the gravity group entered, it would take the form

$$N_{Gr}^{sph} = \frac{2}{9} \frac{(\rho - \rho_{fl})ga^2}{\mu U A_{sph}} \tag{8.5.18}$$

paralleling Eq. (8.4.28).

In the manner described the filter coefficient λ can be determined theoretically by using the appropriate single collector results. Experimentally λ is obtained by measuring the suspended particle number density n as a function of bed depth, and then determining the slope of plots of ln n versus x.

In our discussion it has been supposed that the filter medium remains "unclogged." However, in reality the situation is even more complicated than has been described, because to represent conditions in the filter we must account for the change in porosity and surface area, which occurs as particles accumulate, and for the effects of particle and medium migration that occur under the influence of increasing pressure gradient. The complexity of the situation makes the problem difficult to analyze and is beyond the scope of this book. For details see Scheidegger (1960), Weber (1972), and Ives (1970).

Cell models have also been applied to the approximate determination of the properties of concentrated suspensions. This is touched upon in our discussion of suspension viscosity at high shear rate in Section 9.3.

References

BATCHELOR, G.K. 1967. *An Introduction to Fluid Dynamics.* Cambridge: Cambridge Univ. Press.

CHARLES, G.E. & MASON, S.G. 1960. The coalescence of liquid drops with flat liquid/liquid interfaces. *J. Colloid Sci.* 15, 236–267.

FRIEDLANDER, S.K. 1977. *Smoke, Dust and Haze.* New York: Wiley.

GOREN, S.L. 1970. The normal force exerted by creeping flow on a small sphere touching a plane. *J. Fluid Mech.* 41, 619–625.

HAPPEL, J. 1958. Viscous flow in multiparticle systems: Slow motion of fluids relative to beds of spherical particles. *AIChE J.* 4, 197–201.

HAPPEL, J. 1959. Viscous flow relative to arrays of cylinders. *AIChE J.* 5, 174–177.

HAPPEL, J. & BRENNER, H. 1983. *Low Reynolds Number Hydrodynamics.* The Hague: Martinus Nijhoff.

HIEMENZ, P.C. 1986. *Principles of Colloid and Surface Chemistry*, 2nd edn. New York: Marcel Dekker.

IVES, K.J. 1970. Rapid filtration. *Water Research* 4, 201–223.

LANDAU, L.D. & LIFSHITZ, E.M. 1987. *Fluid Mechanics*, 2nd edn. Oxford: Pergamon Press.

LEVICH, V.G. 1962. *Physicochemical Hydrodynamics.* Englewood Cliffs, N.J.: Prentice-Hall.

OVERBEEK, J.TH.G. 1982. Colloids. A fascinating subject: Introductory lecture. In *Colloidal Dispersions* (ed. J.W. Goodwin), pp. 1–21. Special Publ. No. 43, London: Roy. Soc. Chemistry.

OVERBEEK, J.TH. G. 1972. *Colloid and Surface Chemistry. A Self-Study Course. Part 2. Lyophobic Colloids.* Cambridge, Mass.: M.I.T. Center for Advanced Engg. Study.

PHILIP, J.R. 1970. Flow in porous media. *Ann. Rev. Fluid Mech.* 2, 177–204.

SCHEIDEGGER, A.E. 1960. *The Physics of Flow Through Porous Media.* Toronto: Univ. of Toronto Press.

SCHOWALTER, W.R. 1984. Stability and coagulation of colloids in shear fields. *Ann. Rev. Fluid Mech.* 16, 245–261.

SHAW, D.J. 1980. *Introduction to Colloid and Surface Chemistry*, 3rd edn. London: Butterworths.

SPIELMAN, L.A. 1977. Particle capture from low-speed laminar flows. *Ann. Rev. Fluid Mech.* 9, 297–319.

SPIELMAN, L.A. & FITZPATRICK, J.A. 1973. Theory for particle collection under London and gravity forces. *J. Colloid Interface Sci.* 42, 607–623.

SPIELMAN, L.A. & GOREN, S.L. 1970. Capture of small particles by London forces from low-speed liquid flows. *Environ. Sci. Technol.* 4, 135–140.

VERWEY, E.J.W. & OVERBEEK, J.TH.G. 1948. *Theory of the Stability of Lyophobic Colloids.* Amsterdam: Elsevier.

WEBER, W.J., JR. 1972. *Physicochemical Processes for Water Quality Control.* New York: Wiley-Interscience.

YAO, K.M., HABIBIAN, M.T. & O'MELIA, C.R. 1971. Water and waste water filtration: Concepts and applications. *Environ. Sci. Technol.* 5, 1105–1112.

Problems

8.1 Two circular cylinders of length L and radii a_1 and a_2, respectively, are oriented with their axes of symmetry parallel and with their ends aligned. The cylinder surfaces are separated along their line of centers by a distance h_0, where $h_0 \ll a_1$ and $h_0 \ll a_2$. The surface potential on the cylinders is constant and equal to ϕ_w, and the medium permittivity ϵ is uniform. It is desired to determine the double layer repulsive force between the cylinders. Assume that $zF\phi_w \ll RT$, so the Debye-Hückel approximation is valid, and that the strongest interaction occurs along the line of centers, so the total interaction can be considered to be made up of contributions from infinitesimally small parallel plates.

   a. Employ similar geometrical reasoning that led to Eq. (8.1.13) to derive an analogous integral expression for the cylindrical case considered here and, using Eq. (8.1.9), show that the repulsive force is given by

$$F(h_0) = \frac{2\sqrt{2\pi}\epsilon L \phi_w^2}{\lambda_D^{3/2}} \left(\frac{a_1 a_2}{a_1 + a_2} \right)^{1/2} e^{-h_0/\lambda_D}$$

   where λ_D is the Debye length. Note that

$$\int_0^\infty \frac{e^{-t}}{\sqrt{t}}\, dt = \Gamma(\tfrac{1}{2}) = \sqrt{\pi}$$

   b. Why is it necessary in the derivation of the above result to satisfy the conditions $h_0 \ll a_1$ and $h_0 \ll a_2$?

8.2 a. Verify that the proportionality coefficient in Eq. (8.1.26) for the critical flocculating electrolyte concentration for identical spherical particles with large surface potentials is 3.8×10^{-36} J^2 mol m^{-3}.

   b. What are the theoretical ratios of the critical flocculating concentrations of indifferent electrolytes containing counterions with charge numbers 1, 2, and 3 for identical spherical particles with low surface potentials?

8.3 Consider a colloidal suspension that initially contains equal numbers of small spherical particles of radii a_1 and large spherical particles of radii $a_2 = 10a_1$.

   a. What is the ratio of the rate of collision per unit volume between two small particles, two large particles, and a small particle and a large particle due to Brownian motion, where the diffusion coefficients D_1 and D_2 are given by the Stokes-Einstein equation? Express the answer in whole numbers.

b. Based on the observation that the number of large particles is not affected by collisions between the small particles or between small and large particles, assume that the number density of large particles at any time is given by Eq. (8.2.9). Show using the result of part a that the rate of decrease of the small particle number density n_1 is given by

$$-\frac{dn_1}{dt} = Kn_1^2 + \frac{6Kn_1n_{20}}{1 + t/\tau}$$

where n_{20} = initial number density of large particles
τ = flocculation time for large particles
$K = \frac{4}{3}kT/\mu$

c. Using the fact that the collision process between the small and large particles is the fastest which assumes that n_{20} is not too small; calculate the fraction of small particles remaining after the time τ. Compare this result with that which would be obtained if only small particles were present, supposing equal number densities of large and small particles initially. What can be concluded from the comparison?

8.4 Verify the cylindrical interception efficiency given by Eq. (8.3.25).

8.5 Starting from the convective diffusion boundary layer equation in similarity form (Eq. 8.3.8), derive the general functional relation for the spherical collection efficiency given by Eq. (8.3.26).

8.6 In Section 8.4 a superposition of F_n, F_{St}, and F_{Ad} is carried out, leading to Eq. (8.4.18), which defines the particle velocity with respect to the collector. Explain where these forces come from and describe the superposition of the velocity fields associated with each. Only the governing differential equations and boundary conditions along with simple sketches to indicate the solution domain are required.

8.7 Using a drag model, we want to estimate the permeability of a porous fibrous medium by supposing that the medium is composed of a random distribution of fibers of uniform density, where the fibers are modeled as long circular cylinders of equal lengths L that are very large compared with the radii a. Assume that the random distribution of n fibers per unit volume can be represented by an equipartition of the fibers in three mutually perpendicular directions, one of which is along the direction of the mean flow. The separation between the fibers is large compared with the fiber radius, so the porosity ε is large and it is reasonable to neglect any hydrodynamic interference effects between the fibers. Assuming the flow to be inertia free and the pressure drop necessary to overcome the viscous drag on the fibers to be linearly additive, determine the permeability for the fibrous medium as a function of the porosity ε, fiber radius a, and length L. Note that the form of Eq. (5.1.14) holds for a cylinder moving perpendicular to its axis.

9 Rheology and Concentrated Suspensions

9.1 Rheology

A number of fluids mentioned throughout the text that are of importance in physicochemical hydrodynamics do not behave in the Newtonian fashion outlined in Section 2.2. That is, the stress tensor is not a linear function of the rate of strain tensor. Such "nonlinear" fluids are termed non-Newtonian and the study of their behavior falls under the science of *rheology*, which deals with the study of the deformation and flow of matter. The materials encompassed by this broad subject cover a spectrum from Newtonian fluids at one end to elastic materials at the other with such "fluids" as tars, liquid crystals, and "silly putty" in between. Among the fluids we have discussed in the text that do not exhibit a Newtonian behavior are some polymeric liquids, some protein solutions, and suspensions.

The underlying origin of the non-Newtonian behavior is in the fluid microstructure itself, which can be distorted by the flow because of the relatively long time for the fluid to relax to equilibrium. The principal macroscopic manifestations of this are manyfold and we draw some illustrations from Bird et al. (1987). For example, polymers generally exhibit a viscosity which decreases with increasing shear rate, a behavior that may in fact be their most important characteristic. Thus a polymer liquid with a low shear rate viscosity equal to that of a Newtonian fluid will drain out of a capillary faster than will the Newtonian fluid. Also normal stress effects will arise in simple shear flows. One such effect, strikingly illustrated by rotating a rod in the center of a beaker with a polymeric liquid, is that the liquid climbs up the rod. This contrasts with what is observed with a Newtonian fluid, and would be expected from Newtonian thinking, namely that the liquid would be pushed outward from the rod by centrifugal force and form a dip near the center of the beaker. Other effects that are manifested include an "elastic recoil" resulting from the tendency of polymer molecules which have been stretched by the flow to spring back when the external forces are released. An aluminum soap solution when poured out of a beaker shows this dramatically. Cut in midstream the top half springs back to the beaker and only the bottom half falls down.

An important distinction between polymeric liquids and suspensions arises from their different microstructures and is evidenced by the elastic recoil phenomena that polymers exhibit but suspensions do not. The polymeric or macromolecular system when deformed under stress will recover from very large strains because like an elastic material the restoring force increases with the deformation. With a suspension, however, the forces between the particles decrease with increasing separation so that there is limited mechanism for recovery. There are, however, a variety of rheological properties common to polymeric liquids that suspensions will exhibit including shear rate dependent viscosity and time-dependent behavior. We shall discuss these differences in more detail in the following section.

We may distinguish three general classifications of non-Newtonian fluids in shear flow (Tanner 1988):

1. Time-independent fluids in which the shear stress is a nonlinear and single-valued function of the strain rate.
2. Time-dependent fluids in which the shear stress is not a single-valued function of the strain rate but depends on the shear stress history of the fluid.
3. Viscoelastic fluids that have both viscous and elastic properties in which the shear stress depends upon both the strain rate and strain. The elastic character introduces a time-dependent behavior or "memory" to such fluids.

In what follows, we shall restrict our considerations to the first class of fluid, whose constitutive equation can be expressed in the general form

$$\tau_{ij} = f_{ij}(\varepsilon_{kl}) \tag{9.1.1}$$

where τ_{ij} is the stress tensor, and f_{ij} is a nonlinear tensor function of the rate-of-strain tensor ε_{kl}. This functional form defines the first category of fluid although depending on the flow, it may also be satisfied by a viscoelastic fluid as discussed below. The reason for the lack of uniqueness is because the material functions by which the kinematics is related to the stress field for a non-Newtonian fluid is dependent on the basic flow itself, for example, whether it is a shear flow or a shear-free flow, or if it is steady or unsteady. For purposes of this discussion, we restrict our considerations to a steady incompressible Couette flow defined by Eq. (5.3.1)

$$u = \dot{\gamma}_{yx} y \qquad v = 0 \qquad w = 0 \tag{9.1.2}$$

In a simple shear flow for a Newtonian fluid the only nonzero component of the stress tensor is τ_{yx} and of the rate-of-strain tensor ε_{yx}, where from Eq.

(5.3.2) the shear rate $\dot{\gamma}_{yx} = 2\varepsilon_{yx} = du/dy$. The only material function is the viscosity μ. In general for an incompressible non-Newtonian fluid there will be six independent components of the stress tensor (Eq. 2.2.13). However, assuming the fluid to be isotropic with any directional properties introduced only by the flow itself, it can be shown that it must be invariant to a rotation of coordinate axes and the stress tensor and the rate-of-strain tensor must have the same principal axes. In consequence, for a simple shear flow $\tau_{zx} = \tau_{xz} = \tau_{zy} = \tau_{yz} = 0$. There remain four independent stress components; the three normal stresses τ_{xx}, τ_{yy}, τ_{zz}, and the shear stress $\tau_{yx} = \tau_{xy}$. Because the flow is incompressible, the isotropic pressure cannot be separated from the normal stresses. Of the four stress components, there are only three independent combinations that can be measured, which in turn define three material functions. For a Couette flow, the shear stress τ_{yx} and the two normal stress differences $(\tau_{xx} - \tau_{yy})$ and $(\tau_{yy} - \tau_{zz})$ are the customary choices.

According to Eq. (9.1.1) the stress tensor is a unique function of the rate-of-strain tensor whence for the steady shear flow considered, a function only of $\dot{\gamma}_{yx}$. In analogy with the Newtonian viscosity law, it is customary to express the stress component τ_{yx} through the relation

$$\tau_{yx} = \eta(\dot{\gamma})\dot{\gamma}_{yx} \tag{9.1.3}$$

where η, the most common material function in viscometric flows, is usually referred to as the *apparent viscosity*. The two normal stress differences define two more material functions through the relations

$$\tau_{xx} - \tau_{yy} = \Psi_1(\dot{\gamma})\dot{\gamma}_{yx}^2 \tag{9.1.4}$$

$$\tau_{yy} - \tau_{zz} = \Psi_2(\dot{\gamma})\dot{\gamma}_{yx}^2 \tag{9.1.5}$$

where the functions Ψ_1 and Ψ_2 are termed, respectively, the *first* and *second* *normal stress coefficients*. All of the material functions are even functions of $\dot{\gamma}_{yx}$ so that they are written as a function of $\dot{\gamma} = |\dot{\gamma}_{yx}|$. Note that a viscoelastic fluid will satisfy Eq. (9.1.3) since the fluid is sheared at a constant rate and any time-dependent effects associated with earlier deformations will die out in time.

For a large class of macromolecular fluids, including polymeric liquids, biological fluids, and suspensions, the apparent viscosity generally decreases with increasing shear rate. Finite normal stress coefficients are generally more associated with materials exhibiting a viscoelastic behavior. The first normal stress coefficient is positive and decreases even more sharply with increasing shear rate than the apparent viscosity, while the second normal stress coefficient is much smaller than the first and is generally negative. The rod-climbing experiment described above is a consequence of finite normal stress differences.

We are most concerned in this text with the apparent viscosity, which for the steady Couette flow examined is simply the ratio of the shear stress to shear rate

$$\eta = \tau_{yx}/\dot{\gamma} \tag{9.1.6}$$

Illustrated in Fig. 9.1.1, relative to a Newtonian fluid, are the behaviors of the shear stress versus shear rate in a Couette flow for three principal types of non-Newtonian fluids that can be characterized by the form of the apparent viscosity function in Eq. (9.1.3). A number of empirical functions have been widely employed to characterize the apparent viscosities for these classes of fluids. One termed a Bingham plastic behaves like a solid until a yield stress τ_0 is exceeded subsequent to which it behaves like a Newtonian fluid with a "plastic" viscosity μ_p. The apparent viscosity for this fluid may be written

$$\eta = \mu_p + \frac{\tau_0}{\dot{\gamma}} \qquad \tau_{yx} > \tau_0 \qquad\qquad (9.1.7a)$$

$$\eta = \infty \qquad\qquad \tau_{yx} \leq \tau_0 \qquad\qquad (9.1.7b)$$

Many fluids including colloidal suspensions, slurries, and paints exhibit this type of behavior.

The other two classes of fluids depicted in Fig. 9.1.1 are the *shear thinning* or *pseudoplastic fluid* and the *shear thickening* or *dilatant fluid*. The most

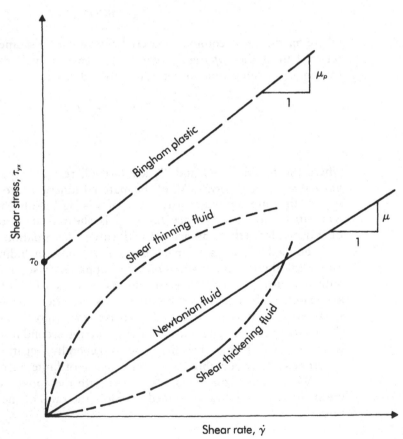

Figure 9.1.1 Shear stress versus shear rate in Couette flow for model non-Newtonian fluids.

widely employed empirical characterization in engineering for these fluids is the power-law model introduced independently by W. Ostwald and A. de Waele

$$\eta = m\dot{\gamma}^{n-1} \qquad (9.1.8)$$

It is seen to contain two parameters; m referred to as the *consistency index* with units of Pa s^n and the dimensionless power law exponent n. If $n = 1$, the fluid is Newtonian and $m \equiv \mu$. For $n < 1$, the fluid is shear thinning and the shear stress–shear rate slope decreases monotonically with increasing shear rate; for $n > 1$, the fluid is shear thickening and the stress–shear rate slope increases with increasing shear rate. We have already indicated that most macromolecular non-Newtonian fluids are shear thinning, however, shear thickening behavior is also observed over some ranges of shear rate with a number of polymer solutions and concentrated particle suspensions (Barnes et al. 1989).

The concept of an apparent viscosity introduced for a Couette flow has been applied empirically to a variety of incompressible inertia free steady shear flows through the generalized relation

$$\tau = \eta\dot{\gamma} \qquad (9.1.9)$$

where τ is the magnitude of the stress tensor, $\dot{\gamma}$ the magnitude of the rate-of-strain tensor (shear rate), and η is a scalar function of $\dot{\gamma}$. A fluid satisfying Eq. (9.1.9) is termed a *generalized Newtonian fluid* (Bird et al. 1987). Equations (9.1.7) and (9.1.8) are among two of the most frequently used empirical expressions for the apparent viscosity.

To illustrate the application of the generalized Newtonian fluid relation, we consider the steady fully developed flow of a non-Newtonian fluid in a circular pipe (Fig. 9.1.2). From an elemental force balance on a cylindrical fluid element of radius r and length Δx, we have on equating the pressure force to the shear force

$$\pi r^2 [p(x + \Delta x) - p(x)] = 2\pi r \Delta x \tau \qquad (9.1.10)$$

with $\tau \equiv \tau_{rx}$, or in the limit as $\Delta x \to 0$

$$\tau = \frac{r}{2} \frac{dp}{dx} = -\frac{1}{2} Gr \qquad (9.1.11)$$

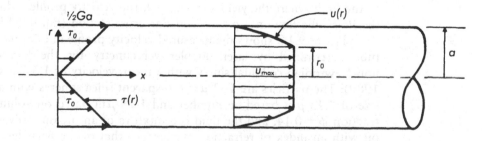

Figure 9.1.2   Non-Newtonian flow in a circular pipe.

Here, we denote by G the negative of the constant pressure gradient, whence a positive value of G causes a flow in the positive x-direction.

Let us choose a Bingham plastic (Eq. 9.1.7) for our generalized Newtonian fluid. From Eq. (9.1.11), the shear stress is seen to vary linearly from 0 at the pipe axis to $\frac{1}{2} Ga$ at the pipe wall, where following convention (Section 2.2), we take the shear stress to be positive. In the pipe core $r \leq r_0$, the shear stress τ is less than the yield stress τ_0, the apparent viscosity is infinite, and the fluid moves as a solid plug. On the other hand, in the annular region near the wall $r_0 < r < a$, where $\tau > \tau_0$, the flow will be quasi-Newtonian (Fig. 9.1.2).

The equation of motion in the wall region from Eqs. (9.1.7a) and (9.1.11) with the shear stress taken positive is

$$\tau_0 - \mu_p \frac{du}{dr} = \frac{r}{2} G \tag{9.1.12}$$

Assuming the flow to be in the positive x-direction, we have set $\dot{\gamma} = -du/dr$ since $\dot{\gamma}$ must be positive. The boundary conditions with no slip at the wall and with the shear continuous at the plug flow boundary $r = r_0$ are

$$u = 0 \quad \text{at } r = a \tag{9.1.13a}$$

$$\frac{du}{dr} = 0, \quad u = u_{max} \quad \text{at } r = r_0 \tag{9.1.13b}$$

Integrating Eq. (9.1.12) and evaluating the constant of integration from the no-slip condition, r_0 from continuity of shear, and u_{max} from the velocity at r_0, we find

$$u = \frac{G}{4\mu_p}(a^2 - r^2) - \frac{\tau_0}{\mu_p}(a - r) \quad r_0 < r < a \tag{9.1.14a}$$

$$u = u_{max} = \frac{\tau_0^2}{\mu_p G}\left(\frac{a}{r_0} - 1\right)^2 \quad r_0 > r > a \tag{9.1.14b}$$

where

$$r_0 = \frac{2\tau_0}{G} \tag{9.1.15}$$

In the limit where the yield stress $\tau_0 = 0$, the velocity profile reduces to that for the Poiseuille flow in a pipe of circular cross-section (Eq. 4.2.14).

Figure 9.1.3 shows the measured velocity profile in a 51 mm circular glass tube determined by laser doppler velocimetry for the laminar flow of a non-Newtonian colloidal slurry with a mean velocity of 1.37 m s^{-1} (Park et al. 1989). The particles are 1–2 μm transparent silica spheres with a mean particle size of 1.13 μm based on number and 1.79 μm based on volume, the volume fraction $\phi = 0.14$, and the fluid is a mixture of an organic solvent and mineral oil with an index of refraction matched to that of the particles.

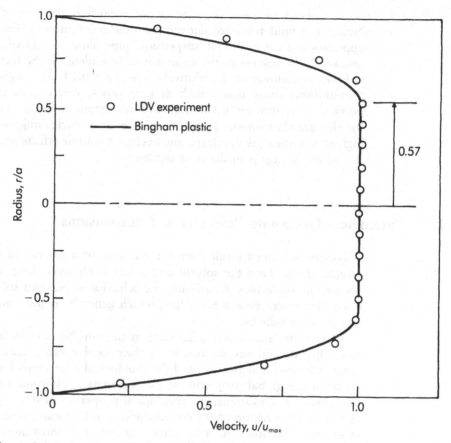

Figure 9.1.3 Laminar velocity profile for the non-Newtonian flow of a transparent colloidal slurry in a 51 mm diameter glass tube at a mean velocity of 1.37 m s^{-1} (data from Park et al. 1989).

From viscometer measurements, the fluid was found to have a so-called yield-power law behavior with an apparent viscosity of the form

$$\eta = \frac{\tau_0}{\dot{\gamma}} + m\dot{\gamma}^{n-1} \qquad (9.1.16)$$

The values of the constants were measured to be $\tau_0 = 10$ Pa, $n = 0.630$, and $m = 0.167$ Pa s^n. This relation is seen to be a combination of the Bingham plastic and power law behavior and is found to fit the measurements to within an accuracy of 1–2%. In Fig. 9.1.3 we have drawn in the velocity distribution for a Bingham plastic fluid using the measured value of the yield stress and a measured value of $r_0/a = 0.57$ from which we calculate $G = 1.4 \times 10^3$ Pa m^{-1}, $\mu_p = 2.28 \times 10^{-2}$ Pa s, and $u_{max} = 1.84$ m s^{-1}. The agreement between the theory and measurement, although not as excellent as for the yield-power law behavior, is nevertheless seen to be quite good and shows clearly the nature of the non-Newtonian behavior associated with the flow of a colloidal suspension.

A word of caution is necessary regarding the use of the generalized Newtonian fluid approach derived from the constant shear Couette flow and its application to a colloidal suspension pipe flow as above. In the example discussed, the success of the approach is dependent on the fact that the particle volume concentration is relatively low ($\phi = 0.14$). In higher concentration non-uniform shear flows, such as pipe flows, where ϕ is say near 0.5, the particle concentration distribution will not remain uniform. Among the reasons for this are shear-induced migration, wherein particles migrate from regions of high to low shear, wall effects, and excluded volume effects (cf. Section 5.7). We return to this again in the next section.

9.2   Parameters Governing Polymers and Suspensions

Polymeric solutions result from the addition of a solvent to the polymer. The interaction between the solvent and solute is relatively large compared to that for smaller molecules. As a result, the behavior of polymer solutions may be far from Newtonian even when dilute, which generally is the condition we emphasize in what follows.

The similarities and differences in the non-Newtonian behaviors of polymeric liquids and suspensions lie in their local microstructures. We, however, shall examine the dependence of the rheological characteristics of these fluids in terms of the global properties of the polymers in solution and the particles in suspension. For example, the rheological properties of dilute polymer solutions are mainly a function of the concentration, temperature, molecular weight, and structure of the long chain molecule, but the single most important factor is the length of the polymer chain. Suspension rheology on the other hand is, in general, a function of the solids loading, particle shape and size distribution, and the chemical state of the system, although the single most important parameter is the particle volume fraction. Of course, both the polymer and suspension rheology depend on the hydrodynamic state of the system, which in the preceding section we characterized in a steady shear flow by the shear rate.

Before considering the relevant forces governing polymer and suspension rheology, we first discuss some general behaviors of polymeric liquids, many of the important features of suspensions having already been brought out in earlier chapters. Synthetic polymers are macromolecules made up of repeating units to form a long flexible chain. The example of polyethylene, which is a linear chain built up from the ethylene group, was cited in Section 1.3. The number of repeating structural units N or *monomers* in one chain is called the *degree of polymerization*, a number which can be very large and which is a measure of the chain length. Symbolically, we may write

$$[-CH_2-]_N \qquad \text{for polyethylene}$$

Another common linear polymer made up from the styrene monomer is polystyrene where we may write

$$\left[-CH_2-CH-\right]_N \quad \text{for polystyrene}$$

The number of structural units N is the molecular weight of the polymer divided by the molecular weight of the individual structural unit. A typical polyethylene molecule will have $N \sim 10^3$ with a molecular weight $M \sim 1.5 \times 10^4$. Typical polystyrene molecules may have $N \sim 5 \times 10^3$ to 5×10^4 with the corresponding molecular weights $M \sim 5 \times 10^5$ to 5×10^6. Generally, flexible linear polymers will have average molecular weights ranging from 10^3 to 10^7 with values of N from around 10 to 10^5. The lower limit on N indicates that polymeric behavior will be observed for more than about 10 monomer units. It should also be noted that N can be larger with entanglement or strong coupling between the molecules, that is, with chemical cross-linking.

Generally, for dilute polymer solutions the fluid is relatively mobile, while for highly concentrated ones it is relatively stiff. The configuration that the polymer takes in solution depends on the solvent. A "good" solvent is one where a stronger interaction occurs between the solvent and polymer than between the solvent and solvent or between various segments of the polymer. In a good solvent, the polymer stretches out in the solution and uncoils, while in a "poor" solvent it coils. This behavior is shown in Fig. 9.2.1. As shown in Fig. 1.3.1, typical lengths for extended polymers (say, polystyrene) are in the 5–10 μm range, and typical diameters for coiled polymers are in the 0.08–0.5 μm range.

The random motions of a flexible chain floating in a solvent has been studied extensively and one widely used model is the "bead-spring" model in which spherical beads are separated by equal lengths of polymer chain. The chains are considered frictionless springs that may be thought of physically as a sequence of monomers long enough to obey Gaussian statistics. Since the molecules are small, the viscous force in the velocity field is Stokesian and is assumed to act only on the beads, that is, at discrete points. Physically the beads may be thought of as groups of impenetrable monomer units. As pointed out in Section 5.1, each spherical bead as it moves through the fluid perturbs the velocity distribution of the fluid nearby and this perturbation is felt by the other

(A)   (B)

Figure 9.2.1 Polymer configurations in solvents: (A) uncoiled molecule in good solvent; (B) coiled molecule in poor solvent.

spheres. The other widely applied model is the "dumbbell" model, which is just a two-bead model in which the viscous drag acts at the spheres located at the ends of a frictionless spring.

Kirkwood (de Gennes 1979) has shown that a single polymer chain behaves hydrodynamically like a solid sphere of a radius R that is proportional to its radius of gyration. A characteristic relaxation time for the chain deformation may then be associated with the rotation of a Brownian sphere of radius R (Eq. 5.2.25)

$$\tau_s \sim \frac{\eta_s R^3}{kT} \qquad (9.2.1)$$

where η_s is the solvent viscosity.

The radius R is dependent on the solvent. Where the polymer is represented as an ideal, freely jointed chain, no account is taken of the effect of the solvent. This represents the borderline case between a poor and good solvent. Here, each link is in a random direction with respect to its neighbors and the end-to-end distance is proportional to $aN^{1/2}$, where a is the effective monomer length. This distance is much smaller than the molecular length aN and is characteristic of the molecular radius R. For the more common case of a chain in a good solvent, excluded volume effects must be considered, and where this volume is of the order of a^3 it can be shown that (de Gennes 1979, 1990)

$$R \approx R_F = aN^{3/5} \qquad (9.2.2)$$

where R_F is called the *Flory radius*. By way of example, for $R_F = 20$ nm and $\eta_s = 10^{-1}$ Pa s, the relaxation time at 300 K is about 2×10^{-4} s.

The characteristic flow time in a steady shear flow is simply given by the reciprocal of the shear rate. Comparing the relaxation time to the flow time, as with hard spheres, the ratio is given by the Peclet number (Eq. 5.3.25)

$$\mathrm{Pe} = \frac{\eta_s \dot{\gamma} R_F^3}{kT} \qquad (9.2.3)$$

The Peclet number is seen to increase as the cube of the Flory radius showing the relative increase in importance of viscous forces with increasing polymer length.

It should be noted here that in polymer rheology, for viscoelastic fluids the commonly used dimensionless parameter to characterize the ratio of elastic force to viscous force is the *Deborah number* denoted by the symbol De. This parameter is essentially just the Peclet number. In terms of characteristic times, it is equal to the ratio of the largest time constant of the molecular motions or other appropriate relaxation time of the fluid compared to the characteristic flow time.

In polymer rheology, the results for viscosity are usually expressed in terms of the intrinsic viscosity $[\eta]_p$ defined by the relation

$$[\eta]_\rho = \lim_{\rho \to 0} \left(\frac{\eta - \eta_s}{\rho \eta_s} \right) \qquad (9.2.4)$$

where ρ is the polymer mass concentration (density) and $[\eta]_\rho$ is dimensional, having the units of inverse concentration. The subscript ρ is generally not appended but intrinsic viscosity is also used in suspension rheology and because it is defined there with volume fraction in place of density, we shall distinguish between them by appropriate subscripts.

The functional dependence of $[\eta]_\rho = f(\dot{\gamma})$ for a dilute solution can be shown by employing the Einstein relation (Eq. 5.3.23) in the form

$$\eta = \eta_s[1 + 2.5\,\phi] \qquad (9.2.5)$$

where ϕ is taken here to be the volume fraction occupied by impenetrable spheres having the chain hydrodynamic radius R_F. Denoting by n_c the number of chains (polymer molecules) per unit volume, the volume fraction occupied by the effective spheres of radius R_F is

$$\phi = n_c \frac{4}{3}\,\pi R_F^3 = \frac{\rho N_A}{M} \frac{4}{3}\,\pi R_F^3 \qquad (9.2.6)$$

where M is the molar mass of each chain and N_A is Avagadro's number. Combining Eqs. (9.2.4)–(9.2.6)

$$[\eta]_\rho \sim \frac{R_F^3 N_A}{M} \qquad (9.2.7)$$

Using this result for the intrinsic viscosity, we may rewrite the relaxation time given by Eq. (9.2.1) in terms of readily measurable quantities as

$$\tau \sim \frac{[\eta]_\rho \eta_s M}{RT} \qquad (9.2.8)$$

where $R = kN_A$ is the gas constant. It has been noted by de Gennes (1990) that this expression has been widely verified.

Using Eq. (9.2.8), we may also rewrite the Peclet number in the form

$$\text{Pe} = \frac{[\eta]_\rho^0 \eta_s M \dot{\gamma}}{RT} \qquad (9.2.9)$$

where for dilute polymer solutions, we have taken the characteristic intrinsic viscosity to be $[\eta]_\rho^0$ defined with respect to the *zero-shear-rate viscosity*, η^0. The zero-shear-rate viscosity is just the constant value the viscosity approaches at low shear rates where the shear stress is proportional to $\dot{\gamma}$. The reduced intrinsic viscosity $[\eta]_\rho/[\eta]_\rho^0$ has been shown to scale with the Peclet number of Eq. (9.2.9) (Fig. 9.2.2). At higher shear rates, over several decades of Peclet number, the power law type of behavior is seen in Fig. 9.2.2. At very high Peclet

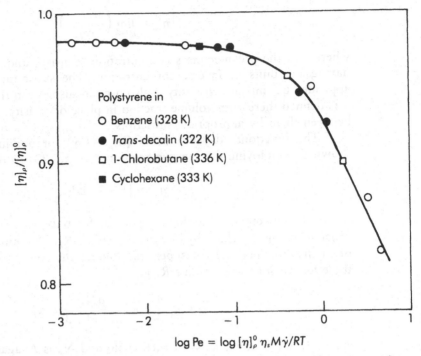

Figure 9.2.2 Intrinsic viscosity of polystyrene solutions with various solvents as a function of reduced shear rate Pe. [After Bird, R.B. et al. 1987. *Dynamics of Polymeric Liquids, vol. 1. Fluid Mechanics,* 2nd edn. New York: Copyright © 1987 by John Wiley & Sons, Inc. With permission.]

numbers, the viscosity "may again become independent of shear rate and approach η^∞, the *infinite-shear-rate-viscosity*" (Bird et al. 1987). The high Peclet numbers may not be achieved, however, because of polymer degradation at high shear rates.

The polymers considered above have been uncharged. Another class of polymers are polyelectrolytes whose chains carry a fixed charge. For Debye lengths small in comparison with the chain size, polyelectrolytes take on a coiled structure. Despite the presence of charge, the description for a polyelectrolyte fluid is similar to that which is obtained for a neutral polymer corresponding to the ideal chain regime (de Gennes 1990).

We next consider the behavior and forces acting on concentrated suspensions, with emphasis on colloidal particles. By concentrated suspension, we refer to one in which the average separation between two similar size particles is close to or less than the particle size. As noted previously, and as first shown by Einstein for dilute suspensions, the solids volume fraction is probably the single most important dimensionless parameter in that an increase in volume fraction enhances the energy dissipation and hence increases the viscosity. The dependence of viscosity on particle fraction is more pronounced at high concentrations, generally above 50%. Particle shape and size distribution also play important roles in the rheological behavior of suspensions, while particle size by itself generally characterizes the magnitude of the forces relative to each other.

Following the descriptions of Probstein and Sengun (1987) and Sengun and Probstein (1989b), we shall examine the forces in concentrated suspensions. We reserve our discussion of the effect of volume fraction for the sections that follow.

We group the forces that control the suspension rheology into two main categories: colloidal forces and viscous forces. The colloidal forces include Brownian diffusion forces and the surface forces of electrostatic repulsion and van der Waals attraction. In order to define the dimensionless scaling parameters that characterize the relative magnitude of these forces we assume the particles are separated by a distance of the order of the particle radius a, which is in turn assumed to be close to the smallest particle separation h_0.

Under the Stokes flow and particle separation assumptions, the viscous force between two approaching particles should scale as $\mu a U$, with μ the Newtonian viscosity of the medium and U the approach velocity. With $U \sim a\dot{\gamma}$, where $\dot{\gamma}$ is the applied shear rate, the energy dissipation within the gap between the particles scales as $\mu a^3 \dot{\gamma}$. We have here assumed that the interaction frequency between the particles is of the order of $\dot{\gamma}$. This will be true so long as the particle concentration is not so high that we are close to the *maximum packing fraction* for which flow can occur, a point which is discussed in greater detail in the following section.

The thermal energy of a particle scales as kT, while with $a \sim h_0$ the van der Waals attractive energy scales as the Hamaker constant A (Eq. 8.1.20). Finally from Eq. (8.1.15), the energy of repulsion is seen to scale as $a\epsilon\zeta^2$ for small zeta potentials, say ζ less than the Nernst potential at standard temperature (26 mV), and for small Debye length to radius ratio.

Comparing the viscous energy dissipation to the thermal energy leads again to the Peclet number

$$Pe = \frac{\mu\dot{\gamma}a^3}{kT} \tag{9.2.10}$$

This is exactly the Peclet number defined by Eq. (5.3.25), which measures the characteristic rotational Brownian diffusion time to the time scale defined by the reciprocal of the shear rate. It is the same measure found for dilute polymer solutions with the particle radius here replacing the Flory radius for the polymer.

A second dimensionless group is termed the *shear-repulsion number*, N_{SR}, defined as the measure of convective force to electrostatic repulsive force

$$N_{SR} = \frac{\mu a^2 \dot{\gamma}}{\epsilon\zeta^2} \tag{9.2.11}$$

Finally, we introduce the *shear-attraction number*, N_{SA}, which is that group characterizing the magnitude of the viscous energy to the van der Waals attractive energy

$$N_{SA} = \frac{\mu a^3 \dot{\gamma}}{A} \tag{9.2.12}$$

The dimensionless groups defined by Eqs. (9.2.10)–(9.2.12) can be used to illustrate the essential features of suspension rheology, in particular they make clear the important role of particle size. The shear-repulsion number increases as the square of the particle size, while both the Peclet number and the shear-attraction number increase as the cube. The parameters therefore increase rapidly with increasing particle size, implying a rapid increase in the relative importance of shear forces. The limit of very large values of these parameters corresponds to what is generally termed the *high shear limit*. This limit does not require high shear rates and more generally it attains as a result of large, non-colloidal particle sizes.

To numerically illustrate the effect of particle size on the dimensionless groups, we note that in water at standard temperature for $a = 100$ μm, a ζ potential of 10 mV, and $A = 10^{-20}$ J, we have $Pe = 10^6 \, \dot{\gamma}$, $N_{SR} = 10^2 \, \dot{\gamma}$, and $N_{SA} = 10^5 \, \dot{\gamma}$. Clearly for 100 μm particles, all of the dimensionless groups are very large compared to unity even at a shear rate of $1 \, s^{-1}$. In the high shear limit, non-Newtonian behavior should vanish and the viscosity should attain a stationary value independent of the shear rate. We note here the analogue between the high shear limit for suspensions and the infinite-shear-rate-viscosity limit for polymers discussed above.

For colloidal particles, the dimensionless parameters are generally small and non-Newtonian effects dominate. Considering the same example as above, but with particles of radius $a = 1$ μm, the parameters take on the values $Pe = \dot{\gamma}$, $N_{SR} = 10^{-2} \, \dot{\gamma}$, and $N_{SA} = 10^{-1} \, \dot{\gamma}$ so that for shear rates of $0.1 \, s^{-1}$ or less they are all small compared to unity. The limit where the values of the dimensionless forces groups are very small compared to unity is termed the *low shear limit*. Here the applied shear forces are unimportant and the structure of the suspension results from a competition between viscous forces, Brownian forces, and interparticle surface forces (Russel et al. 1989). If only equilibrium viscous forces and Brownian forces are important, then there is well defined stationary asymptotic limit. In this case, there is an analogue between suspensions and polymers which is similar to that for the high shear limit, wherein the low shear limit for suspensions is analogous to the zero-shear-rate viscosity limit for polymers.

With surface forces absent, in the limit of $Pe \ll 1$, the distribution of particles is only slightly altered from the Einstein limit. To order ϕ^2, which takes into account two-particle interactions, Batchelor (1977) calculated the effect of Brownian motion on the stress field in a suspension of hard spheres and determined the low shear limit relative viscosity to be given by the Einstein relation with an added term equal to $6.2\phi^2$. This result is found to agree satisfactorily with experiment for $\phi \lesssim 0.1$. Because of the general complexity of the low shear limit with interparticle surface forces, including questions as to the existence of a uniquely defined asymptotic limit, we choose not to discuss this case further, instead referring the reader to Russel et al. (1989) and van de Ven (1989).

To demonstrate the effect of Peclet number, Krieger (1972), in a series of classic experiments, measured the relative viscosity of suspensions of monomodal spheres with sizes from about 0.1 to 0.5 μm. By adjusting the solution electrolyte concentration he was able to minimize any surface force effects. He found that when the relative viscosity was plotted against the reduced shear stress $\tau_r = \tau a^3 / kT$, where τ is the actual shear stress, all of the data for the different size particles and suspension media fell on a single curve for a given particle volume fraction. Some of his data for polystyrene spheres are shown in Fig. 9.2.3.

The reduced shear stress is recognized to be essentially our Peclet number defined by Eq. (9.2.10). As can be seen from Fig. 9.2.3, as the Peclet number is increased, the viscosity reaches a stationary value and in this limit the suspension behaves as a Newtonian fluid. In the opposite limit, as the Peclet number tends to zero, the relative viscosity approaches a higher stationary value. The transition is seen to take place in the neighborhood of a Peclet number close to unity, consistent with our earlier discussion.

For rigid particle suspensions and dilute polymer fluids, where Peclet number is the governing parameter, we have seen that in a steady simple shear flow, there are analogous behaviors of viscosity with Peclet number, including the existence of low and high Peclet number stationary limits. The dependence of the polymer viscosity on Peclet number is, however, much stronger than that of the suspension viscosity. This may be attributed to the fact that for the dilute polymer solution, the polymer molecules are deformable so that with increasing shear rate the individual polymers will elongate with an effective decrease in viscosity. On the other hand, for a rigid particle suspension, the Peclet number effect is smaller since the principle consequence of increasing shear rate is the loss of the Brownian motion forces. If the suspension is dilute, interparticle forces are absent because of the large separation distance, that is, the particle separation distance is irrelevant. It is only when the volume fraction is increased

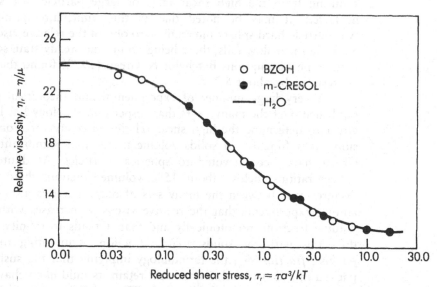

Figure 9.2.3 Relative viscosity as a function of reduced shear stress for monodisperse colloidal polystyrene spheres of different sizes in different fluids (benzyl alcohol, meta-cresol, and water) at 50% volume fraction (after Krieger 1972).

to the point where two-body and higher order interactions become important that the Peclet number effect manifests itself for spherical particles.

The practical consequence is that a polymeric fluid will evidence a far stronger non-Newtonian shear thinning behavior than will a suspension. This can result in a fairly wide range in the "power law" exponent with the exponent n in Eq. (9.1.8) ranging between 0.1 and 0.6 for typical polymeric liquids (Bird et al. 1987), a behavior not found in suspension rheology.

Another significant distinction, which has previously been noted, is that polymeric liquids will generally maintain uniform concentrations, whereas suspensions with volume fractions greater than, say, 20–25% develop concentration gradients in non-uniform shear flows. Under these conditions a general approach for specifying the problem, even in the simple case of pipe flow, is not at hand.

9.3   High Shear Limit Behavior of Suspensions

In the last section we introduced the concept of two "asymptotic" viscosity limits for shear thinning colloidal suspensions as a function of shear rate. One is the high shear limit which corresponds to high values of the Peclet number where viscous forces dominate over Brownian and interparticle surface forces. Generally this limit is attained with non-colloidal size particles since to achieve large Peclet numbers by increase in shear rate alone requires very large values for colloidal size particles. In this limit, non-Newtonian effects are negligible for colloidal as well as non-colloidal particles.

In discussing the high shear limit, it is assumed that inertial effects resulting from the high shear rates or large particles are small and can be neglected. It may be noted that in this limit, attempting to account for two-particle hard-sphere interactions to obtain the relative viscosity to $O(\phi^2)$ in a simple shear flow fails, there being no unique steady state solution. However, with some assumptions Batchelor & Green (1972) found the coefficient of the ϕ^2 term to be about 5.2.

A very large number of experimental and theoretical papers have been published over the many years that suspension rheology has been studied in an effort to determine the high shear relative viscosity for monodisperse suspensions as a function of solids volume fraction. A significant portion of these efforts have been devoted to spherical particles. At dilute and semi-dilute concentrations, below about 15% volume fraction, there is relatively little disagreement between the many sets of data. There is also general agreement among experiments that the relative viscosity increases with increasing solids volume fraction monotonically and that it tends to infinity asymptotically. It does this with the solids volume fraction asymptoting to some *maximum packing fraction* ϕ_m, the terminology implying that the suspension cannot be packed in a denser fashion and still retain its fluid-like behavior. Alternatively, this limit is termed the *fluidity limit*. The maximum packing fraction is derived from experiment by extrapolation, which may involve the use of semi-empirical models. This is necessitated by the fact that for practical reasons, it is not possible to obtain a steady viscosity measurement closer than about 5% to the maximum packing fraction.

For suspensions of uniform spherical particles, there is considerable variation in the reported high shear maximum packing fraction, ranging from about 0.53 to 0.71. The reader should note that the maximum attainable solid fraction for spherical particles is 0.74, corresponding to a face centered cubic structure. In Fig. 9.3.1 are shown the relative viscosity data that yield the maximum packing fraction bounds quoted. The curve through the data that gives the asymptote of 0.53 is obtained from a semi-empirical equation valid for $\phi \gtrsim 0.25$ that is discussed below. The curve through the data that gives the asymptote of 0.71 is from another semi-empirical model that is also discussed below. Shapiro and Probstein (1992) conclude that there is in fact not a single value for the maximum packing fraction but rather a range, with the higher values ascribable to a greater degree of ordering of the particles. Their results suggest that the extent of the ordering in a given viscometric experiment is not a uniquely definable quantity but rather depends upon the system and the initial and boundary conditions imposed.

Shapiro & Probstein following earlier suggestions (Sengun & Probstein 1989a, Onoda & Liniger 1990) showed indirectly, from both viscosity experiments of the type illustrated in Fig. 9.3.1 and measurements of the dry random

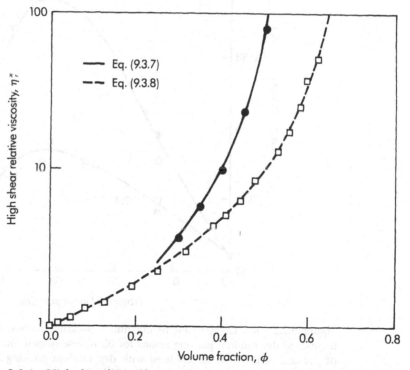

Figure 9.3.1 High shear limit relative viscosity for monodisperse spherical particles as a function of solids volume fraction. Circles are data of Shapiro & Probstein (1992), squares are data of de Kruif et al. (1986), curves are semi-empirical equations.

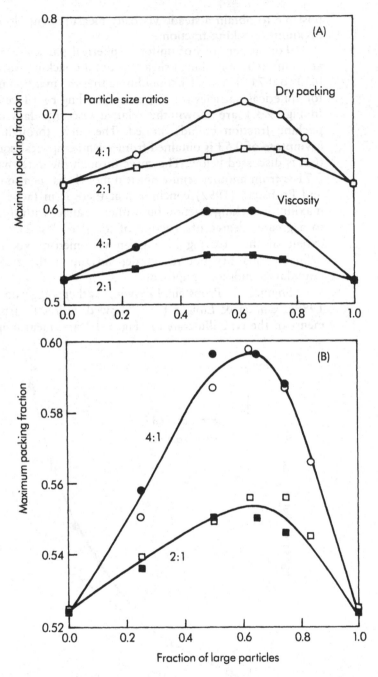

Figure 9.3.2 (A) Comparison of maximum packing fraction from viscosity measurements and dry random packing results for bidisperse suspensions with particle size ratios of 2:1 and 4:1; (B) data replotted with dry random packing results divided by 1.19. [After Shapiro, A.P. & Probstein, R.F. 1992. Random packings of spheres and fluidity limits of monodisperse and bidisperse suspensions. *Phys. Rev. Lett.* **68**, 1422–1425. Copyright 1992 by the Am. Phys. Soc. With permission.]

close packing of spheres, that the lower bound on the maximum packing fraction (~0.53) corresponds to a random suspension microstructure.

The justification for the above conclusions is based on the experimental results shown in Fig. 9.3.2 for bidisperse noncolloidal suspensions with particle size ratios of 2:1 and 4:1 for different fractions of the large particles. When the large particle fraction is either 0 or 1, this corresponds to a monomodal suspension. The figure shows the lower bound maximum packing fraction derived from Couette viscometer experiments through extrapolations of the type shown in Fig. 9.3.1, together with the dry random close packings of the spherical beads. The dry random packings are determined by pouring beads into graduated cylinders which are then vibrated for several hours after which the total volume is measured and divided into the known particle volume. It is well known from experiment that the dry random close packing for monomodal spherical particles is 0.63.

The shapes of the curves in Fig. 9.3.2A are seen to be strikingly similar. Figure 9.3.2B shows the same data except that the random close packing concentrations have been scaled by dividing by the ratio of the dry random packing to the lower bound viscometric maximum packing fraction for mono-disperse spheres; 0.63/0.53 = 1.19. The value 1.19 is termed the *filler dilatancy factor*. The good agreement between the scaled dry packing fraction and the lower bound maximum packing fraction suggests that the suspension near the lower bound maximum packing fraction is of random structure. Moreover, this correlation allows the lower bound maximum packing fraction to be determined independently of viscometry experiments. These results have also been extended to high values of the particle size ratio and to polydisperse suspensions of noncolloidal spherical particles. Similar results were again found with the same value of the filler dilatancy factor (Probstein et al. 1994).

A rational and fundamental theoretical determination of the high shear limit relative viscosity as a function of volume fraction for concentrated suspensions, even for monomodal spherical particles, is a difficult matter as we have seen from our brief discussion of Batchelor & Green's efforts for a semi-dilute suspension. Most approaches to the problem have been semi-empirical and make use of an assumed knowledge of the maximum packing fraction. If the appropriate maximum packing fraction is the lower bound maximum packing fraction, which is obtainable from simple dry packing experiments independent of viscosity experiments, this makes the evaluation of such semi-empirical models both simpler and more reliable. van de Ven (1989) has pointed out that there are at least 100 semi-empirical relations between η_r^∞ and ϕ in the literature, where we use the ∞ superscript to denote the high shear limit. In what follows we discuss two of the basic forms.

We first present the derivation, as outlined by van de Van (1989), of the widely used Krieger–Dougherty expression for the high shear limit relative viscosity. In deriving this relation, the concept of intrinsic viscosity is used although here, in contrast with the definition in polymer rheology, it is defined with volume fraction in place of density (Eq. 9.2.12) thereby making it dimensionless. To distinguish these two intrinsic viscosities, here we use the symbol $[\eta]_\phi$ so that

$$[\eta]_\phi = \lim_{\phi \to 0} \left(\frac{\eta - \mu}{\phi \mu} \right) \tag{9.3.1a}$$

or

$$\eta = \mu[1 + [\eta]_\phi \phi] \tag{9.3.1b}$$

where μ is the viscosity of the suspending medium. Note that in the limit of small ϕ with $[\eta]_\phi = 2.5$, this is just the Einstein relation.

Assume a suspension having a suspended volume fraction ϕ_1 to which is added a small number of particles of volume fraction $\phi_2 \ll 1$. From Eq. (9.3.1b) the viscosity of the new suspension is given approximately by

$$\eta(\phi_1 + \phi_2) \approx \eta(\phi_1)[1 + [\eta]_\phi \phi_2] \tag{9.3.2}$$

There is, however, an excluded volume effect since when the particle fraction ϕ_2 was added, not all the fluid volume was available, part being already taken up by the particle fraction ϕ_1. Krieger and Dougherty therefore assumed that the volume fraction available to the new particles was only $1 - k\phi_1$, where k is a constant termed a *crowding factor*. It follows that Eq. (9.3.2) should be written

$$\eta(\phi_1 + \phi_2) = \eta(\phi_1)\left[1 + \frac{[\eta]_\phi \phi_2}{1 - k\phi_1}\right] \tag{9.3.3}$$

With $\Delta\eta$ the change in viscosity associated with the increase in volume fraction $\Delta\phi \equiv \phi_2$, we write

$$\eta + \Delta\eta = \eta\left[1 + \frac{[\eta]_\phi \Delta\phi}{1 - k\phi}\right] \tag{9.3.4}$$

or in the limit as $\Delta\phi \to 0$

$$\frac{d\eta}{\eta} = [\eta]_\phi \frac{d\phi}{1 - k\phi} \tag{9.3.5}$$

which can be integrated to give

$$\eta = \mu(1 - k\phi)^{-[\eta]_\phi / k} \tag{9.3.6}$$

where μ is the Newtonian viscosity of the particle free fluid.

Since the apparent viscosity $\eta \to \infty$ as $\phi \to 1/k$, it follows that $1/k$ may be interpreted as the maximum packing fraction ϕ_m. The high shear limit relative viscosity may therefore be written

$$\eta_r^\infty = \left(1 - \frac{\phi}{\phi_m}\right)^{-[\eta]_\phi \phi_m} \tag{9.3.7}$$

This is one form of the Krieger–Dougherty relation, which for $\phi \to 0$ and

$[\eta]_\phi = 2.5$ goes to the Einstein limit. Numerous modifications to this relation have been made including those by van de Ven (1989) who approximately has taken account of the fact that $k = k(\phi)$. It is evident that, except at very small volume fractions, the expression for the high shear viscosity is critically dependent on the value of the maximum packing fraction.

The equation used to fit the data of the high shear limit relative viscosity giving $\phi_m = 0.71$ in Fig. 9.3.1 is Eq. (9.3.7) with the exponent $[\eta]_\phi \phi_m = 1.93$ (de Kruif et al. 1985). This corresponds to a value of $[\eta]_\phi = 2.72$ which, although not equal to, nevertheless closely approximates the Einstein limit for small ϕ.

Another semi-empirical expression derives from the widely used work of Frankel & Acrivos (1967) who assumed a dense suspension of monomodal spherical particles, where the particle separation is small compared to the particle radius. In this limit the dissipation is taken to be governed by lubrication forces associated with the relative velocity of the particles along their line of centers. In the low volume fraction or dilute limit, this effect vanishes and the dissipation is close to that associated with the externally applied shear rate. Sengun & Probstein (1989a) calculated the dissipation in a unit cell assuming the total dissipation at any volume fraction to be given by the linear sum of the two dissipations.

The result for the relative viscosity is shown to be given by

$$\eta_r^\infty = 1 + C\left(\frac{3\pi}{8}\right)\left(\frac{\beta}{\beta+1}\right)\left\{\frac{3 + 4.5\beta + \beta^2}{\beta+1} - 3\left(\frac{\beta+1}{\beta}\right)\ln(\beta+1)\right\} \tag{9.3.8}$$

where C is an undetermined proportionality constant of $O(1)$ and β is the ratio of the particle diameter to the minimum separation distance between the particle surfaces. Here, β is a key parameter because it characterizes the average effective surface separation which is critical in any lubrication type flow. To express β in terms of ϕ requires the selection of a specific configuration for the spherical particles. This problem is sidestepped by introducing the maximum packing fraction, ϕ_m, and writing

$$\beta = \frac{(\phi/\phi_m)^{1/3}}{1 - (\phi/\phi_m)^{1/3}} \tag{9.3.9}$$

In this way, β is related to the particle volume fraction in terms of the maximum packing fraction such that the separation between the particle surfaces approaches zero in the limit $\phi \to \phi_m$. It is to be noted that Eq. (9.3.9) is obtained if it is assumed that the effective microstructure of a flowing suspension is a simple cubic ($\phi_m = 0.52$), or body centered cubic ($\phi_m = 0.68$), or face centered cubic ($\phi_m = 0.74$). It is therefore assumed that Eq. (9.3.9) is also applicable to other effective suspension microstructures such as the random microstructure. Equation (9.3.8) is appropriate only for high solid volume fractions ($\phi \gtrsim 0.25$) since it was developed for concentrated suspensions for which the average separation distance between two similar size particles is close to or less than the particle size.

The equation used to fit the data of the high shear limit relative viscosity giving $\phi_m = 0.53$ in Fig. 9.3.1 is Eq. (9.3.8) with $C = 1.5$ found to give the best fit to the viscosity data (Shapiro & Probstein 1992). This equation was also used for determining ϕ_m for the bidisperse data of Fig. 9.3.2.

As the separation between particle surfaces approaches zero with increasing solids volume fraction, particle surface roughness and deformity may be expected to play an important role. The value of the volume fraction above which such effects are significant should depend on the properties of the suspension and may possibly be located by an abrupt change in the slope of the viscosity versus volume fraction curve. Above this value, high shear limit viscosity expressions such as Eqs. (9.3.7) or (9.3.8) are not expected to be applicable. In this regard, from an engineering point of view, the maximum packing fraction should be considered to be a parameter that characterizes the suspension viscosity through semi-empirical relations, as those given here, up to a value close to but somewhat smaller than ϕ_m, say, to within a couple of percent.

It has been shown (Adler et al. 1990) from a rigorous asymptotic, lubrication-theory analysis that lubrication concepts cannot lead to a singular behavior of the viscosity of a spatially periodic suspension in which layers of particles slide past one another. This means that the use of Eq. (9.3.8), for example, which employs lubrication concepts to characterize suspension viscosity is limited to suspensions where particle layering does not take place, for example, where the microstructure is random.

Cell models akin to those discussed in Section 8.5 have also been applied to the determination of the properties of concentrated suspensions (Happel & Brenner 1983, van de Ven 1989). Although it is another method which has been used to obtaining approximate expressions for the high shear relative viscosity, we choose not to expand upon it here, instead referring the reader to the references cited. One of the difficulties is that the determination of the boundary conditions at the cell surface is somewhat arbitrary. Furthermore, expressions obtained by this approach indicate that the cell model is inappropriate for highly concentrated suspensions and is most satisfactory only at low to moderate concentrations.

Although the discussion of the high shear limit viscosity relations has centered on monomodal spherical suspensions, Probstein et al. (1994) have shown the applicability of Eq. (9.3.8) to bidisperse and polydisperse suspensions experimentally and on theoretical grounds.

9.4   Bimodal Model for Suspension Viscosity

In this section we turn our attention to bidisperse suspensions, following which we will briefly discuss polydisperse suspensions. It is known that for a bidisperse suspension the relative viscosity decreases significantly in comparison to that of a monodisperse suspension of the same material and with the same solids volume fraction.

Chong et al. (1971) measured the high shear limit viscosity of bimodal suspensions of spherical glass particles. Their results showed that at a given total solids volume fraction, the viscosity decreased as the ratio of the large to small particle size increased. However, above a ratio of 10 : 1 the viscosity remained relatively insensitive to any further increase in this parameter. These results confirmed the pioneering study of Fidleris and Whitmore (1961), who observed that when a large particle is dropped in a suspension of much smaller particles, such that the large particle size is at least ten times bigger than the small particle size, the drag experienced by the large particle is the same as for the motion of the large particle through a pure liquid having the same density and viscosity as the suspension.

From the above observations and from the discussion of the forces governing suspension rheology, Probstein & Sengun (1987) (see also Sengun & Probstein 1989b) introduced a model to characterize the viscous behavior of concentrated suspensions. In their model the suspension is bimodal, wherein it is made up of a colloidal fine fraction and a coarse fraction of noncolloidal particles.

Truly bimodal suspensions of colloidal and noncolloidal particles are of considerable practical interest. For such suspensions at low shear rates, the viscosity is high so that, for example, during storage, settling is reduced. On the other hand, because the mixture is shear thinning, at higher shear rates when the suspension is pumped the viscosity decreases, thereby enabling the mixture to be pumped at a lower pressure drop.

The important feature of the bimodal model is that the colloidal fraction is assumed to act independently of the coarse fraction and to impart to a stable suspension most of its important non-Newtonian characteristics such as its shear thinning behavior. The large particles are unaffected by the colloidal forces and essentially are unaware of the existence of the colloidal particles. Instead, the large particles see a "stiffened" single phase fluid with the same rheological behavior as the colloidal suspension, and contribute to the viscosity rise solely through hydrodynamic dissipation. We note that the two fractions are not completely, but are largely, independent. For example, as the volume fraction of the coarse particles increases, the colloidal fraction that is squeezed out from between the coarse particles will experience a shear higher than that applied by the viscometer walls and this must be accounted for (Sengun & Probstein 1989c). These nonindependence effects are not considered here.

The relevant independent variables of the problem are the various particle volume fractions. In the bimodal model, the total volume fraction of the solid particles, for example, is

$$\phi_t = \frac{V_c + V_f}{V_c + V_f + V_l} \tag{9.4.1}$$

where V_c is the volume of the coarse particles, V_f the volume of the fine particles, and V_l the volume of the carrier liquid.

According to the model, the behavior of the colloidal particles plus liquid mixture is independent of V_c and is governed by what is termed the *fine filler* volume fraction defined by

$$\phi_{ff} = \frac{V_f}{V_f + V_l} \qquad (9.4.2)$$

while the viscosity rise due to the coarse particles is characterized by the coarse volume fraction defined by

$$\phi_c = \frac{V_c}{V_c + V_f + V_l} \qquad (9.4.3)$$

According to the bimodal concept, the *net relative viscosity*, η_{nr}, of a bimodal suspension is given by

$$\eta_{nr} = \eta_{fr} \eta_{cr} = \left(\frac{\eta_f}{\mu}\right) \left(\frac{\eta_c}{\eta_f}\right) \qquad (9.4.4)$$

Here, η_{fr}, is termed the *fine relative viscosity* and represents the contribution of the colloidal size particles. It is defined by the ratio of the apparent viscosity of the mixture of suspending liquid plus fine particles, η_f, to the viscosity of the suspending liquid, μ. The quantity, η_{cr}, is termed the *coarse relative viscosity* and is the contribution of the coarse particles to the net relative viscosity. It is defined by the ratio of the apparent viscosity of the coarse suspension, η_c, to the apparent viscosity of the fine fraction, η_f.

Because of the large characteristic size of the coarse particles, as discussed in Section 9.2, even at relatively low shear rates, say, $1 \, s^{-1}$, the dimensionless force groups Eqs. (9.2.10)–(9.2.12) are all very large compared to unity and η_{cr} will exhibit a shear rate independent behavior characteristic of the high shear limit.

Equation (9.4.4) may be usefully written in the form

$$\log \eta_{nr} = \log \eta_{fr} + \log \eta_{cr} \qquad (9.4.5)$$

According to the bimodal model, η_{cr} is independent of the shear rate so that if $\log \eta_{nr}$ and $\log \eta_{fr}$ were plotted against the logarithm of the shear rate on the same graph, the curves should be parallel and the difference between the curves should be equal to $\log \eta_{cr}$.

An application of the bimodal model to truly bimodal coal-water suspensions was carried out by Sengun and Probstein (1989b,c). First they measured the viscosity versus shear rate of a colloidal coal-water suspension with an average particle size of 2.3 μm and fine filler volume fraction $\phi_{ff} = 0.30$. Their experimental results are represented by the solid line in Fig. 9.4.1, where the colloidal effects are evidenced by the strongly non-Newtonian shear thinning behavior. Next they added to the same suspension coarse coal particles in the size range 200–300 μm and their viscosity measurements for a coarse volume fraction $\phi_c = 0.52$ are represented by the triangles in Fig. 9.4.1. The dashed line drawn through these experimental points is just the fine relative viscosity curve shifted upward so that it is parallel to the initial curve, with the amount of the

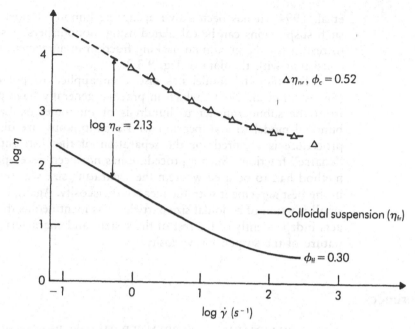

Figure 9.4.1 Relative viscosity of a bidisperse coal slurry made up of a colloidal fine fraction of mean diameter 2.3 μm and a noncolloidal coarse fraction of 200–300 μm particles of mean diameter about 250 μm as a function of shear rate. The volume fraction of the colloidal particles $\phi_{ff} = 0.30$ and of the coarse particles $\phi_c = 0.52$. The solid line is a mean curve through the measured viscosities of the colloidal fraction. The triangles are the experimental points for the measured viscosity for the fine plus coarse mixture. The dashed line is the fine relative viscosity experimental curve redrawn through the data points to illustrate the parallelism. The upward shift of this curve corresponds to a coarse relative viscosity $\log \eta_{cr} = 2.13$. [After Sengun, M.Z. & Probstein, R.F. 1989. Bimodal model of slurry viscosity with application to coal-slurries. Part 2. High shear limit behavior. *Rheol. Acta* **28**, 394–401. Steinkopff Darmstadt. With permission.]

shift given by $\log \eta = 2.13$. This contribution is just the contribution of the coarse relative viscosity. The near perfect parallelism between the data for the net relative viscosity and the data for the fine relative viscosity is a striking confirmation of the validity of the bimodal model. It indicates that the colloidal and the coarse fractions do behave independently of each other and the shear-dependent behavior is caused solely by the colloidal fraction. Independent high shear limit viscosity measurements of suspensions of only the coarse particles showed close agreement with the values obtained by differencing curves typical of Fig. 9.4.1 (Sengun & Probstein 1989c).

It should be noted that Farris (1968) developed a bimodal model for polydisperse suspensions in which the fine and the coarse fractions were also assumed to behave independently of each other. However, the arguments were purely geometric and the issues related to the non-Newtonian character of the viscosity were not treated.

The bimodal model just described has been successfully applied in the high shear limit to disperse suspensions with a very large particle size ratio (Probstein et al. 1994). It has been shown using the bimodal model that the viscosity of such suspensions can be calculated using monodisperse viscosity data. This led to results for the maximum packing fraction in agreement with experiment and consistent with the data of Fig. 9.3.2.

The bimodal model has also been applied to polydisperse suspensions (Probstein et al. 1994), which in practice generally have particle sizes ranging from the submicrometer to hundreds of micrometers. In order to apply the bimodal model to a suspension with a continuous size distribution, a rational procedure is required for the separation of the distribution into "fine" and "coarse" fractions. Such a procedure has not been developed so that an inverse method had to be used wherein the separating size was selected which resulted in the best agreement with the measured viscosity. Again, however, the relatively small fraction of colloidal size particles was identified as the principal agent that acts independently of the rest of the system and characterizes the shear thinning nature of the suspension viscosity.

References

ADLER, P.M., NADIM, A. & BRENNER, H. 1990. Rheological models of suspensions. *Adv. Chem. Eng.* **15**, 1–72.

BARNES, H.A., HUTTON, J.F. & WALTERS, K. 1989. *An Introduction to Rheology.* Amsterdam: Elsevier.

BATCHELOR, G.K. 1977. The effect of Brownian motion on the bulk stress in a suspension of spherical particles. *J. Fluid Mech.* **83**, 97–117.

BATCHELOR, G.K. & GREEN, J.T. 1972. The determination of the bulk stress in a suspension of spherical particles to order c^2. *J. Fluid Mech.* **56**, 401–427.

BIRD, R.B., ARMSTRONG, R.C. & HASSAGER, O. 1987. *Dynamics of Polymeric Liquids, vol. 1. Fluid Mechanics*, 2nd edn. New York: Wiley.

CHONG, J.S., CHRISTIANSEN, E.B. & BAER, A.D. 1971. Rheology of concentrated suspensions. *J. Appl. Polym. Sci.* **15**, 2007–2021.

DE GENNES, P.G. 1979. *Scaling Concepts in Polymer Physics.* Ithaca: Cornell Univ. Press.

DE GENNES, P.G. 1990. *Introduction to Polymer Dynamics.* Cambridge: Cambridge Univ. Press.

DE KRUIF, C.G., VAN LERSEL, E.M.F., VRIJ, A. & RUSSEL, W.B. 1985. Hard sphere colloidal dispersions: Viscosity as a function of shear rate and volume fraction. *J. Chem. Phys.* **83**, 4717–4725.

FARRIS, R.J. 1968. Prediction of the viscosity of multimodal suspensions from unimodal viscosity data. *Trans. Soc. Rheol.* **12**, 281–301.

FIDLERIS, V. & WHITMORE, R.N. 1961. The physical interactions of spherical particles in suspensions. *Rheol. Acta* **1**, 573–580.

FRANKEL, N.A. & ACRIVOS, A. 1967. On the viscosity of a concentrated suspension of solid spheres. *Chem. Eng. Sci.* **22**, 847–853.

HAPPEL, J. & BRENNER, H. 1983. *Low Reynolds Number Hydrodynamics.* The Hague: Martinus Nijhoff.

KRIEGER, I.M. 1972. Rheology of monodisperse lattices. *Adv. Colloid Interface Sci.* **3**, 111–136.

ONODA, G.Y. & LINIGER, E.G. 1990. Random loose packings of uniform spheres and the dilatancy onset. *Phys. Rev. Lett.* **64**, 2727–2730.

PARK, J.T., MANNHEIMER, R.J., GRIMLEY, T.A. & MORROW, T.B. 1989. Pipe flow measurements of a transparent non-Newtonian slurry. *J. Fluids Eng. (Trans. ASME)* **111**, 331–336.

PROBSTEIN, R.F. & SENGUN, M.Z. 1987. Dense slurry rheology with application to coal slurries. *PhysicoChem. Hydrodynamics* **9**, 299–313.

PROBSTEIN, R.F., SENGUN, M.Z. & TSENG, T-C. 1994. Bimodal model of concentrated suspension viscosity for distributed particle sizes. *J. Rheol.* **38**, No. 4 (1994).

RUSSEL, W.B., SAVILLE, D.A. & SCHOWALTER, W.R. 1989. *Colloidal Dispersions.* Cambridge: Cambridge Univ. Press.

SENGUN, M.Z. & PROBSTEIN, R.F. 1989a. High-shear-limit viscosity and the maximum packing fraction in concentrated monomodal suspensions. *PhysicoChem. Hydrodynamics* **11**, 229–241.

SENGUN, M.Z. & PROBSTEIN, R.F. 1989b. Bimodal model of slurry viscosity with application to coal-slurries. Part 1. Theory and experiment. *Rheol. Acta* **28**, 382–393.

SENGUN, M.Z. & PROBSTEIN, R.F. 1989c. Bimodal model of slurry viscosity with application to coal-slurries. Part 2. High shear limit behavior. *Rheol. Acta* **28**, 394–401.

SHAPIRO, A.P. & PROBSTEIN, R.F. 1992. Random packings of spheres and fluidity limits of monodisperse and bidisperse suspensions. *Phys. Rev. Lett.* **68**, 1422–1425.

TANNER, R.I. 1988. *Engineering Rheology*, revised edn. Oxford: Clarendon Press.

VAN DE VEN, T.G.M. 1989. *Colloidal Hydrodynamics.* San Diego: Academic.

Problems

9.1   a.   Consider the fully developed flow in a channel of a non-Newtonian fluid with a viscosity that obeys the power-law model of Eq. (9.1.8). The channel width is $2h$ and the flow is in the x-direction. Show that the volumetric flow rate Q is given by

$$Q = \left(\frac{2n}{2n+1}\right)\left(\frac{G}{m}\right)^{1/n} h^{(2n+1)/n}$$

b.   Assume the fluid to behave as a Bingham plastic (Eq. 9.1.7) instead of a power-law fluid. With $2z_0$ the width of the core or plug region, show that the volumetric flow rate is now given by

$$Q = \frac{\tau_0}{3\mu_p z_0}(2h^3 - 3h^2 z_0 + z_0^3)$$

9.2   The power-law fluid of Problem 9.1.1a is squeezed out from between two circular disks of radius R, by the top disk approaching the stationary bottom disk at a speed V. When the distance separating the disks, $2h$, is small compared to R, show that the normal force F on the disks is given by

$$F = \left(\frac{\pi m}{n+3}\right) R^{n+3}\left(\frac{2n+1}{2n}\right)^n \frac{(-\dot{h})^n}{h^{2n+1}}$$

9.3 In the dumbbell model, a polymer chain in a solvent is pictured as two massless spheres of equal size connected by a frictionless spring. The spheres experience a hydrodynamic drag proportional to their size, characterized by the Flory radius. Assume that the displacement of the spring generated by the thermal energy is also characterized by the Flory radius. Write the equation of motion for the dumbbell and show that the characteristic relaxation time for the chain deformation is that given by Eq. (9.2.1).

9.4 Assume that the crowding factor, k, in Eq. (9.3.5) varies linearly with ϕ from the value k_0 at $\phi = 0$ to $1/\phi_m$ at $\phi = \phi_m$. What form does the viscosity expression take under this assumption?

9.5 Consider the fully developed flow of a stable suspension of spherical particles through a pipe of radius $a = 5$ mm. The suspending medium is water with a total solids volume fraction of 0.26. The applied pressure gradient, G, along the pipe is 1750 Pa m^{-1}.

a. Assume that the particles are colloidal and monomodal in size. Determine the volumetric flow rate, Q, for the following experimentally determined shear viscosity behavior

for $\dot{\gamma} < 10^2$ s^{-1} Newtonian behavior with $\ln \eta_r$
$$= \frac{2.18\phi}{1 - 3.14\phi}$$

for 10^2 s$^{-1} < \dot{\gamma} < 10^4$ s^{-1} power-law behavior with $\eta_r = (m/\mu)\dot{\gamma}^{n-1}$

for $\dot{\gamma} > 10^4$ s^{-1} Newtonian behavior with η_r
$$= (1 - \phi/0.71)^{-1.93}$$

where $\dot{\gamma}$ is the magnitude of the shear rate, and ϕ the solids volume fraction.

b. Assume now that the suspension is bimodal with equal volumes of colloidal and coarse particles. The total solids volume fraction is the same as in Part a, and the colloidal fraction has the same viscosity behavior. The coarse particles are very large compared to the colloidal particles, so that they are not affected by colloidal forces. Determine the volumetric flow rate, Q, and compare the result with that obtained in Part a.

10 Surface Tension

<hr style="border-top: 3px solid;" />

10.1 Physics of Surface Tension

The effects of surface and interfacial tensions give rise to so many commonplace phenomena observed in liquid behavior that we often take for granted the complex physical-chemical interactions involved, not all of which are understood even today. Among the many familiar examples of surface tension effects are the formation of soap bubbles that float gently upward until they break, or the thin capillary tube in which a liquid will rise to a height greater than the pool in which it is placed. There is also the breakup into drops of a stream of water flowing out of a faucet, the physics of which is the basis of the ink jet printer or gel encapsulation processes to encase everything from monoclonal antibodies to perfume. Or there is the phenomenon of a liquid drop remaining stationary when placed on a solid surface, as well as the opposite situation of the spreading of a drop of water when placed on a clean glass surface. The examples, both observed and applied, that result from interfacial effects between liquids, gases, and solids are indeed numerous.

Interfacial phenomena attracted considerable scientific attention from the 18th century onward. At first the attempts to characterize the different behaviors were mechanical, where the liquids were described as being stretched at their interfacial surface like a membrane, with a state of tension existing there. Today we know that the liquid state itself is composed of molecules in motion that are kept relatively close to each other by attractive van der Waals forces. However, a principal method of analysis of problems of interfacial effects rests upon the assumption that the liquid can be described by a continuum *mean-field approximation* or *mean molecular field*, wherein it is assumed possible to define an element of the liquid that is small compared to the range of the intermolecular force but large enough to contain a sufficient number of molecules. This approximation implies that on average the attractive force on any molecule in the liquid is the same in all directions, giving to the liquid its fluid characteristics.

The above argument cannot hold, for example, at a liquid-gas interface since although the molecules are free to move in the liquid, their motion is far more restricted than in a gas where there is little attraction between the molecules. The attraction between the liquid molecules will prevent but a small fraction of them from escaping (vaporizing) into the gas. Therefore, the liquid molecules at the interface are attracted inward and to the side, but there is no *outward* attraction to balance the pull because, by comparison, there are not many liquid molecules outside in the gas (see Fig. 10.1.1). As a result, the liquid molecules at the surface are attracted inward and normal to the liquid-gas interface, which is equivalent to the tendency of the surface to contract (shrink). The surface of the liquid thus behaves as if it were in tension like a stretched membrane. We emphasize, however, that there really is not a macroscopic smooth meniscus-type surface at which the molecular concentration changes discontinuously from that of the liquid phase to that of the gaseous phase. Rather, this change between the two phases takes place continuously over a small distance of about 100 nm or less.

An alternative description from an energetic point of view follows from the fact that because a liquid molecule at a liquid-gas interface must be attracted to less neighboring molecules than one in the interior of the fluid, the attractive energy per molecule at the surface must then be some fraction of that in the interior. The energy of a surface molecule is therefore higher than that of one in the bulk liquid, so energy must be expended to move a molecule from the interior to the surface. However, since the free energy of the system will tend to a minimum, the surface of the liquid phase will tend to contract. With σ the force per unit length tending to contract the surface, we may therefore write that, at constant temperature and volume for a given number of moles of system,

$$\sigma = \frac{\partial F}{\partial A} \qquad (10.1.1)$$

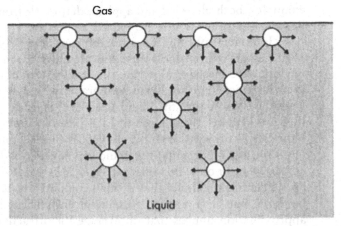

Figure 10.1.1 Attractive forces between molecules at surface and in interior of a liquid for a plane liquid-gas interface.

where

$$F = E - TS \qquad (10.1.2)$$

is the Helmholtz free energy (capital letters indicating the extensive form). Equation (10.1.1) simply states since ∂F will decrease together with a decrease in ∂A (that is, surface contraction), then σ will be positive. The quantity σ is called the *surface tension* and is usually given in units of force per unit length. Table 10.1.1 shows some values of surface tensions for pure liquids interfacing with the vapor phase.

It can readily be shown (Davies & Rideal 1963) that, provided the viscosity of the liquid is not very high, the surface tension is equal to the Helmholtz free energy per unit area of surface, and hence also has the equivalent dimensions of energy per unit area. From the definition of the Helmholtz free energy applied at the surface, it also is easily demonstrated that the surface tension decreases with increasing liquid temperature; that is,

$$\frac{\partial \sigma}{\partial T} < 0 \qquad (10.1.3)$$

In practice, σ decreases very nearly linearly with increasing temperature. A typical value is about $-0.1 \, \text{mN m}^{-1} \, \text{K}^{-1}$.

It is well known that an insoluble adsorbed monolayer at the surface will also lower the surface tension because the adsorbed molecules will tend to "spread" the surface and hence lower the surface tension since it tends to contract the surface. Surface-active materials that are strongly adsorbed at an interface in the form of an oriented *monomolecular layer* (*monolayer*) are termed *surfactants* and are capable of reducing the surface tension of water or aqueous solutions by up to five orders of magnitude. The degree of surface activity of a material will depend on its surface adsorption and its mixing in the aqueous phase. Inorganic ions in water are not surface active because they are "pulled" into the aqueous phase. On the other hand, typical aqueous surfactants such as detergents are organic compounds with a long-chain hydrocarbon tail and a polar head. It is well known that hydrocarbons are relatively insoluble, water being highly polar. An ideal surfactant behavior is shown in Fig. 10.1.2,

Table 10.1.1
Surface Tensions of Liquid-Vapor Interfaces at
20°C [a]

	σ mN m^{-1}
Water	72.88
Nitromethane	32.66
Benzene	28.88
Methanol	22.50
Mercury	486.5

[a] Adamson (1982).

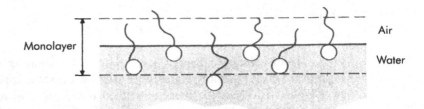

Figure 10.1.2   Schematic of surfactant monolayer.

where the polar molecules adsorb at the surface as a monolayer with the hydrocarbon tails "up" and the polar heads "down." It is not difficult to show that it takes a relatively small number of surfactant molecules to form a monolayer and thereby greatly reduce the surface tension of water, which is relatively high.

The effect of surface-active materials in altering surface tension is expressed quantitatively through the *Gibbs equation*, which relates the change in surface tension to the concentration of surface-active materials. The relation takes a particularly simple form for a two-component dilute solution consisting of a solvent and a single solute. If the *surface excess concentration* (mol m^{-2}) of the adsorbed solute is denoted by Γ and c is the bulk molar concentration, then

$$\Gamma = -\frac{c}{RT}\frac{\partial \sigma}{\partial c} \qquad (10.1.4)$$

The Gibbs equation shows that there is a decrease in surface tension with an increase in concentration of a particular solute that is positively adsorbed at an interface.

Although at low concentrations the surfactant molecules behave independently, at higher concentrations they aggregate to form *micelles*. The micelles are roughly spherical and typically contain about 50 to 100 molecules. An ionic micelle is shown schematically in Fig. 10.1.3. The polar heads are in contact with the water, and surrounded by a double layer shell, while the central core of the micelle is essentially water free, being made up of the hydrocarbon tails. The concentration at which micelle formation begins is called the *critical micelle concentration*. Above this critical value the concentration of free surfactant molecules is essentially unchanged and, therefore, so is the surface tension. Any further addition of surfactant molecules would only go into micelle formation (Hiemenz 1986).

Extending our molecular picture to include kinetic considerations, we can show that the time for a liquid surface to take up its equilibrium value of surface tension is very short. The time may be estimated from the kinetic theory relation $t = l^2/D$, where D is the diffusion coefficient and l is approximately equal to the lattice spacing. This time characterizes the time for molecules at the surface to exchange with those in the immediate bulk below. With $D \sim 10^{-9}$ m^2 s^{-1} and $l \sim 0.3$ nm (the diameter of a water molecule), we find $t \sim 10^{-10}$ s, which is indeed very small. In practice, the time is somewhat larger than this, but nevertheless extremely small.

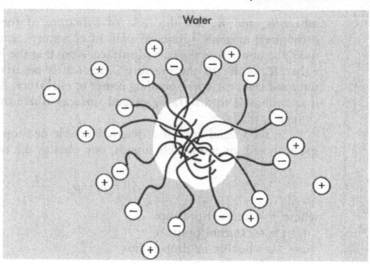

Figure 10.1.3 Schematic of an aqueous ionic micelle.

Another time of interest is the residence time of a molecule at the surface that is in equilibrium with the surrounding vapor, that is, where the number of molecules evaporating are equal to those condensing from the vapor phase. This time is much larger than that to take up the equilibrium surface tension value, but nevertheless is still quite small. Considering water vapor, $D \sim 10^{-6}\,\mathrm{m^2\,s^{-1}}$ and $l \sim 0.1\,\mu\mathrm{m}$ at atmospheric conditions. An estimate of the rate of collision of the vapor molecules with the liquid surface is then D/l^2 or $10^8\,\mathrm{s^{-1}}$. Taking the fraction of molecules that strike the surface and condense on it to be about 0.1 gives a lifetime for a molecule on the surface before it evaporates to be about 10^{-7} s. This rate still indicates a very rapid change and strong agitation at the surface. In actuality mass transfer boundary layer resistance makes the rate considerably lower, particularly in the presence of external hydrodynamic effects.

The above discussion for a liquid-gas interface is also applicable to a liquid-liquid interface between two immiscible liquids with an interfacial tension acting at the interface. As before, there is an imbalance of intermolecular forces, although smaller. The magnitude of the interfacial tension usually lies between the surface tensions of each liquid.

We have thus far restricted our discussion to plane interfaces. However, because of the existence of surface tension, there will be a tendency to curve the interface, as a consequence of which there must be a pressure difference across the surface with the highest pressure on the concave side. The expression relating this pressure difference to the curvature of the surface is usually referred to as the *Young-Laplace equation*. It was published by Young in 1805 and, independently, by Laplace in 1806. From a calculation of the p-V work required to expand the curved surface and so change its surface area, it is relatively straightforward to show that this equation may be written

$$\Delta p = \sigma \left(\frac{1}{R_1} + \frac{1}{R_2} \right) \qquad (10.1.5)$$

where R_1 and R_2 are the radii of curvature of the surface along any two orthogonal tangents (principal radii of curvature), and Δp is the difference in fluid pressure across the curved surface. Note that the individual contribution of either R_1 or R_2 to the pressure difference is negative when moving radially outward from the corresponding center of curvature. As Eq. (10.1.5) is written, it is applicable to arbitrarily shaped surfaces where the radii of curvature may change spatially.

In the special case of a spherical bubble or drop of either a liquid in gas, gas in liquid, or immiscible liquids, one obtains the well-known result

$$\Delta p = p_i - p_e = \frac{2\sigma}{a} \tag{10.1.6}$$

where p_i = internal pressure
$\quad\ p_e$ = external pressure
$\quad\ a$ = bubble or drop radius

This formula is consistent with the fact that in stable equilibrium the energy of the surface must be a minimum for a given value of bubble or drop volume, and a sphere has the least surface area for a given volume. For general curved surfaces the radius a in Eq. (10.1.6) is frequently taken to be the mean radius of curvature defined as half the sum of the inverse principal radii of curvature. For immiscible liquids σ refers to the *interfacial tension*, which, for example, for benzene over water at 20°C is 35 mN m^{-1}. Obviously for a plane interface, where the mean radius tends to infinity, the pressure difference will be zero.

The behavior of liquids on solid surfaces is also of considerable practical importance. However, the molecules or atoms at a solid surface, unlike those at a liquid surface, are essentially immobile. Therefore, a solid-fluid interface will not have the same behavior as a fluid-fluid interface. For a planar interface, as in Fig. 10.1.1, with the gas replaced by a solid, the liquid molecules could be attracted more strongly to the solid surface than between the liquid molecules themselves, a situation representative of water on very clean glass. In this case the liquid molecules would be attracted outward to the interface, but the inward attraction to balance the pull would be less, so the liquid molecules would be attracted outward and normal.

It is almost impossible with solid surfaces to obtain a purely planar surface free from inhomogeneities in contrast to a liquid surface, which can be made to have a high degree of homogeneity. Solid surfaces will therefore almost always be contaminated by impurities, which can have a marked effect on the surface tension. The impurities will manifest themselves most strongly at the surface since, as with a liquid, the surface molecules will be bonded to a smaller number of neighboring molecules than in the interior. For example, for a close-packed arrangement each molecule will be bound to 12 in the interior but only 9 on the surface and, as discussed previously with a liquid, will have a correspondingly lower binding energy than inside the material itself. It is therefore difficult to define a priori either the magnitude *or sign* of the solid-liquid surface tension.

Normally when a liquid drop is placed on a plane solid surface, it will be in contact not only with the surface but often with a gas such as air, as shown in

Fig. 10.1.4. The liquid may spread freely over the surface, or it may remain as a drop with a specific angle of contact with the solid surface. Denote this *static contact angle* by θ. There must be a force component associated with the liquid-gas surface tension σ that acts parallel to the surface and whose magnitude is $\sigma \cos \theta$. If the drop is to remain in static equilibrium without moving along the surface, it has to be balanced by other forces that act at the *contact line*, which is the line delimiting the portion of the surface wetted by the liquid, for example, a circle. *It is assumed* that the surface forces can be represented by surface tensions associated with the solid-gas and solid-liquid interfaces that act along the surface, σ_{sg} and σ_{sl}, respectively. Setting the sum of the forces in the plane of the surface equal to zero, we have

$$\sigma \cos \theta = \sigma_{sg} - \sigma_{sl} \tag{10.1.7}$$

This equation, known as *Young's equation*, was published in 1805, shortly before its independent publication by Laplace.

As derived, Young's equation follows only from an equilibrium force balance in the horizontal direction with no balance in the vertical direction. Since static equilibrium requires a balance of all forces, a normal force (per unit length) $\sigma \sin \theta$ must act vertically downward on the solid at the contact line. Dussan V. (1979) has pointed out that such a force can exist, since the solid is modeled as a rigid body, but she and others have asked the question, "what restricts this force to a vertical component?" Allowing for a horizontal component, all forces could be balanced for any value of the contact angle. At present there is no unambiguous resolution to this question, although Young's equation can be deduced from thermodynamic considerations.

A system is in static equilibrium if it is in a configuration of minimum energy, and it is this principle that Gibbs (1906) applied to deduce Eq. (10.1.7).

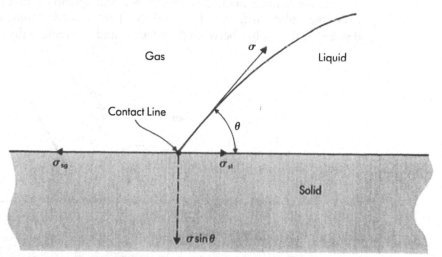

Figure 10.1.4 Static equilibrium of a liquid drop in a gas at line of contact with a horizontal solid surface.

However, as far back as 1829 Gauss provided an alternative derivation based on the inclusion of a surface energy in the energy conservation equation, corresponding to the inclusion of surface tension in the momentum conservation equation (Dussan V. 1979). A simple demonstration of the minimum energy derivation has been given by de Gennes (1985), and we repeat it here.

Interfacial tension is interpreted as free energy per unit area, and the condition is applied that at equilibrium the total system energy must be stationary with respect to any infinitesimal displacement of the contact line. In Fig. 10.1.5 a liquid wedge is shown translated along the plane of the solid by an amount δx. The translation does not affect the bulk energies since the pressure is unchanged in the liquid and the gas. The energy in the region of radius δx is also unaffected since the material is simply translated. However, the interfacial area per unit depth of contact line is changed in an amount $+\delta x$ for the surface-gas area, $-\delta x$ for the surface-liquid area, and $-\delta x \cos \theta$ for the liquid-gas area. The condition that the energy be stationary is then

$$\sigma_{sg}\delta x - \sigma_{sl}\delta x - \sigma\delta x \cos \theta = 0 \qquad (10.1.8)$$

which is just Young's equation in the limit $\delta x \to 0$.

Young's equation may be written in the form

$$k = \frac{\sigma_{sg} - \sigma_{sl}}{\sigma} = \cos \theta \qquad (10.1.9)$$

where k is sometimes termed the *wetting coefficient*. Note that there is no restriction on the magnitude of σ_{sg} or σ, or on the sign or magnitude of σ_{sl}, since their values depend on the appropriate intermolecular forces. The only statement that can be made is that k must lie between -1 and $+1$ (Rowlinson & Widom 1982).

If the contact angle θ is zero, $k = 1$ and spreading takes place completely over the solid surface. The solid is then termed "completely wetted." If $0 < k < 1$, then θ lies between 0 and $\pi/2$, and conventionally the solid is said to

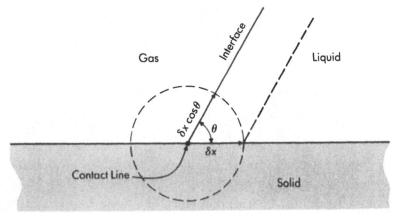

Figure 10.1.5   Translation from left to right of a liquid wedge interface.

be *wetted* by the liquid. However, it is clear that this statement gives an incorrect impression since the surface is actually only "partially wetted."

At the other extreme if $k = -1$, the contact angle θ is π and the solid is completely unwetted. This limit is unrealistic but can be approached physically by assuming the liquid to be supported from the surface by a thin film of vapor. Here, σ_{sl} is very large and positive with a value approaching σ. If $-1 < k < 0$, then θ lies between π and $\pi/2$, and conventionally the solid is said to be *unwetted* by the liquid. Again this statement gives an incorrect impression, since the solid is actually partially wetted, σ_{sl} not being large enough to inhibit solid-liquid contact. The extent of the wetting is less than in the range of $0 < \theta < \pi/2$. A physical example of θ between π and $\pi/2$ is mercury on glass, where $\theta \sim 140°$, a consequence of the very strong forces between the atoms within the liquid.

It is usually pointed out in most texts on the subject of surface tension that no equilibrium position of the contact line exists when

$$S = \sigma_{sg} - \sigma_{sl} - \sigma > 0 \qquad (10.1.10)$$

where S is termed the *spreading coefficient* of the liquid phase at the solid-gas interface. This can be seen more clearly by rewriting Eq. (10.1.9) in terms of S to give

$$k = 1 + \frac{S}{\sigma} = \cos\theta \qquad (10.1.11)$$

Evidently if $S > 0$ then $k > +1$. Were $S > 0$ so that $\sigma_{sg} > \sigma_{sl} + \sigma$, this would imply that the solid-gas interface would immediately coat itself with a layer of the liquid phase and replace the supposedly higher free energy per unit area of direct solid-gas contact, σ_{sg}, by the supposedly lower sum of the free energies per unit area of solid-liquid and liquid-gas contacts, $\sigma_{sl} + \sigma$, thereby lowering the free energy of the system. However, in thermodynamic equilibrium this cannot be realized (Gibbs 1906, Rowlinson & Widom 1982). Therefore, for a spreading film in thermodynamic equilibrium $k = +1$ ($S = 0$), and locally there is a state of mechanical equilibrium at the contact line between the three phases.

In the absence of equilibrium, say, over molecular length scales at the interfaces, it is possible to have $S > 0$, although it is difficult on a macroscopic scale to differentiate between whether $S = 0$ or $S > 0$. This distinction is important since experiment and theory indicate that the larger positive S is, the better the spreading over a solid surface is (de Gennes 1985). We discuss the case of complete wetting ($\theta = 0$) here since we shall be concerned with it in Sections 10.3 and 10.5.

We conclude this discussion by alerting the reader to the concept of the *dynamic contact angle* (and line), which appears in the literature of flows governed by surface tension (Dussan V. 1979). In a flow field where the contact line moves, it is necessary to know the contact angle as a boundary condition for determining the meniscus shape. If this angle is a function of the speed of the contact line relative to the solid surface, then the force balance inherent in

Young's equation is no longer satisfied because, among other things, viscous stresses would deform the interface. In fact, the viscous stresses lead to singular stresses at the contact line. The point to recognize is that the contact angle concept is a continuum one, and, although quite useful, it does break down near the surface. However, the concept of the moving contact line may be applied by assuming a fluid continuum with artifices such as relaxing the "no-slip" boundary condition to permit a "slip velocity" wherein the velocity jumps discontinuously at the surface, a condition that can remove the singular nature of the stresses. Ultimately, however, the understanding of the contact line behavior can only be resolved through an understanding of the physics at the molecular scales, including the deviations from thermodynamic equilibrium (see de Gennes 1985).

10.2   Capillarity and Capillary Motion

Capillarity may be defined as the phenomena resulting from the fact that a free liquid surface has a finite or zero contact angle with a solid wall and will attain this angle when placed in contact with the wall. It is commonly thought of as the rise (or fall) of liquids in small tubes or finely porous media. More generally, *capillary motion* can be said to be any flow that is governed in some measure by the forces associated with surface tension. "Ordinary capillarity" is observed in a fine tube open at both ends that is placed vertically in a pool of liquid exposed to the atmosphere, with the liquid seen to attain a level in the tube above the level of the pool. The actual rise velocity of the free surface of the liquid in the tube from the level of the pool is one simple example of "capillary motion."

Capillarity and capillary motion are widely seen in nature and are applied extensively in a broad range of technical processes. Capillarity has been observed by anyone who has ever used a blotter. It plays a major role in soil hydrology. Coating flows in which thin uniform liquid films are deposited on solid surfaces frequently employ the surface tension characteristics of the film. This fact is usually not thought of in the common task of painting a wall. Surface tension may not be uniform due to adsorbed material, temperature gradients, or varying electric charge, and this may give rise to unbalanced forces, resulting in fluid motions. A child's example of this is the small boat with a piece of camphor at the back that propels itself on a dish of water. At another extreme, surface tension gradients can be used for crystal growth to produce semiconductors by imposing large temperature gradients along a molten melt in the low-gravity environment of a spacecraft.

The forces that give rise to the phenomena spoken of appear because of the alteration in stresses at the interface between two immiscible fluid phases. For a curved interface there is a difference in pressure between the two fluids given by the Young-Laplace equation. This pressure difference is termed the *capillary pressure*, and since the normal stress component at the interface must be continuous, then that pressure added to the hydrostatic pressure must balance. A balance can always be achieved under static conditions. In addition, the tangential stress must also be continuous at the interface. However, if there is a gradient in surface tension in the tangential direction, this leads to an additional tangential stress that adds an unbalanced force that will set the fluid into motion; that is, the interface cannot be balanced when the fluids are stationary. On the other hand, the conditions of continuity of tangential velocity and no normal velocity across the immiscible interface are unaltered by surface tension.

The subject of capillarity and capillary motion can perhaps best be introduced by using the classical example of the rise of a liquid in a circular capillary tube of radius a, that sits vertically in a pool of the liquid open to the atmosphere, as shown in Fig. 10.2.1. It is assumed that the surface tension at the liquid-air interface is uniform with any possible gradients neglected, leaving the discussion of the effects arising from such gradients to Section 10.4.

If the capillary is circular in cross section, the meniscus will be approximately hemispherical with a constant radius of curvature $a/(\cos\theta)$, where θ is taken to be the static contact angle (Fig. 10.2.1). Departure from hemisphericity is associated with the variation in liquid pressure over the surface due to the difference in gravitational force over the meniscus height h. A measure of the hydrostatic gravitational force to surface tension force is given by the *Bond number*

$$\text{Bo} = \frac{\text{gravitational force}}{\text{surface tension force}} = \frac{\rho g L^2}{\sigma} \qquad (10.2.1)$$

where L is the characteristic length scale. When the Bond number is large, the capillary pressure effect can usually be neglected in a liquid at rest. For the meniscus we may identify L with its height h; then the criterion for hemis-

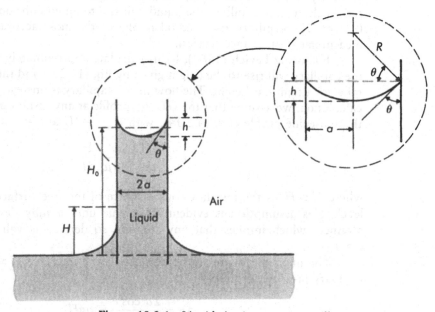

Figure 10.2.1 Liquid rise in an open capillary.

phericity can be written

$$h \ll \left(\frac{\sigma}{\rho g}\right)^{1/2} \equiv \Delta_c \tag{10.2.2}$$

where Δ_c is termed the *capillary length*.

We apply the Young-Laplace equation (10.1.5) to determine the equilibrium height H_0 that the column of liquid will attain, recognizing it is the pressure under the meniscus that controls the height. If the pressure variation along the interface is neglected, $R_1 = R_2 = a/(\cos\theta)$, and the pressure difference across the interface Δp is $p_{atm} - p_{liq}$, with p_{liq}, say, the mean pressure in the liquid at the interface. The hydrostatic pressure at the pool level is $p_{liq} + \rho g H_0$, which is approximately equal to p_{atm}. It follows that the approximate condition for hydrostatic equilibrium is given by

$$\rho g H_0 = \frac{2\sigma\cos\theta}{a} \tag{10.2.3}$$

from which the equilibrium column height is

$$H_0 = 2\left(\frac{\sigma}{\rho g}\right)\frac{\cos\theta}{a} = 2\frac{\Delta_c^2}{a}\cos\theta \tag{10.2.4}$$

Depending on the contact angle, the height can take on any value in the interval $-2\Delta_c^2/a \to 2\Delta_c^2/a$. Thus for mercury, where $\theta \sim 140°$, the capillary will fall, not rise.

From Eq. (10.2.4) it is evident that for small capillaries H_0 can become relatively large. For example, for water with $\sigma = 73\ mN\ m^{-1}$ in a 0.1-mm-radius clean glass capillary the liquid will rise to an equilibrium height of about 0.15 m. The capillary rise method is one of the most accurate means for the measurement of surface tension.

Following Levich (1962), let us calculate approximately the rate at which the capillary will rise to the height given by Eq. (10.2.4) and the length of time it takes to attain that height. The flow in the capillary is unsteady. To simplify the calculation, we assume that the velocity profile at any instant of time is given by the Poiseuille profile (Eq. (4.2.14), with $u_{max} = 2U$ and $h = a/\sqrt{2}$)

$$U = \frac{dH}{dt} = \frac{a^2}{8\mu}\frac{\Delta p}{H} \tag{10.2.5}$$

where $H \le H_0$ is the instantaneous distance of the free surface above the pool level. This assumption is evidently not true until a fully developed profile is attained, which implies that any solution so derived is valid only for times $t \gg a^2/\nu$.

The unbalanced pressure difference at the interface $H \le H_0$, that is, the capillary pressure, is simply

$$\Delta p = \frac{2\sigma\cos\theta}{a} - \rho g H \tag{10.2.6}$$

Eliminating the pressure difference between the above two equations gives the following expression for the liquid velocity:

$$\frac{\mu}{\sigma}\frac{dH}{dt} = \frac{1}{8}\left[2\left(\frac{a}{H}\right)\cos\theta - \frac{\rho g a^2}{\sigma}\right] \tag{10.2.7}$$

where σ/μ is a characteristic speed. The parameter $\rho g a^2/\sigma$ is simply the Bond number based on the capillary radius.

A second dimensionless quantity is represented by the left side of Eq. (10.2.7). However, this quantity is a variable, not a parameter, since dH/dt is not an imposed velocity. When there is an imposed characteristic velocity, call it U, then the group represented by the form of the left side is called the *capillary number*. This number measures the ratio of the viscous force to surface tension force and is defined by

$$Ca = \frac{\text{viscous force}}{\text{surface tension force}} = \frac{\mu U}{\sigma} \tag{10.2.8}$$

We note in passing that the ratio of the capillary number to the Bond number is termed the *Stokes number*. It is a measure of the viscous force to gravity force and is defined by

$$N_{St} = \frac{\text{viscous force}}{\text{gravitational force}} = \frac{\mu U}{\rho g L^2} \tag{10.2.9}$$

We can write Eq. (10.2.7) in a form more suitable for integration, using the equilibrium result for H_0 (Eq. 10.2.4) to give

$$dt = \frac{8\mu H_0}{\rho g a^2}\left(\frac{H/H_0}{1 - H/H_0}\right)d(H/H_0) \tag{10.2.10}$$

where the characteristic time to attain the equilibrium height is

$$\tau = \frac{8\mu H_0}{\rho g a^2} = 8\frac{\mu}{\sigma}\frac{H_0}{Bo} \tag{10.2.11}$$

Upon integrating, we find the actual rise time to be

$$t = \tau\left(\ln\frac{1}{1 - H/H_0} - \frac{H}{H_0}\right) \tag{10.2.12}$$

This relation shows that $H \to H_0$ only as $t \to \infty$, a consequence of the approximations made.

Taking account of the largeness of the log term as $H/H_0 \to 1$, we can expand and rewrite Eq. (10.2.12) in the form

$$\frac{H}{H_0} \approx 1 - \exp\left(-\frac{t}{\tau}\right) \tag{10.2.13}$$

showing the characteristic time τ given by Eq. (10.2.11) to be the $1/e$ time to reach H_0. That this time is relatively rapid may be seen from the capillary example cited in connection with Eq. (10.2.4). For that case, where the equilibrium height is about 0.15 m, Eq. (10.2.11) gives $\tau \approx 12$ s. Except at the beginning of the rise the condition of $\tau \gg a^2/\nu$ ($\sim10^{-2}$ s) is clearly satisfied. The Bond number, which measures the ratio of the gravitational force to surface tension force, is approximately 1.3×10^{-3}.

The capillary flow example given is seen to be a one-parameter problem in which the parameter is the Bond number (Eq. 10.2.1). An important class of technical problems is "coating flows," in which a uniformly thin liquid film is made to cover a substrate. For such flows a characteristic velocity is usually imposed on the substrate or the fluid. As a consequence, the behavior of the coating process generally depends on the capillary number and may depend on the Stokes number. Because of this distinction from the present example, and because of the technical importance of such flows, we shall treat them separately in the next section.

10.3   Coating Flows

A coating flow is a fluid flow in which a large surface area is covered with one or more thin, uniform liquid layers. A summary of work in this field may be found in the review article of Ruschak (1985). In general, though not always, the film laid down is subsequently dried or cured. Examples of technical importance include the manufacture of synthetic membranes, photographic film, and the more prosaic application of painting.

A fundamental key in coating is to spread the liquid over a relatively large substrate by means of viscous forces while maintaining a thin layer of uniform thickness at the substrate by the action of surface tension. Although surface tension stresses are quite small compared with the pressure differences associated with the spreading of the viscous flow, they are important because they fix the amount of fluid laid down.

The hydrodynamic problem in coating flows is to define the steady solution giving the relation between the film-forming geometry, the liquid properties, the speed of application of the film, and a desired coating thickness. The analysis is complicated by several factors, one of which is that such flows are essentially two-dimensional. More importantly, they are free surface–type flows, which are basically nonlinear in that the region occupied by the liquid is not known initially but is defined by the solution. Moreover, such free surface flows must be analyzed to determine if they are stable.

We take as our first example the well-known procedure of withdrawal from a pool of liquid of a surface to be coated. This technique, termed *dip coating*, is, as we shall discuss below, an analogue of a number of other widely used coating procedures. Landau & Levich (1942; see also Levich 1962) were the first to provide a hydrodynamic analysis of dip coating by considering the simplified example of the vertically upward withdrawal of an infinite flat plate from a liquid pool (Fig. 10.3.1). They derived a simple approximate relationship

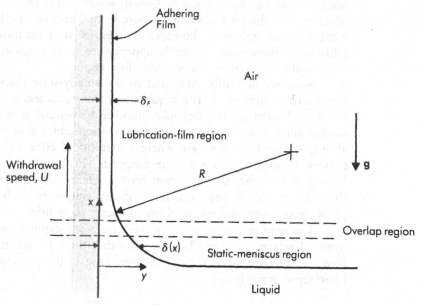

Figure 10.3.1   Vertical withdrawal of an infinite flat plate from a liquid pool (dip coating).

for the film thickness and flow rate at which the liquid is dragged out of the bath by the plate in the limit of small values of the ratio of viscous forces to surface tension forces (capillary number), where gravity drainage down the plate may be neglected. Because of the difficulty of analyzing two-dimensional, film-forming flows for arbitrary values of the capillary number, we shall restrict our considerations to low capillary numbers both here and in what follows.

As pointed out by Ruschak (1985), the approximate analysis of Landau & Levich rests upon the fact that the final film thickness δ_f is small compared with the overall length scale of the flow field when the capillary number is small. In this case viscous effects are important over the length scale δ_f of the *lubrication-film region*, where the film adheres to the plate, but they are negligible outside this region in the *static-meniscus region*, where the shape of the meniscus is controlled by surface tension and the hydrostatic pressure field (Fig. 10.3.1). Because the film is relatively thin, this static meniscus appears to be tangent to the moving plate corresponding to complete wetting, that is, zero contact angle. Ruschak points out that if the radius of curvature of the static meniscus at its apparent point of tangency is R, then the overall scale of the flow field can be taken as R, and the assumption is that $\delta_f \ll R$. In the lubrication-film region the flow is very nearly rectilinear because of the near-zero contact angle at which the meniscus approaches the wall, and viscosity and surface tension act to define the film characteristics.

We first analyze the lubrication-film region and then the matching of this region with the static-meniscus region. Although this problem can be more elegantly treated by a formal matched asymptotic expansion procedure (Ruschak 1976, Wilson 1982), we shall follow the physically intuitive approach originally set out by Landau & Levich, which yields the same analytic solution obtainable in the limit $Ca \to 0$ by more formal means. The basic equation is the Navier-Stokes equation; however, numerous simplifications characteristic of lubrication theory can be made appropriate to the region of film adherence, whence the designation "lubrication-film region."

Since we are only interested in the steady-state thickness, any unsteady terms can be neglected. The requirement for steadiness will be met so long as $t \gg R^2/\nu$. Moreover, the Reynolds number $Re = \rho U R/\mu$ is also supposed sufficiently small that any inertial terms are negligible; that is, $Re \ll 1$. As noted above, we consider the case where gravitational effects are also small, so the gravity drainage term ρg may be dropped. This requires that the Bond number $Bo = \rho g R^2/\sigma$ satisfy the condition $Bo \ll 1$. The result of these approximations is that the Navier-Stokes equation reduces simply to a balance between the pressure gradient and the gradient in shear, as in lubrication theory.

The appropriate pressure in this case is the capillary pressure given by the Young-Laplace equation. This assumes that there is essentially no effect of the flow on the interfacial pressure change. One of the principal curvatures of the interface is zero; hence

$$-\Delta p = \frac{\sigma}{R} = \frac{\sigma \delta''}{(1 + \delta'^2)^{3/2}} \approx \sigma \delta'' \tag{10.3.1}$$

Here $\delta(x)$ is the film thickness with x measured vertically upward from the pool level; primes denote differentiation with respect to x. Because of the quasi-one-dimensional character of the flow in the lubrication-film region, we have taken $\delta'^2 \ll 1$. It follows that the Navier-Stokes equation can be written as a balance between viscous and surface tension forces:

$$\sigma \frac{d^3\delta}{dx^3} + \mu \frac{\partial^2 u}{\partial y^2} = 0 \tag{10.3.2}$$

where u is the vertical velocity of the fluid in the film and y is measured perpendicular to the plate.

Equation (10.3.2) may be integrated with respect to y at constant x subject to the no-slip condition

$$u = U \qquad \text{at } y = 0 \tag{10.3.3a}$$

and the condition that the stress at the free surface of the film is vanishingly small:

$$\mu \frac{\partial u}{\partial y} = 0 \qquad \text{at } y = \delta(x) \tag{10.3.3b}$$

The parabolic velocity profile so obtained is

$$u = U - \frac{\sigma}{\mu} \frac{d^3\delta}{dx^3} \left(\frac{y^2}{2} - \delta y \right) \tag{10.3.4}$$

The volume flow rate per unit width of plate of "lubricant" is given by the quadrature

$$\tilde{Q} = \int_0^{\delta(x)} u \, dy = U\delta + \frac{\sigma}{\mu} \frac{d^3\delta}{dx^3} \frac{\delta^3}{3} \qquad (10.3.5)$$

showing the characteristic lubrication behavior for $\tilde{Q}$ as a function of δ. Far above the static-meniscus region the film is of constant thickness and parallel to the plate (complete wetting), whence

$$\tilde{Q} = U\delta_f \qquad (10.3.6)$$

Eliminating $\tilde{Q}$ from the last two expressions gives the following ordinary nonlinear differential equation for $\delta(x)$:

$$\delta^3 \frac{d^3\delta}{dx^3} + \left(\frac{3\mu U}{\sigma}\right)\delta = \left(\frac{3\mu U}{\sigma}\right)\delta_f \qquad (10.3.7)$$

where $\mu U/\sigma$ is the capillary number Ca.

Examination of Eq. (10.3.7) shows it may be written in dimensionless form in terms of the reduced variables

$$\eta = \frac{\delta}{\delta_f} \qquad \xi = \frac{x}{\delta_f}\left(\frac{3\mu U}{\sigma}\right)^{1/3} \qquad (10.3.8)$$

to give the universal form

$$\eta^3 \frac{d^3\eta}{d\xi^3} = 1 - \eta \qquad (10.3.9)$$

The equation is of third order, so four conditions are required to fix both the solution and define the unknown film thickness δ_f. One of the requirements is that far above the meniscus region ($x \to \infty$) the film thickness approaches δ_f or

$$\eta \to 1 \qquad \text{as } \xi \to \infty \qquad (10.3.10)$$

Now in the neighborhood of $\eta = 1$, Eq. (10.3.9) reduces to $d^3\eta/d\xi^3 = 1 - \eta$, and the particular solution there may be written $\eta = 1 + Ae^{-\xi}$ by ruling out the two exponentially growing solutions. This supplies a second condition. A third condition comes from the observation that the constant A may be arbitrarily chosen because the differential equation (10.3.9) is invariant to a shift in the origin of ξ, so the constant A may, for example, be set equal to 1.

The fourth condition must be obtained by somehow smoothly merging the solution of Eq. (10.3.9) valid in the lubrication-film region into that for the static-meniscus region. Landau & Levich's intuitive argument was that the meniscus curvature must be the same for the lubrication film where it overlaps with the static-meniscus layer. This may be stated as an asymptotic equivalence in terms of the reduced variables we have introduced as

$$\left(\frac{d^2\eta}{d\xi^2}\right)^{menis}_{\eta\to 0} = \left(\frac{d^2\eta}{d\xi^2}\right)^{lube}_{\eta\to\infty} = \alpha \qquad (10.3.11)$$

where α is a constant found from the numerical integration of Eq. (10.3.9). This overlap region is thus one of constant mean curvature, which implies a quadratic behavior in ξ.

The static-meniscus curvature is readily calculated from the free surface interface shape for a liquid meeting a plane rigid wall (Fig. 10.3.1). Applying the condition of static equilibrium in a constant-density fluid, we get $p - \rho g x =$ constant, and it follows from Eq. (10.3.1) that

$$\frac{\sigma\delta''}{(1+\delta'^2)^{3/2}} - \rho g x = 0 \qquad (10.3.12)$$

Integrating once gives

$$\frac{\delta'}{(1+\delta'^2)^{1/2}} = \frac{\rho g x^2}{2\sigma} - 1 \qquad (10.3.13)$$

Here, the constant of integration has been set to -1 based on the requirement that the liquid far from the pool surface, where $x \to 0$, must be horizontal; that is, $\delta' \to -\infty$.

Following Landau & Levich, the transition to the lubrication regime must begin when the liquid film is almost parallel to the plate; that is, when $\delta' \to 0$ or when, according to Eq. (10.3.13), $x \to (2\sigma/\rho g)^{1/2}$. In this limit the curvature is given from Eq. (10.3.12) by the asymptotic relation

$$\delta'' \to \left(\frac{2\rho g}{\sigma}\right)^{1/2} = \frac{\sqrt{2}}{\Delta_c} \qquad (10.3.14)$$

where Δ_c is the capillary length (Eq. 10.2.2). In terms of the reduced variables defined by Eq. (10.3.8),

$$\left(\frac{d^2\eta}{d\xi^2}\right)^{menis}_{\eta\to 0} = \sqrt{2}\,\frac{\delta_f}{\Delta_c}\,(3Ca)^{-2/3} \qquad (10.3.15)$$

where according to the asymptotic equivalence of Eq. (10.3.11) the dimensionless curvature is equal to the constant α.

The constant α is determined by a stepwise numerical integration of Eq. (10.3.9) in the direction of decreasing ξ, starting from an origin near $\eta = 1$. This has been carried out by several authors, and the result reported, which is slightly different from that originally given by Landau & Levich (see Levich 1962), is $\alpha = 0.643$. From Eq. (10.3.15) the limiting thickness of the film adhering to the plate is

$$\frac{\delta_f}{\Delta_c} = 0.946 Ca^{2/3} \qquad (10.3.16)$$

Alternatively, with R the meniscus radius of curvature, Eqs. (10.3.1) and (10.3.16) give

$$\frac{\delta_f}{R} = 0.643(3\mathrm{Ca})^{2/3} \qquad (10.3.17)$$

where we have expressed the result in this form for later comparison with that from another problem.

We recall from our introductory physical argument that $\delta_f \ll R$ was implicitly assumed, from which it follows that $\mathrm{Ca} \ll 1$. A formal asymptotic matching (Wilson 1982) shows the solution given by Eqs. (10.3.16) or (10.3.17) to be valid strictly only as $\mathrm{Ca} \to 0$, although in practice it is only necessary that $\mathrm{Ca} \lesssim 0.01$. In this regard note that in the small capillary number limit the gravitational field does not affect the lubrication film. The effect of gravity enters at about the same capillary number for which the assumption of nearly parallel flow in the film region breaks down (Ruschak 1985). Accounting for the first-order effect of finite capillary number, Wilson has shown that the correction to Eq. (10.3.16) is given by $(1 - 0.113\mathrm{Ca}^{1/3} + \cdots)$. This correction, which is quite small, to the order considered brings in both the effect of gravity and the nonparallel nature of the lubrication film. In the opposite limit to the one discussed here, when surface tension effects are negligible the capillary number is large and the Stokes number is $O(1)$; that is, viscous forces balance gravitational forces, and surface tension is unimportant in the liquid adhesion process.

A closely related film-forming procedure to dip coating is roll coating. In one embodiment of this method, illustrated in Fig. 10.3.2, two cylinders of equal radii with a viscous liquid on one side are counterrotated at the same

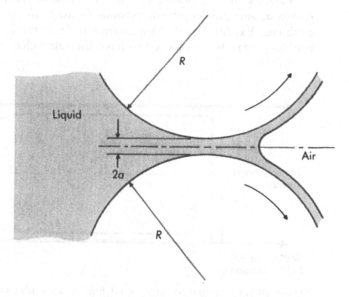

Figure 10.3.2   Roll coating.

speed so as to drag the liquid through the gap between them. The liquid then divides and forms a coating film over each roller. The roller surfaces are close together with the separation distance at the "nip" or point of closest approach small compared with the radii so that the flow there is a nearly parallel lubrication-type flow.

In connection with a study of cavitation in lubrication bearings, Taylor (1963) pointed out that the problem he considered, which we observe is like roll coating, is analogous to the deposition of a thin film on the inside of a capillary tube by blowing a viscous liquid out of the tube with air (Fig. 10.3.3). The analogy is readily seen in a reference frame in which the bubble moving into the tube with a velocity $-U$ is stationary, with the wall moving to the left at a speed U. The bubble air-liquid interface forms itself into a round meniscus at its front end that travels down the tube until it reaches the end. After the meniscus has passed any point, it leaves behind a liquid film that is essentially at rest with a constant pressure along its length. In the case of small enough Reynolds number and a very small capillary radius, gravitational and inertial forces may be neglected, and the flow depends only on a balance between viscous and surface tension forces. Moreover, as Taylor noted, for any given length of liquid the rate of outflow depends essentially on the applied pressure, whereas the amount of fluid left in the tube after the air has reached the end depends essentially on surface tension.

Taylor (1961) studied experimentally the bubble problem described while, at about the same time, Bretherton (1961) analyzed both theoretically and experimentally the slow motion in a capillary of a long bubble, where differently shaped interfaces at the front and rear menisci have to be considered. If, however, only the front meniscus and the film thickness to a point sufficiently far back of the front are examined, then the configuration is essentially the same as Taylor's. In the bubble problem the characteristic length scale is the tube radius a, and this length corresponds to the half-nip distance in the roll coater problem. Except for the appearance of a natural length scale, all of these problems may be recognized to have the same character as the Landau-Levich

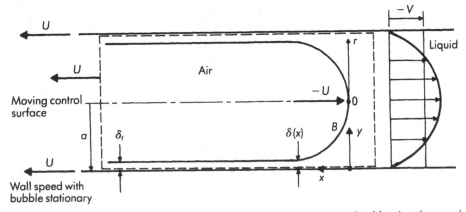

Figure 10.3.3 Formation of a liquid film in a circular capillary by blowing in an air bubble.

dip coating problem in the limit of small capillary number, where the surface tension force dominates the viscous force. Indeed, Bretherton's analytical description concentrated on small capillary numbers and employed an approach that closely paralleled the Landau-Levich work of almost 20 years earlier, although it was done independently. It is also interesting that some 20 more years passed when Wilson (1982), unaware of Taylor's and Bretherton's work, showed by matched asymptotic expansions that the low capillary number flow problem of a rotating cylinder in a liquid whose top is close to but somewhat above the free surface of the liquid is also the analogue of dip coating in the film region where the fluid is dragged up.

In the bubble problem for small capillary number the meniscus in the neighborhood of the front end of the bubble will approach a hemispherical shape of mean curvature $2/a$ (cf. Eq. 10.1.6), changing to the curvature $1/a$ (cf. Eq. 10.3.1) where the film approaches the constant thickness δ_f. Point B in Fig. 10.3.3 represents the position at which the hemispherical meniscus region starting from the origin overlaps smoothly with the lubrication-film layer, with the curvatures of the two regions matched and equal to $2/a$. Although done with greater justification of the approximations made, Bretherton's analysis was essentially the same as that of Landau & Levich. He derived the same governing differential equation for the interface shape (Eq. 10.3.9) and found the same result for the limiting film thickness δ_f as given by Eq. (10.3.17), with R replaced by the tube radius a.

Bretherton did write the expression for the interface shape in the region of overlap as

$$\delta = \frac{1}{2}\left(\frac{x^2}{a}\right) + 1.79(3\text{Ca})^{2/3}a \qquad (10.3.18)$$

This relation is seen to be consistent with the approximate curvature $\delta'' + a^{-1}$ being equal to $2/a$. Note that the parabolic shape in x follows from integrating the matching condition Eq. (10.3.11), setting the coefficient of the linear term in x equal to zero, and determining the coefficient of the constant term from the numerical solution of Eq. (10.3.9). Dropping the coefficient of the linear term comes from the disposable constant associated with the ability to arbitrarily shift the origin of ξ, in accordance with our discussion of the conditions to fix the solution of Eq. (10.3.9).

A point made by Bretherton related to his own, Taylor's, and earlier experimental observations was that the volume flow rate of liquid swept out by the bubble moving with the speed of the air-liquid interface relative to the wall must equal the average speed of the liquid in front of it, $-V$, multiplied by the tube cross-sectional area. Applying conservation of mass to a control surface in uniform motion with a velocity $+U$ so that the bubble is stationary with respect to the surface (Fig. 10.3.3), we get

$$(0 + U)A_{\text{film}} - (-V + U)A_{\text{tube}} = 0 \qquad (10.3.19)$$

Here, A_{film} is the cross-sectional area of the lubrication film where it attains the uniform thickness δ_f, and A_{tube} is the cross-sectional area of the tube whose radius is a. It follows that

$$\frac{V}{U} = \frac{A_{bubble}}{A_{tube}} = \frac{\pi(a - \delta_f)^2}{\pi a^2} \qquad (10.3.20)$$

with A_{bubble} the bubble cross-sectional area where the film thickness has the limiting value δ_f. Denoting the speed by which the bubble exceeds the average speed of the liquid in the tube $U - V$ by UW, we may write

$$\frac{V}{U} = 1 - W = \left(1 - \frac{\delta_f}{a}\right)^2 \qquad (10.3.21)$$

With $\delta_f \ll a$ the fractional velocity change to the bubble speed W is given by

$$W \approx 2\frac{\delta_f}{a} \qquad (10.3.22)$$

The importance of the fractional velocity change W is that it can be estimated directly from experiment by measuring the volume of fluid ejected from the tube when the bubble moves a known distance. The measurements can then be compared with the theoretical result for the film thickness by using Eq. (10.3.22). Experiments by Bretherton did in fact show good agreement with Eq. (10.3.17) for capillary numbers in the range $10^{-4} < Ca < 10^{-2}$. However, below about $Ca \sim 10^{-4}$ there is an unexplained but systematic divergence from theory. Other experimental work concerned with dip coating of vertical plates (White & Tallmadge 1965, Spiers et al. 1974) also show good agreement with the Landau-Levich result for capillary numbers greater than about 10^{-4}, and up to about 10^{-2}, when the small capillary number assumption begins to break down. Comparison between theory and experiment will be discussed further in the last section of this chapter where the effect of nonuniform surface tension is examined.

In this section we have considered the equilibrium configurations for a number of thin-film coating flows. It must be emphasized that because these are free surface flows subject to interfacial and gravity forces, it is not evident when the film deforms due to a disturbance that the interface is stable and will return to its equilibrium shape. In roll coating flows, for example, an important type of instability that occurs is termed ribbing-line instability in which the flow becomes spanwise periodic and the coating becomes ribbed. In this instability the lower the capillary number is and the greater the divergence between the roller surfaces where the meniscus is located, the more likely the flow will be stable. There are a variety of different instabilities that can cause the layer thickness of a thin film to become nonuniform, and an understanding of them is of considerable importance in any coating process (Ruschak 1985, 1987). Closely related to these interfacial instabilities is the phenomenon of interfacial wave motion and the conditions under which surface tension dominates in the formation, growth, or decay of the waves called capillary waves. We choose not to discuss further the instability of thin films. Instead, in the following section, we illustrate the general problem of interfacial instability and its relation to the breakup of a circular liquid jet into drops.

10.4   Surface Waves and Jet Breakup

A commonly observed phenomenon seen with a slow-moving cylindrical stream of water issuing from a faucet is that at some distance below the faucet the stream becomes undulated and then further down breaks up into drops. The problem of the breakup of a liquid jet issuing into a gaseous, and also a liquid medium, is of considerable practical importance in connection with such diverse procedures as microencapsulation, ink jet printing, emulsification, and many other problems of practical importance.

In 1879 Rayleigh (see Rayleigh 1894) was the first to demonstrate by a hydrodynamic stability analysis that a liquid jet is unstable to small perturbations and breaks up into segments that, under the action of surface tension, form into individual drops. The disturbance that gives rise to the instability may be random or forced. The current procedure of choice for the production of a uniformly sized droplet stream with uniform droplet spacing is forced longitudinal vibrations in the direction of the jet flow.

To better appreciate the nature of jet instability and jet breakup, we first consider the closely related phenomenon of horizontally propagating sinusoidal waves on the free surface of a "deep" liquid, say, water. When the waves are driven by a balance between the fluid's inertia and the restoring force of gravity, they are termed *surface gravity waves*. If instead of gravity the restoring force to bring the surface flat is surface tension, the waves are referred to as *capillary waves* or *ripples*. Both of these wave types are confined to a distance of about a wavelength from the surface. Although the wave amplitudes may be small, such waves are dispersive; that is, the speed of the wave varies with the wavelength.

Below, we examine the propagation of unchanging small disturbances on a liquid surface and then investigate the conditions under which they remain unchanged, become attenuated, or grow exponentially in time so as to give rise to unstable motions. It is this latter condition that is of interest in connection with jet breakup.

Consider a small-amplitude sinusoidal plane wave of length λ and amplitude a propagating along an air-water interface in the positive x direction with wave speed c, as sketched in Fig. 10.4.1. The water is taken to be incompressible with viscosity and other dissipative effects neglected so that the wave amplitude at the interface remains unchanged. The vertical displacement of the disturbed surface, ζ, may be written

$$z = \zeta(x, t) = a \sin(\omega t - kx) \qquad (10.4.1)$$

where the wave number k and radian frequency ω are defined in terms of the period τ and wavelength λ by

$$\omega = \frac{2\pi}{\tau} \qquad (10.4.2a)$$

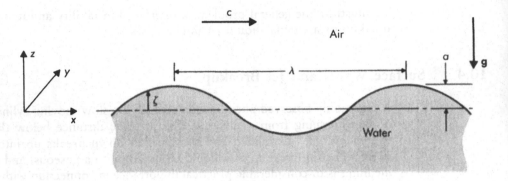

Figure 10.4.1   Sinusoidal wave propagating on surface of air–water interface.

$$k = \frac{\omega}{c} = \frac{2\pi}{\lambda} \qquad (10.4.2b)$$

with

$$\lambda = c\tau \qquad (10.4.2c)$$

Although there is no damping or growth of the wave, it is dispersive, so the frequency depends on wave number. To determine the dispersion relation, we simplify the momentum equation for a wave whose disturbance amplitude is small compared with the wavelength. Because of the small amplitude, terms involving squares and products may be neglected and the linearized inviscid, constant-density form becomes (Lighthill 1978)

$$\rho \frac{\partial \mathbf{u}}{\partial t} = -\nabla p_e \qquad (10.4.3)$$

where $\mathbf{u}$ is the disturbance velocity and p_e is the *excess pressure* over the undisturbed value due to the disturbance. The variation of the excess pressure over any cross section (specified by its x coordinate) may be neglected. This derives from the condition that for longitudinal waves the longitudinal gradients of p_e can be large compared with those resulting from transverse gradients only if the variation of p_e is negligible across each cross section. With the restoring force of gravity and surface tension taken into account, the excess pressure at any point is given by the sum of the capillary pressure due to the wave curvature and the hydrostatic pressure due to gravity.

The undisturbed hydrostatic pressure distribution, with z measured upward from the undisturbed free surface, is

$$p_0 = p_a - \rho g z \qquad (10.4.4)$$

where p_a is the atmospheric pressure. The pressure at any point in the disturbed liquid is

$$p = p_0 + (p_e)_g \tag{10.4.5}$$

with $(p_e)_g$ the excess pressure from gravity due to the disturbance. Since $p = p_a$ at the free surface $z = \zeta$, it follows that

$$(p_e)_g = \rho g \zeta \tag{10.4.6}$$

As noted, there is also an excess pressure from surface tension that is associated with the interface curvature. From the Young-Laplace equation for the plane case considered, this excess pressure is $(p_e)_\sigma = \sigma/R$, where R is the local radius of curvature of the wave. For small displacements $R^{-1} = -\partial^2 \zeta/\partial x^2$ (Eq. 10.3.1), from which

$$(p_e)_\sigma = -\sigma \frac{\partial^2 \zeta}{\partial x^2} \tag{10.4.7}$$

For the sinusoidal wave described by Eq. (10.4.1), $\partial^2 \zeta/\partial x^2 = -k^2 \zeta$, so

$$(p_e)_\sigma = \sigma k^2 \zeta \tag{10.4.8}$$

The total excess pressure, equal to the sum of the capillary and hydrostatic pressures, is

$$p_e = \rho \left(g + \frac{k^2 \sigma}{\rho} \right) \zeta \tag{10.4.9}$$

or

$$p_e = \rho g_{\text{eff}} \zeta \tag{10.4.10}$$

In the case of sinusoidal waves, surface tension is thus seen to manifest itself simply as an increase in g by an amount $k^2 \sigma/\rho$ to a new effective value g_{eff}.

From the above arguments it may be seen that the waves will become pure capillary waves with surface tension forces dominant when

$$\lambda \ll 2\pi \left(\frac{\sigma}{\rho g} \right)^{1/2} \tag{10.4.11}$$

On the other hand, the waves will be pure gravity waves with gravitational forces dominant when

$$\lambda \gg 2\pi \left(\frac{\sigma}{\rho g} \right)^{1/2} \tag{10.4.12}$$

The capillary length $\Delta_c = (\sigma/\rho g)^{1/2}$ is the characteristic length scale governing the change from one regime to the other. For water with $\sigma = 73$ mN m^{-1} and $\rho = 1000$ kg m^{-3} the wavelength $2\pi(\sigma/\rho g)^{1/2}$ equals 1.7×10^{-2} m. This shows capillary waves to be short-wavelength, high-frequency waves, whereas gravity waves are long wavelength.

In addition to the momentum equation, it is also necessary to satisfy the equation of continuity, which for the incompressible fluid considered is $\nabla \cdot \mathbf{u} = 0$. The disturbance velocity is also irrotational; hence, in addition, $\nabla \times \mathbf{u} = 0$ (Lighthill 1978). Therefore the disturbance velocity field is derivable from a gradient of a velocity potential; that is, $\mathbf{u} = \nabla \phi$ with the potential satisfying Laplace's equation $\nabla^2 \phi = 0$. This may seem surprising at first, but Laplace's equation can describe a wave motion when boundary conditions are satisfied at a free surface.

The excess pressure field associated with the velocity field is given from the integral of the momentum equation (Eq. 10.4.3) by

$$p_e = -\rho \, \frac{\partial \phi}{\partial t} \tag{10.4.13}$$

which may be recognized as the unsteady Bernoulli equation. From Eq. (10.4.10) the dynamic boundary condition for the free surface is then

$$\left(\frac{\partial \phi}{\partial t} \right)_{z=0} = -g_{\text{eff}} \zeta \tag{10.4.14}$$

Note that consistent with the linearization of the flow equations, $\partial \phi / \partial t$ is evaluated at the undisturbed free surface, the difference between the values of the derivatives at $z = \zeta$ and $z = 0$ being of higher order.

A second boundary condition is provided by the kinematic requirement that each particle on the surface remain there. With w the vertical velocity component this can be expressed through

$$\frac{D\zeta}{Dt} = w \qquad \text{at } z = \zeta \tag{10.4.15}$$

Linearizing, we can neglect the convection term $\mathbf{u} \cdot \nabla \zeta$ in $D\zeta/Dt$ and the normal velocity $\partial \phi / \partial z$ at $z = \zeta$ can be evaluated at $z = 0$, whence

$$\frac{\partial \zeta}{\partial t} = \left(\frac{\partial \phi}{\partial z} \right)_{z=0} \tag{10.4.16}$$

The velocity potential ϕ satisfies the Laplace equation and the free surface boundary condition on ϕ is obtained by differentiating Eq. (10.4.14) with respect to t and eliminating $\partial \zeta / \partial t$ from Eq. (10.4.16). For waves on deep water the second boundary condition on the potential is supplied by the requirement that there are no disturbances deep in the water, or that $\phi = \text{constant}$ as $z \rightarrow -\infty$.

A sinusoidal wave solution of Laplace's equation satisfying both the free surface condition and the deep water condition is

$$\phi = A e^{kz} \cos(\omega t - kx) \tag{10.4.17}$$

with A a constant. The solution shows the disturbance to die off exponentially with distance down from the surface, with the $1/e$ "decay depth" equal to $\lambda/2\pi$. The solution given for ϕ satisfies both the conditions of Eq. (10.4.14) and Eq. (10.4.16), as well as the postulated sinusoidal free surface shape of Eq. (10.4.1), provided

$$\omega^2 = k g_{\mathrm{eff}} \qquad (10.4.18)$$

We note here that although the free surface shape was set down at the outset, it could have been found directly by solving for the velocity potential ϕ as outlined and then determining ζ, say, from Eq. (10.4.14).

Equation (10.4.18) is the sought after dispersion relation for surface waves on deep water. It may also be written as a dependence of wave speed on wavelength:

$$c = \frac{\omega}{k} = \left(\frac{2\pi\sigma}{\rho\lambda} + \frac{g\lambda}{2\pi} \right)^{1/2} \qquad (10.4.19)$$

The wave speed is readily shown to have a minimum value as a function of wavelength, which is expressible in the form

$$c_{\min} = \left(\frac{g\lambda_{\min}}{\pi} \right)^{1/2} \qquad (10.4.20)$$

where the wavelength corresponding to the minimum wave speed is

$$\lambda_{\min} = 2\pi \left(\frac{\sigma}{\rho g} \right)^{1/2} \qquad (10.4.21)$$

The wavelength $\lambda_{\min}$ is seen from Eqs. (10.4.11) and (10.4.12) to be that length associated with the transition from the dominance of capillary waves to gravity waves, or vice versa. For water, as noted earlier, $\lambda_{\min} = 1.7 \times 10^{-2}$ m to which the corresponding minimum wave speed is 0.23 m s^{-1}.

With gravity neglected, the surface tension is related to the wave speed by

$$\sigma = \frac{c^2 \rho \lambda}{2\pi} \qquad (10.4.22)$$

This expression for capillary waves has been used as the basis for the measurement of surface tension by measuring the propagation velocity of capillary waves generated at a predetermined frequency (see Levich 1962 and, for a more recent approach using laser light scattering, Hård et al. 1976).

In the absence of dissipation and with uniform surface tension, plane small-amplitude capillary waves will propagate undamped and unamplified. Viscosity and surface tension gradients lead to the damping of capillary waves. In the following section we shall discuss the damping due to the presence of surface-active substances, which because of the wave shape are not uniformly distributed, giving rise to a spatially nonuniform surface tension. Of interest in connection with our introductory remarks on the breakup of a liquid jet are the conditions under which small-amplitude capillary disturbances on the surface of a cylindrical jet will amplify and lead to jet breakup.

The physical reason why a slow-moving liquid jet breaks up into drops at some distance below the nozzle lies in the interaction between small-amplitude disturbances on the jet and surface tension, with subsequent high-gain amplification of the capillary perturbation. The initial disturbances may be a result of random excitations, such as jet friction or nozzle roughness, or they may be impressed on the jet.

If a cylindrical jet is disturbed into an undulating or "varicose" shape, it will, under the action of surface tension, tend to minimize its surface area to drive the free energy to a minimum and thus release surface energy. That an axisymmetric deformation can decrease the surface area of a jet is simply illustrated by considering the limiting situation of the jet breaking up into spherical drops as a result of the bulbous portions of the jet surface expanding further and the constricted regions becoming narrower, thereby forming a neck that thins and eventually breaks (Fig. 10.4.2A). A simple geometrical calculation will show that if the jet is considered to be a fixed cylindrical volume, its surface area can be reduced if it breaks up into spherical drops of radius greater than 1.5 times the cylinder radius. With n equal to the number of drops into which the jet breaks up, and $n \geq 2$, the uniform spacing between the droplets would be greater than $3n/(n-1)$ times the drop radius, where $n/(n-1)$ will lie between 2 and 1. It is precisely because an instability on a round jet can decrease the surface area that breakup occurs; the decreased surface area gives rise to a release of free energy. It may be recognized that this area reduction is a consequence of the axial symmetry and is not characteristic of plane flow.

There are a number of modes of jet breakup, but we consider only the breakup into spherical drops caused by capillary forces, which implies low jet velocities into an external medium whose density is low by comparison with the jet, say, air. At high velocities the dynamic effect of the surrounding medium on the jet surface will alter the surface pressure, and tangential stresses at the surface due to viscosity may also affect the breakup. It is a well-known observation that at sufficiently high velocities a jet will "atomize" into a large number of small droplets compared with a relatively smaller number of big drops at low velocities.

Even if the jet velocity is low enough that just capillary forces need be accounted for, the hydrodynamic stability problem is relatively simple when only the jet stability to small disturbances is considered, that is, disturbances whose amplitudes are small compared with the jet radius. When this is not the case, nonlinear mechanisms enter, which are manifest in various phenomena, including the formation of satellite droplets, which are small spherules that form between the drops (Fig. 10.4.2B). For literature on these and other nonlinear effects of jet instability, see Bogy (1979).

The problem of the linear stability of an infinitely long, initially stationary, circular, inviscid, incompressible liquid jet in air was first analyzed in 1879 in a classic paper by Rayleigh (see Rayleigh 1894). This paper and his other studies on jet instability remain a pleasure to read for their clarity and insight despite

Upper portion   Upper portion

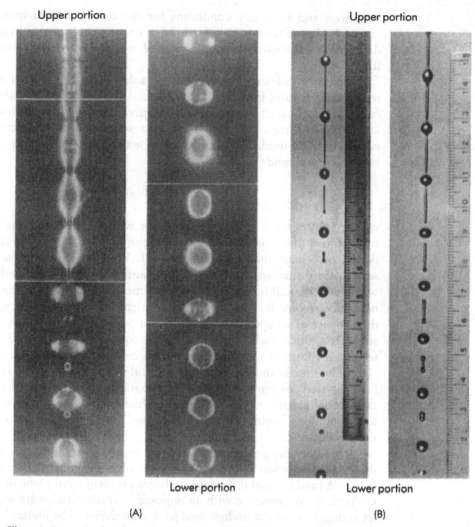

Lower portion   Lower portion

(A)   (B)

Figure 10.4.2 Photographs of liquid jet breakup in air: (A) into spherical drops. [Courtesy of Prof. Richard K. Chang. From Qian, S-X. et al. 1986. Lasing droplets: Highlighting the liquid-air interface by laser emission. *Science* 231, 486–488. Copyright 1986 by the AAAS. With permission.] (B) into spherical drops with satellite droplets. [Courtesy of Prof. M.C. Yuen. From Goedde, E.F. & Yuen, M.C. 1970. Experiments on liquid jet instability. *J. Fluid Mech.* 40, 495–511. Cambridge University Press. With permission.]

numerous subsequent works on the subject by other authors and the passage of well over a century.

All linear hydrodynamic stability analyses of steady laminar flows can be recognized to have essentially four fundamental steps (Lin 1955, Drazin & Reid 1981). The first is the specification of the basic flow through the knowledge of the velocity and other fields such as pressure and temperature at each point in the field. Next the basic flow is assumed to be slightly disturbed, and the equations and boundary conditions for the disturbed field quantities are obtained by linearization. This leads to a linear homogeneous system of partial differential equations, the coefficients of which may vary spatially but not in time.

In the third step, an elementary solution of the system in appropriate mathematical form is chosen for the initial disturbance. Typically the complex form of the Fourier representation of periodic functions, although the more cumbersome form of an expansion in a series of sine and cosine terms may equally well be used. For example, the elementary solution might be chosen to be the *normal mode*

$$\phi(x, z, t) = \Phi(z) \, e^{i(\omega t - kx)} \tag{10.4.23}$$

where, in general, Φ and ω are complex with k real. Here, the real part of ϕ represents a one-dimensional wave traveling in the x direction that may grow, decay, or remain unchanged in time (cf. Eq. 10.4.17). With $\omega = \omega_r + i\omega_i$ for $\omega_i < 0$, the disturbance will be *unstable* and grow with time, and for $\omega_i > 0$ the wave will be *stable* and decay with time, and for $\omega_i = 0$ the wave will be *neutrally stable*. It is generally convenient to represent an arbitrary initial disturbance as a superposition of normal modes, each of which may be treated separately since each satisfies a linear equation. The analysis then involves finding a complete set of normal modes to represent the disturbance.

The fourth and final step in the stability analysis is the reduction of the linear system of partial differential equations to a system of ordinary linear differential equations, the solution of which, subject to the appropriate boundary conditions, yields the "eigenfunction" $\Phi(z)$ and the associated complex wave velocity c.

Much of the procedure for the analysis of jet stability has already been set down in connection with the discussion of undamped surface waves on deep water. A fundamental difference in the jet problem from plane deep water waves is that it is axisymmetric with an imposed characteristic length scale equal to the jet radius a. Since the undisturbed jet is considered to be inviscid and in uniform flow, it can be reduced to a state of rest simply by a Galilean transformation. With gravity neglected and only surface tension forces acting, the pressure at any point within the jet is $p_a + \sigma/a$. This then describes the basic flow needed for the first step of the stability analysis.

The second step is the linearization of the governing flow equations and boundary conditions assuming the flow to be only slightly disturbed. The momentum equation so linearized is given by Eq. (10.4.3), with p_e the excess or "disturbance" pressure and $\mathbf{u}$ the disturbance velocity. The continuity equation for the disturbance velocity is as before, $\nabla^2 \phi = 0$. The natural coordinates for the problem are the cylindrical coordinates (r, θ, z) in which Laplace's equation takes the form

$$\frac{1}{r} \frac{\partial}{\partial r} \left(r \frac{\partial \phi}{\partial r} \right) + \frac{1}{r^2} \frac{\partial^2 \phi}{\partial \theta^2} + \frac{\partial^2 \phi}{\partial z^2} = 0 \tag{10.4.24}$$

where the radial displacement of the disturbed jet surface ζ may be expressed functionally as

$$r = \zeta(z, \theta, t) \tag{10.4.25}$$

The linearized dynamic boundary condition is physically the same as given by Eq. (10.4.7) for the plane surface wave, which with allowance for the cylindrical symmetry of the jet problem may, from the Young-Laplace equation, be written

$$p_e = -\sigma\left(\frac{\partial^2\zeta}{\partial z^2} + \frac{1}{a^2}\frac{\partial^2\zeta}{\partial\theta^2} + \frac{\zeta}{a^2}\right)_{r=a} \tag{10.4.26}$$

Similarly, from Eq. (10.4.16), the kinematic boundary condition that each particle remain on the surface is

$$\frac{\partial\zeta}{\partial t} = \left(\frac{\partial\phi}{\partial r}\right)_{r=a} \tag{10.4.27}$$

The linearized partial differential equations for the flow and the boundary conditions having been defined, we next specify the form of the disturbance. We avoid generality and take the disturbance potential ϕ to be represented by the typical wave component

$$\phi(r, \theta, z, t) = \Phi(r)\, e^{\beta t}\cos(kz + n\theta) \tag{10.4.28}$$

where k is the real axial wave number and n is an integer. The *amplification factor* β may be positive real or imaginary, the latter case corresponding to an unchanging wave component and the former to amplification. Note that Eq. (10.4.28) cannot satisfy viscous boundary conditions, and, in the absence of viscosity or other dissipative mechanism, there can be no damping. With real β restricted to positive values, the potential represents both a longitudinal and azimuthal perturbation that may remain unchanged or grow exponentially in time.

In the fourth and last step, by substituting the wave form Eq. (10.4.28) into Laplace's equation and separating variables, one arrives at Bessel's modified differential equation of order n for $\Phi(r)$:

$$\frac{d^2\Phi}{dr^2} + \frac{1}{r}\frac{d\Phi}{dr} - \left(\frac{n^2}{r^2} + k^2\right)\Phi = 0 \tag{10.4.29}$$

The general solution for Φ is $AI_n(kr) + BK_n(kr)$, where A and B are arbitrary constants, and I_n and K_n are the modified Bessel functions of the first and second kind. Since $K_n \to \infty$ as $r \to 0$, we have, on requiring ϕ be bounded,

$$\Phi = AI_n(kr) \tag{10.4.30}$$

The determination of the equation for the free surface shape and the "eigenvalue" relation between the amplification factor β, the wave number k, and the integer n is carried out just as in the evaluation of the dispersion relation for plane waves. In particular, from the kinematic boundary condition Eq. (10.4.27) and the solution for the disturbance potential ϕ given by Eqs. (10.4.28) and (10.4.30), we can immediately obtain the expression for the radial displacement of the disturbed surface as

$$\zeta = \frac{AkI_n'(ka)}{\beta} \, e^{\beta t} \cos(kz + n\theta) \qquad (10.4.31)$$

The excess pressure p_e is then found from the dynamic boundary condition Eq. (10.4.26). Using the unsteady Bernoulli equation $p_e + \partial\phi/\partial t = 0$, we can eliminate the unknown constant A to give the eigenvalue relation

$$\beta^2 = \left(\frac{\sigma}{\rho a^3}\right) \frac{\alpha I_n'(\alpha)}{I_n(\alpha)} (1 - \alpha^2 - n^2) \qquad (10.4.32a)$$

where

$$\alpha = ak \qquad (10.4.32b)$$

The linear stability characteristics of the jet are specified by Eq. (10.4.32), where we note that $\beta^2 \sim \sigma/\rho a^3$, which may be compared with the plane capillary wave result where $\omega^2 \sim \sigma/\rho\lambda^3$. This behavior is not surprising and can be deduced from dimensional arguments. Indeed, for the jet when $\alpha \gg 1$, that is, when the wavelengths are small compared with the jet radius, we have from the properties of the Bessel function that $I_n'(\alpha)/I_n(\alpha) \approx 1$. With $\beta = i\omega$, Eq. (10.4.32) reduces to the dispersion relation $\omega^2 = k^3\sigma/\rho$ for stable, sustained surface capillary waves on deep water (Eq. 10.4.19).

Of greater interest here are the general conditions where $\beta^2 > 0$, leading to an instability. Since $\alpha I_n'(\alpha)/I_n(\alpha)$ is positive for all nonzero real α, it follows from Eq. (10.4.32) that $\beta^2 < 0$ for all α if $n \neq 0$ or for $\alpha \geq 1$ or $\alpha \leq -1$ if $n = 0$. Under these conditions β is an imaginary number, and as with surface capillary waves on deep water the oscillations sustain themselves without growth or decay. The case $n \neq 0$ corresponds to nonaxisymmetric disturbances, and $n = 0$ to axisymmetric disturbances, so we may conclude that if the disturbances are nonaxisymmetric the jet is always stable. On the other hand, $\beta^2 > 0$ for $-1 < \alpha < 1$ if $n = 0$. Since $\alpha = ak$, this says the jet is unstable to axisymmetric disturbances whose wavelength $\lambda = 2\pi/k$ is greater than the undisturbed jet circumference $2\pi a$.

The wave number of the fastest-growing disturbance can be obtained by differentiating Eq. (10.4.32) with respect to α for $n = 0$ and setting the result to zero. This calculation was made by Rayleigh, and he found the amplification factor to have a maximum value as a function of α given by

$$\beta_{\max} = 0.343\left(\frac{\sigma}{\rho a^3}\right)^{1/2} \qquad (10.4.33)$$

The wavelength corresponding to the maximum amplification factor occurs for $\alpha = 0.697$ or

$$\lambda_{max} = 9.02a \tag{10.4.34}$$

Rayleigh then postulated that this fastest-growing mode would dominate the instability and lead to the jet breakup. Drazin and Reid (1981) note this may not necessarily be correct because all disturbances might not have the same initial amplitude and because nonlinear effects may be important, though they do observe that it is a "good working rule."

Although these results apply to the instability of a stationary jet, they can be used to estimate the length at which a circular jet of uniform initial velocity U_j will break up. We estimate the breakup length as

$$L_B = U_j t_B \tag{10.4.35}$$

where t_B is the time for the fastest-growing mode to increase from its initial value to a value on the order of the jet circumference. The maximum dimensionless logarithmic growth rate $\beta_{max}(\rho a^3/\sigma)^{1/2}$ will amplify the initial disturbance amplitude by the factor e in a time given by $\beta_{max} t_B = 1$ or

$$t_B = 2.92\left(\frac{\rho a^3}{\sigma}\right)^{1/2} \tag{10.4.36a}$$

or to 100 times its initial amplitude ($\beta_{max} t_B = 4.61$) in a time

$$t_B = 13.4\left(\frac{\rho a^3}{\sigma}\right)^{1/2} \tag{10.4.36b}$$

For a 5-mm-diameter water jet the characteristic capillary time $(\rho a^3/\sigma)^{1/2}$ is 4.14×10^{-2} s, so we may expect such a slow-moving jet to break up very quickly, in distances on the order of a centimeter for speeds approximately 0.1 m s^{-1}. The jet breakup length is predicted remarkably well by Rayleigh's linear result over a wide range of disturbance amplitudes even though the breakup process may be strongly nonlinear.

If the interval between drops is λ_{max} and the spherical drops that form at this spacing have a diameter d and the same volume as a cylinder of length λ_{max} with a radius a equal to that of the jet, then

$$\frac{\pi d^3}{6} = \pi a^2 \lambda_{max} \tag{10.4.37}$$

Inserting the value of λ_{max} from Eq. (10.4.34), we find

$$d = 3.78a \tag{10.4.38}$$

Our earlier, simple, nonflow, geometrical argument showed that the surface area of a jet can be reduced if it breaks up into spherical drops with $d > 3a$, a remarkably close bound to the result given by Eq. (10.4.38). From Eqs. (10.4.34) and (10.4.38) it is readily shown that

$$\lambda_{max} = 2.38d \tag{10.4.39}$$

Our nonflow geometrical calculation showed that $\lambda > 3d$ for the cylinder breaking into two equally spaced spherical drops with $d > 3a$, again a surprisingly close bound. We recall that these bounds were derived from the argument that a decrease in the jet surface area would tend to drive the free energy to a minimum and thereby release surface energy.

Examination of Fig. 10.4.2A shows that in the breakup of the jet before the drops become spherical they undergo an oscillation about a spherical shape. This oscillation is associated with capillary waves on the drop surface and from dimensional considerations the characteristic oscillation frequency must be $(\sigma/\rho d^3)^{1/2}$, with d the drop diameter. Rayleigh (1894) (see also Levich 1962) showed this estimate to be exactly the minimum natural oscillation frequency from which the length to form the uniformly spaced spherical drops can be estimated.

We conclude this section with reference to Fig. 10.4.2B, where it can be seen that smaller satellite drops are interspersed between the main drops. These drops form when the ligaments separate from the main drops at both ends. In many practical applications of drop formation, including drug microencapsulation and ink jet printing, uniform drop sizes are required and satellite drops are an impediment. Analytical study of satellite drop formation requires both temporal and spatial nonlinear analyses. Such treatments will not be considered here because they are beyond the scope of the text, but for further information see Bogy (1979).

10.5   Flows Driven by Surface Tension Gradients

Spatial variations in surface tension at a liquid-gas interface result in added tangential stresses at the interface and hence a surface tractive force that acts on the adjoining fluid, giving rise to fluid motions in the underlying bulk liquid. This force is in addition to any arising from viscous tangential stresses at the interface and can lead to interfacial dissipation within a surface boundary layer that can exceed the dissipation in the bulk of the fluid. The motion induced by tangential gradients of surface tension is usually termed the *Marangoni effect*, after C. Marangoni whose initial work on the subject appeared in 1871. Drazin & Reid (1981) have pointed out, however, that the phenomenon was actually first described by James Thomson, the elder brother of Lord Kelvin, in 1882.

Gradients in surface tension can also lead to an instability, with subsequent cellular-type flows. These unstable flows are similar in character to the unstable convection that results when a density gradient is parallel to, but opposite, a body force, such as gravity. In this case the fluid is in unstable equilibrium with the heavier fluid on top of the lighter fluid. When a critical density gradient is exceeded, the flow will assume a steady cellular or vortex roll configuration (Ostrach 1980). We shall discuss this surface tension induced instability in the following section.

Spatial gradients in surface tension may arise from a variety of causes, including spatial variations at the interface in temperature (Eq. 10.1.3), in surface concentrations of an impurity or additive (Eq. 10.1.4), or in electric charge or surface potential. The resulting flows are termed, respectively, thermocapillary flows, diffusocapillary flows, and electrocapillary flows. We shall limit our discussion of electrocapillary phenomena because of space restrictions but instead refer the reader to Levich (1962) and Newman (1991).

Many examples of stable flows driven by tangential stresses derived from surface tension gradients are commonplace: the camphor ball that will "dance" on a water surface, the ripples that form on the skinned surface of chocolate pudding near the cup center, and the calming effect of "oil on troubled waters," as phrased in Plutarch's question "Why does pouring oil on the sea make it clear and calm?" We shall attempt to answer this question in part by showing how insoluble surface-active substances can lead to strong damping of plane capillary waves on deep water. Following Herbolzheimer (1988), we shall also show how surfactants can strongly modify the pressure drop required to push through a bubble in a fine capillary at a given speed, a problem examined for constant surface tension in Section 10.3 in relation to coating flows. This particular problem is of considerable practical importance in connection with the displacement of oil from a porous strata by the technique of foam flooding. Surface tension driven flows arise in many other technically important fields, including metals processing (Szekely 1979) and crystal growth (Ostrach 1983, Langlois 1985). Finally, as an indication of the diversity of phenomena resulting from flows driven by surface tension gradients, we note the biological model of cell cleavage (cytokinesis) proposed by Greenspan (1977) in which the cell splitting results from a difference in the equatorial and polar surface tensions that produces an unstable contraction of the spherical surface toward the equator.

Clearly, then, we must know the concentration, temperature, and charge distributions at the interface in order to define the surface tension variation required to solve the hydrodynamic problem. However, these distributions are themselves coupled to the equations of conservation of mass, energy, and charge through the appropriate interfacial boundary conditions. The boundary conditions are obtained from the requirement that the forces at the interface must balance. This implies that the tangential shear stress must be continuous across the interface, and the net normal force component must balance the interfacial pressure difference due to surface tension.

If the surface tension varies along the interface, a tangential force per unit area will exist on the interface, given by

$$\mathbf{f}_s = \nabla_s \sigma \qquad (10.5.1)$$

where ∇_s denotes the surface gradient and $\mathbf{f}_s$ is the force component in the surface. The positive sign on $\nabla_s \sigma$ indicates that the liquid tends to move in a direction from lower to higher surface tension.

Let us denote the force per unit area exerted on the interface from the viscous stresses and pressures associated with the boundary fluids as f^α and f^β. The superscripts α and β refer to the two different fluids on each side of the interface. With $\mathbf{n}$ the unit normal vector into the fluid β, the forces may be written (Newman 1991, Edwards et al. 1991)

$$f^\alpha = \mathbf{n} \cdot \mathbf{\mathfrak{T}}^\alpha + \mathbf{n}p^\alpha \qquad (10.5.2a)$$

$$f^\beta = -\mathbf{n} \cdot \mathbf{\mathfrak{T}}^\beta - \mathbf{n}p^\beta \qquad (10.5.2b)$$

We now can make a tangential and normal force balance. In the tangential direction on the interface the forces f^α and f^β are purely viscous, and from Eqs. (10.5.1) and (10.5.2) we have

$$f_s^\alpha + f_s^\beta + \mathbf{\nabla}_s\sigma = 0 \qquad (10.5.3)$$

Thus the shear stress depends on the local surface tension gradient, in the absence of which Eq. (10.5.3) simply reduces to the usual fluid dynamic boundary condition that the tangential viscous stress is continuous at the interface of two different fluids. The normal force balance simply gives the scalar equation

$$f_n^\alpha + f_n^\beta = \sigma\left(\frac{1}{R_1} + \frac{1}{R_2}\right) \qquad (10.5.4)$$

where we have used the Young-Laplace equation (10.1.5). The normal forces f_n include both the thermodynamic pressures as well as the normal viscous stresses.

It is perhaps clearer now from Eqs. (10.5.3) and (10.5.4) that if the surface tension terms are not small compared with the shear terms that the velocity distribution will be affected in each of the phases. However, σ depends on the concentration, temperature, and charge at the interface, the determination of which is coupled to the solution of the appropriate equation of change. Therein lies the difficulty in solving this type of problem.

To illustrate a particularly simple example of how a surface tension gradient can give rise to a bulk fluid motion, we consider the thermocapillary motion generated in an open rectangular shallow pan with a very thin liquid layer at the bottom (Fig. 10.5.1). The variation of surface tension is brought about by maintaining one of the side walls at a higher constant temperature than the other side wall, which is at a different constant temperature. The difference in side wall temperatures results in a temperature gradient along the surface and a corresponding surface temperature gradient. We recall, as shown in Fig. 10.5.1, that for liquids $\partial\sigma/\partial T < 0$.

The pan is considered to be much deeper (into the paper) than the liquid height in the pan, h. Moreover, the pan itself is much longer than it is wide, so $l/h \gg 1$, where l is the pan length. Under these conditions, any flow nonuniformities at the side walls are small and the flow is essentially two-dimensional. Since $h \ll l$, the flow is nearly lateral, but, as we shall show, the liquid height

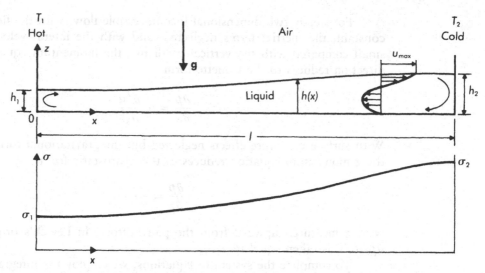

Figure 10.5.1 Thermocapillary motion in a shallow pan.

does vary along the length, so there is a small but finite vertical flow. However, except at the side walls, which are far removed from the bulk flow, the vertical velocity component is very much smaller than the horizontal component and any effects of the free surface curvature may be neglected.

The essentials of this problem were first suggested by Levich (1962), although his solution contained some simplifying assumptions, one of which we shall retain, and some inconsistencies pointed out by Yih (1968), who gave a more general solution. This work was later generalized further to unsteady flows by Pimputkar & Ostrach (1980).

In his analysis Levich assumed the liquid layer was thin enough that inertial effects were negligible, implying that the appropriate Reynolds number was sufficiently small. In light of the "shallow water" theory approximation made, we may choose for the characteristic length the initial liquid film height h_1. There is no specified characteristic velocity, and we take it to be the maximum lateral velocity at the interface, u_{max}, due to the driving force $d\sigma/dx$. We emphasize, however, that this quantity is not given but is defined by the solution. In order of magnitude, u_{max} is specified by the balance between the shear force at the interface and the tractive force due to the surface tension gradient, whence

$$u_{max} \sim \frac{h_1}{\mu} \frac{d\sigma}{dx} \sim \frac{h_1}{\mu} \frac{\sigma_2 - \sigma_1}{l} \qquad (10.5.5)$$

The criterion for the smallness of the Reynolds number $\mathrm{Re}_h = u_{max} h_1/\nu$ can therefore be written

$$\mathrm{Re}_h = \frac{\sigma_2 - \sigma_1}{\rho \nu^2} \left(\frac{h_1}{l}\right)^2 l \ll 1 \qquad (10.5.6)$$

For steady two-dimensional incompressible flow, with the liquid viscosity constant, the inertial terms neglected, and with the lateral velocity gradients small compared with the vertical gradients, the momentum equation in the x direction reduces to the Couette form

$$\frac{\partial p}{\partial x} = \mu \frac{\partial^2 u}{\partial z^2} \qquad (10.5.7)$$

With surface curvature effects neglected but the gravitational force considered, the z momentum equation reduces to the hydrostatic form

$$\frac{\partial p}{\partial z} = -\rho g \qquad (10.5.8)$$

with z measured upward from the pan bottom. In Levich's original analysis $\partial p/\partial z$ was taken equal to zero.

To complete the system of equations, we employ the integral form of the continuity equation, recognizing that the liquid surface layer set in motion by the surface tension force (Eq. 10.5.1) must be accompanied by a motion of the fluid in the opposite direction below the surface, as sketched in Fig. 10.5.1. With no net flux across any cross section, the continuity equation for the fully developed flow is

$$\int_0^{h(x)} u(z)\, dz = 0 \qquad (10.5.9)$$

Yih pointed out that this is a special case and that, depending on the geometry of the end conditions, the net flux need not be zero, but we retain this assumption as appropriate to the particular physical problem considered.

Next, the boundary conditions must be specified, and at the pan surface it is simply the no-slip condition

$$u = 0 \qquad \text{at } z = 0 \qquad (10.5.10)$$

At the interface, with curvature neglected, the tangential stress is continuous, so from Eq. (10.5.3)

$$\mu \frac{\partial u}{\partial z} = \frac{d\sigma}{dx} \qquad \text{at } z = h(x) \qquad (10.5.11)$$

there being no normal gradients and the air viscosity being negligibly small compared with that of the liquid. The last boundary condition is supplied by continuity of pressure at the surface

$$p = p_a \qquad \text{at } z = h(x) \qquad (10.5.12)$$

where p_a is the atmospheric pressure.

Levich integrated Eq. (10.5.7) and applied the boundary conditions (10.5.10) and (10.5.11) to obtain for the velocity profile

$$\mu u = \left(\frac{d\sigma}{dx} - h\,\frac{\partial p}{\partial x}\right)z + \frac{1}{2}\,\frac{\partial p}{\partial x}\,z^2 \qquad (10.5.13)$$

Levich did not, however, specify the means for evaluating $\partial p/\partial x$. This may have resulted from his assumption that h was constant. However, as Yih noted, from integrating the z momentum equation, the pressure is given by the local hydrostatic condition

$$p = p_a + \rho g(h - z) \qquad (10.5.14)$$

This specifies the relation between the pressure gradient and variation of free surface height in the x direction as

$$\frac{\partial p}{\partial x} = \rho g\,\frac{dh}{dx} \qquad (10.5.15)$$

We can now use the continuity relation (Eq. 10.5.9) together with the above equation to obtain the variation in surface tension gradient with the gradient in free surface height:

$$\frac{d\sigma}{dx} = \frac{2}{3}\,\rho g h\,\frac{dh}{dx} \qquad (10.5.16)$$

This can be immediately integrated to give

$$\sigma - \sigma_1 = \frac{\rho g}{3}\,(h^2 - h_1^2) \qquad (10.5.17)$$

where the constant of integration has been determined by the requirement that $\sigma = \sigma_1$ and $h = h_1$ at $x = 0$. Note that since we have taken the net flux across any cross section equal to zero, there is no freedom in specifying h_2 at $x = l$ if σ_2 is given. Both σ_2 and h_2 could be specified at $x = l$, but then it would not be possible to specify h_1 at $x = 0$ with σ_1 given. The important point to be seen in Eq. (10.5.17) is that a variation in σ automatically requires a corresponding variation in h. The parameter measuring the relative change in h^2 for a given change in σ is the Bond number $\rho g h_1^2/\sigma_1$.

From the solution for the velocity profile (Eq. 10.5.13) and the relation between $d\sigma/dx$ and $\partial p/\partial x$, we have

$$u = \frac{z}{2\mu}\left(\frac{3}{2}\,\frac{z}{h} - 1\right)\frac{d\sigma}{dx} \qquad (10.5.18)$$

As sketched in Fig. 10.5.1, the liquid velocity has a maximum value at the interface in the direction of positive x given by

$$u_{max} = \frac{h}{4\mu}\,\frac{d\sigma}{dx} \qquad (10.5.19)$$

which is in accord with our earlier order-of-magnitude estimate. The velocity reverses direction at $y = 2h/3$, attaining a maximum negative value at $y = h/3$, and then decreasing monotonically to the pan bottom.

The requirement that the Reynolds number $\mathrm{Re}_h = h_1 u_{max}/\nu$ be small compared with unity may be expressed as

$$h_1^2 \leq \frac{0.4\rho\nu^2}{d\sigma/dx} \qquad (10.5.20)$$

where we have interpreted "small compared with unity" as less than or equal to 0.1. As discussed in Section 10.1, the variation of σ with temperature for liquids is close to linear. For water $\partial\sigma/\partial T \approx -0.15\,\mathrm{mN\,m^{-1}K^{-1}}$; thus with a temperature gradient $dT/dx = -100\,\mathrm{K\,m^{-1}}$ this would give a value of $d\sigma/dx = 15\,\mathrm{mN\,m^{-2}}$. With $\nu = 10^{-6}\,\mathrm{m^2\,s^{-1}}$, $\rho = 10^3\,\mathrm{kg\,m^{-3}}$, and h_1^2 defined by the equality in Eq. (10.5.20), the Bond number is 3.6×10^{-3} for $\sigma_1 = 73\,\mathrm{mN\,m^{-1}}$ (the surface tension of water at 20°C). The small value of the Bond number shows the strong effect of the surface tractive force relative to the force of gravity. We conclude that surface tension gradients can indeed be important in such quasi-one-dimensional examples as considered, with very thin liquid layers of mm size or less, or in a reduced gravity environment (termed *microgravity*). For example, a crystal grown from its melt under reduced gravity is governed by convection driven by thermally induced surface tension gradients rather than buoyancy forces (Ostrach 1983, Xu & Davis 1983). The reader should be cautioned, however, that steady-state solutions are not always achievable for arbitrarily imposed physical conditions such as a specified free surface temperature distribution (Yih 1968, Pimputkar & Ostrach 1980, Xu & Davis 1984). An example to be discussed in the next section is cellular convection induced by surface tension gradients.

As we have shown, the surface forces at an interface depend upon the surface tension gradients there. If adsorbed surface-active materials are distributed at an interface, then this distribution must be known to determine the surface forces, since the surface tension gradients depend on the local surface concentration of adsorbed material. The surface mass concentration of the adsorbed substance follows from an interfacial mass balance.

Because the interface is generally in motion, it is convenient to derive the mass conservation relation at the interface, employing a moving control surface. Typically a "pillbox" control volume straddling the interface is used. Any net convective outflow is then governed by the relative velocity $\mathbf{u}_{rel} = \mathbf{u} - \mathbf{u}_{cs}$ with respect to each area element $\mathbf{n}\,dA$. Thus the outflows are calculated with respect to an observer moving with the velocity $\mathbf{u}_{cs}$ of an element of the control surface.

The derivation of the overall interfacial mass conservation equation is carried out in a fashion similar to the bulk mass conservation derivation in Section 3.1. The result will be of the same form as Eq. (3.1.4), except for the addition of a term corresponding to the "jump" in $\rho\mathbf{u}_{rel}$ across the interface. With $\tilde{\Gamma}$ the interface mass density ($\mathrm{kg\,m^{-2}}$), the equation is

$$\frac{\partial\tilde{\Gamma}}{\partial t} + \nabla_s \cdot \tilde{\Gamma}\mathbf{u}_s + [\rho\mathbf{u}_{rel}] \cdot \mathbf{n} = 0 \qquad (10.5.21)$$

where the subscript s refers to the surface, $\mathbf{n}$ is the outward unit normal, and the brackets denote the jump in $\rho\mathbf{u}_{rel}$ across the interface.

If we consider the system as a binary one with a surface-active material and bulk liquid, it is physically instructive to write the individual material balance relation for the surface excess concentration Γ (mol m^{-2}). The procedure for this is exactly as was carried out for the bulk binary system treated in Section 3.3. No chemical reaction at the interface is assumed, the system is considered to be dilute, the multicomponent mass flux is assumed to follow Fick's law, and the diffusion coefficients are taken to be constant. The expression for the surface concentration then becomes

$$\frac{\partial \Gamma}{\partial t} + \boldsymbol{\nabla}_s \cdot (\Gamma \mathbf{u}_s) = D_s \boldsymbol{\nabla}_s^2 \Gamma + [D\boldsymbol{\nabla}c] \cdot \mathbf{n} \qquad (10.5.22)$$

where $\mathbf{u}_s$ = molar average velocity
         = the mass average velocity for the dilute system considered
    D_s = binary interfacial diffusion coefficient
     D = binary diffusion coefficient of the dissolved surface-active material
     c = bulk molar concentration of the surface-active material

Equation (10.5.22) shows that the local interfacial concentration varies as a result of local and convective acceleration along the interface and surface diffusion, and the "jump" term represents the difference in diffusion of dissolved surface-active material to or from the adjacent bulk solution. An assumption made in writing the jump term in the way done is that the adsorption-desorption kinetics are assumed to be rapid compared with the diffusion rate; that is, the surface concentration is always assumed to be in equilibrium with the concentration of surface-active material in the liquid immediately adjacent to the interface. This need not necessarily be true. It is clear from Eq. (10.5.22) that the velocity distribution in the liquid must be known to define the interface distribution of surface-active material. This distribution in turn defines the surface forces, which couple to the velocity distribution, thereby making, in general, a relatively difficult closure problem.

One important enhanced oil recovery procedure is the displacement of oil in porous strata by foam flooding. Such flows are exceedingly complicated, including the factors influencing bubble size and the tendency for the bubbles to block a large percentage of the flow paths in the complex porous geometry. Herbolzheimer (1988), in an effort to better understand the phenomena, reexamined the Bretherton problem discussed in Section 10.3, wherein an air bubble is used to blow a viscous liquid out of a circular capillary, leaving a thin film deposited on the inside wall. In the analysis of Bretherton (1961), where the motion of a long closed bubble was considered, the interfacial tension was assumed everywhere constant. To relate to the flooding problem in which surfactants are employed, Herbolzheimer assumed that surfactant was added to the liquid surrounding the bubble. As described, for example in Section 4.3, capillary flows are often used to model porous media flow.

Herbolzheimer observed that as the bubble travels down the capillary, the flow in the surrounding liquid causes a nonuniform distribution of surfactant to develop on the bubble surface. Thus the mechanical boundary condition of Eq. (10.5.3) by means of Eq. (10.5.22) becomes coupled to the flow in the surrounding liquid through the transfer of the surfactant between the liquid and the interface. In an analysis that we outline below, he showed that the surfactant addition strongly alters the flow pattern, resulting in pressure drops to push the bubble through the capillary with a given velocity, which can be from two to four orders of magnitude larger than those in the absence of surfactant.

The "clean" capillary bubble problem previously analyzed is sketched in Fig. 10.3.3. In treating this problem in the presence of a surfactant, for convenience we modify the geometry slightly and suppose the bubble to be closed and long, as Bretherton originally did. We employ a reference frame in which the bubble is stationary, as shown in Fig. 10.5.2, so that the fluid is seen as flowing from the front to the rear of the bubble. As in Section 10.3, the interfaces at the front and rear menisci are shaped differently. The radius of the rear cap is somewhat larger than that of the front cap, and the transition of the lubrication-film layer to the rear meniscus is inflected. For the simplified analysis to be presented we neglect these differences, taking both the front and rear bubble caps to be hemispherical with radii equal to the capillary radius a. The length of the constant thickness lubrication-film layer is l.

With surface-active material present the convection of the liquid, from right to left in the reference frame in which the bubble is stationary, results in a nonuniform surfactant distribution on the bubble surface. The surfactant is swept toward the rear of the bubble where it accumulates. As a consequence, the surface tension varies along the bubble with its lowest value at the rear end. The surface tension gradient exerts a tractive force on the bubble and increases its resistance to motion under a driving pressure gradient $p_2 - p_1$.

At this point it is useful to recall some results from the earlier Bretherton analysis for bubble motion with no surfactant present. We note first that in this case the boundary condition at the interface is that the shear stress vanishes at the bubble surface, the gas viscosity being so much smaller than that of the liquid. From Bretherton's result for the interface shape at the front end in the overlap region between the hemispherical cap and the film layer (Eq. 10.3.18),

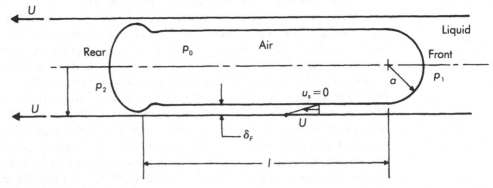

Figure 10.5.2   Air bubble moving in capillary filled with a viscous liquid as viewed in a reference frame with the bubble fixed and the capillary wall moving from right to left.

the true mean curvature for small capillary number of the central portion of the front meniscus region is not quite $2/a$ but is $(2/a)[1 + 1.79(3\text{Ca})^{2/3}]$. The first term in brackets is the static contribution to the curvature, and the second term is the change in curvature resulting from the fluid motion in the central region of the meniscus. It follows from the Young-Laplace equation that the dynamic portion of the pressure drop at the front meniscus is $3.58(\sigma/a)(3\text{Ca})^{2/3}$. The curvature increase at the rear meniscus is somewhat different, and the dynamic pressure difference there is $-0.94(\sigma/a)(3\text{Ca})^{2/3}$, which gives a total pressure drop to drive the bubble of

$$p_2 - p_1 = 9.40\text{Ca}^{2/3}\,\frac{\sigma}{a} \qquad (10.5.23)$$

An important feature of Eq. (10.5.23) is that it shows pressure drop is independent of bubble length. This is a consequence of the fact that the zero shear stress condition at the interface results in the fluid in the film layer slipping by with no drag force, with all the pressure drop taking place in the front and rear overlap regions of the meniscus. Since $\text{Ca} \ll 1$, not only is the pressure drop relatively small, but from Eq. (10.5.23) the drop goes as $\sigma^{1/3}$, indicating that with a lower surface tension the driving pressure will decrease. Herbolzheimer has pointed out that this prediction "fails spectacularly," since experimentally measured pressure drops increase by orders of magnitude when surfactant is added. However, there is no error in the analytical approach of Bretherton. The error, as Herbolzheimer recognized, is in trying to apply Bretherton's result to a physical situation for which it is not applicable, namely to the different type of flow condition prevailing at a bubble interface when surface-active materials are present.

To explain the seeming paradox discussed, we consider the limiting case in which the transfer of surfactant takes place only by surface convection. We assume the capillary number is small, and essentially we follow Herbolzheimer's approach. If we make Eq. (10.5.22) appropriately dimensionless, choosing U as the characteristic velocity, a as the characteristic length, and D as the diffusion coefficient, then both terms on the right side of Eq. (10.5.22) are on the order of Pe^{-1}, where $\text{Pe} = Ua/D$. For the steady flow configuration examined with $\text{Pe} \gg 1$, surface convection is large compared with diffusive transfer, and Eq. (10.5.22) reduces to $\nabla_s \cdot \Gamma \mathbf{u}_s = 0$. From continuity $\nabla_s \cdot \mathbf{u}_s = 0$, whence the expression for the excess surface concentration simply becomes

$$(\nabla_s \Gamma) \cdot \mathbf{u}_s = 0 \qquad (10.5.24)$$

Since what is of interest here is finite $\nabla_s \Gamma$, and since $\nabla_s \Gamma$ is parallel to $\mathbf{u}_s$, we must have $\mathbf{u}_s = 0$. The condition $\mathbf{u}_s = 0$ at the interface implies that the surface-active substance on the interface is insoluble, and the surfactant film behaves like an incompressible thin solid membrane at which the liquid velocity drops to zero (in a reference frame where the bubble is fixed). With the tangential velocity zero at the bubble surface, the change in velocity across the lubrication film is much sharper than in the absence of surfactant, where the shear stress is zero at the bubble surface.

The zero-velocity condition at the bubble surface dramatically increases the shear stress in the film, which remains essentially of uniform thickness. Employing a simple force balance on the moving fluid in the lubrication layer, we get, corresponding to Eq. (9.1.10),

$$\pi a^2 (p_2 - p_1) = 2\pi a l \tau_s = \frac{2\pi a l \mu U}{\delta_f} \tag{10.5.25}$$

where τ_s is the constant shear stress in the film. In order to evaluate the pressure drop to drive the bubble, we must determine the film thickness δ_f. Although we have indicated that the pressure change with surface-active material present is large compared with the clean case, this actually results from only a relatively small fractional change in surface tension along the bubble, a result that we shall show a posteriori.

With the assumption that σ is a slowly varying function of x, we can again use the dip coating momentum equation (Eq. 10.3.2) to calculate the lubrication film thickness δ_f, the term $\delta'' \, d\sigma/dx$ being neglected as small. The boundary condition at the bubble surface (Eq. 10.3.3b) changes to

$$u = 0 \qquad \text{at } y = \delta(x) \tag{10.5.26}$$

and for the assumed profile in the film the volume flow rate of "lubricant" per unit width is half that given by Eq. (10.3.6), or

$$\tilde{Q} = \tfrac{1}{2} U \delta_f \tag{10.5.27}$$

Carrying out the integration of Eq. (10.3.2) and applying the new conditions given above, we obtain the same equation as Eq. (10.3.7), except that the factor 3 is replaced by 6. The resulting equation can thus be transformed to the universal form of Eq. (10.3.8) with the 3 in the definition of ξ replaced by 6. The solution procedure follows exactly as in the dip coating and clean bubble problem, giving for the lubrication film thickness, in place of Eq. (10.3.17),

$$\frac{\delta_f}{a} = 0.643(6\mathrm{Ca})^{2/3} \tag{10.5.28}$$

Substituting this value of the film thickness into Eq. (10.5.25), we find the relation for the pressure drop driving the bubble to be

$$p_2 - p_1 = 0.942 \mathrm{Ca}^{1/3} \frac{\sigma}{a} \frac{l}{a} \tag{10.5.29}$$

Now the total pressure drop driving the bubble is also given by the difference in pressure drops at the front and rear menisci. In particular, $p_2 - p_1 = (p_2 - p_0) - (p_1 - p_0)$, whence from the Young-Laplace equation

$$p_2 - p_1 = \frac{2}{a} (\sigma_{\text{front}} - \sigma_{\text{rear}}) \tag{10.5.30}$$

Here, we have invoked the assumption made earlier that the front and rear bubble caps are spherical and each of radius a. Inserting the value of $p_2 - p_1$ from Eq. (10.5.29), we arrive at the expression for the maximum difference in surface tension between the front and rear of the bubble:

$$\frac{\sigma_{\text{front}} - \sigma_{\text{rear}}}{\sigma} = 0.471 \text{Ca}^{1/3} \frac{l}{a} \qquad (10.5.31)$$

This relation shows that for Ca $\ll$ 1, as assumed, the fractional change in σ along the bubble surface is relatively small, allowing terms in $d\sigma/dx$ in the momentum equation to be neglected.

The film thickness as given by Eq. (10.5.28) is seen to increase by only about 59% from the constant σ result of Eq. (10.3.17), and both have the same behavior with capillary number. On the other hand, the pressure drop is increased markedly from the clean bubble value of Eq. (10.5.23). With surfactant present the pressure drop is proportional to $\text{Ca}^{1/3} l/a$, whereas for constant surface tension it goes as $\text{Ca}^{2/3}$. For small values of Ca, around 10^{-6}, the difference in driving pressure with surfactant is about two orders of magnitude higher than it is in the absence of surface-active material. The pressure drop with surfactant more closely approaches the constant surface tension value as Ca increases, the difference being relatively small at Ca $\sim 10^{-1}$. This may explain the experimental observations of Bretherton (1961), Schwartz et al. (1986), and others who have shown in "clean" systems for capillary numbers less than 10^{-3} to 10^{-4} a systematic increase with decreasing Ca, roughly power law in character, in the measured fractional change in bubble speed W over that predicted theoretically (Eq. 10.3.22). Since $W \approx 2\delta_f/a$, this result says that the experimental film thickness correspondingly increases with a decrease in Ca over the value predicted by Eq. (10.3.17), in the same fashion as the fractional change in bubble speed. Herbolzheimer argues that this and other differences noted at very low Ca in systems assumed to be "clean" may derive from the presence of trace amounts of surface-active impurities, which could lead to a change in the bubble motion dynamics in the direction illustrated, consistent with the surfactant analysis presented.

A phenomenologically closely related problem to the one just examined is that of the "calming effect of oil on troubled waters." Damping of waves on liquids by surface-active materials is complicated by the close coupling between the boundary conditions at the wave surface and the fluid motion below. In Section 10.4 we examined clean, undamped, and growing capillary waves in the absence of viscosity. We propose here to introduce viscosity and briefly sketch the strong viscous damping effect on capillary waves that results from the presence of a layer of insoluble surface-active material on the wave surface. If a surfactant is present on a wave traveling along a liquid surface, the surfactant concentration will vary with position along the interface, giving rise to surface tension gradients. These gradients affect the surface force balance and lead to a viscous surface boundary layer in which the tangential stresses can increase markedly from their value in the bulk fluid. A graphic example of this is the large increase in shear stress that we saw to take place in the lubrication film adjacent to an air bubble in a capillary when surfactant is added at the interface. Such enhanced shearing can greatly increase the viscous dissipation at the wave surface, causing strong wave damping far in excess of that which might result solely from viscous stresses acting throughout the wave.

Before showing the relatively strong damping effect that a surfactant can exert on plane capillary waves, we first consider the pure viscous damping of deep water capillary waves, with the flow taken to be laminar. The liquid viscosity is assumed low enough that it does not appreciably affect the fluid motion or wave frequency. In this case the irrotational wave motion solution given by Laplace's equation, which was discussed in the preceding section, satisfies the Navier-Stokes equations but does not satisfy the viscous boundary conditions at the wave surface, particularly the vanishing of the tangential viscous stress, leading to a boundary layer-type behavior there. The solution to this problem is to be found in numerous texts (Levich 1962, Lighthill 1978, Miller & Neogi 1985). It has exactly the same form as given for the ideal fluid except that there is an exponential attenuation of the wave amplitude with time $\exp(-\beta_{u_y=0}t)$, where

$$\beta_{u_y=0} = (2k)k\nu \qquad (10.5.32)$$

with k the wave number and ν the kinematic viscosity. What this shows is that the viscous attenuation of the irrotational fluid motion in time takes place approximately over the depth k^{-1} below the surface, which is that region in which the fluid motion associated with the undamped wave dies off exponentially.

Let us now look at the same capillary wave with surfactant present on the wave surface. As in the bubble problem, we again consider the limiting case of a concentrated monolayer of surfactant in which the surfactant transfer takes place only by convection and Eq. (10.5.24) holds. For small-amplitude surface waves the condition at the wave surface may be evaluated at the undisturbed liquid level (cf. discussion leading to Eq. 10.4.16), whence

$$u_s = 0 \qquad \text{at } z = 0 \qquad (10.5.33)$$

With this condition, as for the bubble problem, the change in velocity through the wave becomes much sharper than for a clean surface, where the viscous boundary condition at the surface is that of zero tangential stress.

Equation (10.5.33) implies, as noted in connection with Eq. (10.5.24), that the surfactant film is incompressible and behaves like a membrane or thin metal sheet that bends as the surface deforms but which is inextensible; that is, it neither contracts at the wave trough or expands at its crest (Levich 1962). The effect is the inhibition of longitudinal motion, which is replaced by a transverse wave motion—in other words, a laminar viscous wave that propagates into the fluid in the $-z$ direction because of the surface deflection.

At this point we digress slightly to recall the classic Stokes oscillating boundary layer problem (Lighthill 1978, Landau & Lifshitz 1987). In this problem a large volume of fluid, say in the region $z < 0$, is bounded by a solid plane wall at $z = 0$ and is subjected to a spatially uniform pressure gradient $\partial p_e / \partial x$, which oscillates sinusoidally in time with frequency ω. The solution is quite straightforward, and it shows the development of a velocity profile $u(z, t)$ in the fluid below the wall, which can be characterized as a damped transverse wave of wave number $k = (\omega / 2\nu)^{1/2}$ propagating in the $-z$ direction. The transition to the irrotational solution, where the velocity oscillations are in phase with those of the external motion to within about 5%, takes place in a boundary layer thickness

$$\delta_{\text{Stokes}} = 2 \left(\frac{2\nu}{\omega} \right)^{1/2} \qquad (10.5.34)$$

From the description of the Stokes problem, we conclude that the incompressible surfactant film causes an oscillating Stokes-type boundary layer to develop, which is the predominant damping mechanism. By the same argument as before, the amplitude of the wave solution satisfying Laplace's equation is modified by an exponential attenuation $\exp(-\beta_{u_s=0} t)$. Here, however, we take the attenuation length to be given by the viscous boundary layer thickness of Eq. (10.5.34), from which, by comparison with Eq. (10.5.32),

$$\beta_{u_s=0} = \frac{1}{2} \left(\frac{\omega}{2\nu} \right)^{1/2} k\nu \qquad (10.5.35)$$

For $(\omega / \nu k^2)^{1/2} \gg 1$, Eq. (10.5.35) is the result obtained by Levich (1962) through a detailed wave analysis treatment. However, we must admit to a small "fudge" by our selective choice of definition for the Stokes boundary layer thickness. We would also point out that no properties of the surface-active material appear explicitly in our solution because of the limiting case treated.

The quantity $(\omega / \nu k^2)^{1/2}$, which is assumed to be large compared with unity, is proportional to the ratio of the damping coefficient with surfactant (Eq. 10.5.35) to that without (Eq. 10.5.32). This quantity, which we denote by B, may also be written in terms of wave speed c and wavelength λ as

$$B = \left(\frac{\omega}{\nu k^2} \right)^{1/2} = \left(\frac{c\lambda}{2\pi\nu} \right)^{1/2} \qquad (10.5.36)$$

For purposes of estimate let us choose c to be the minimum wave speed associated with the transition from capillary wave to gravity wave dominance in the undamped case (Eq. 10.4.20), and λ the associated wavelength (Eq. 10.4.21). If we take $\lambda_{\min} = 1.7 \times 10^{-2}$ m and $c_{\min} = 0.23$ m s^{-1} for water ($\nu = 10^{-6}$ m^2 s^{-1}), then $B = 25$, showing the criterion of B being large compared with unity to be met. Although this is only a rough estimate, for example σ with surfactant will be lower than that for clean water, it nevertheless shows that capillary wave damping with surfactant is large compared with damping without surfactant. A similar result can also be shown for gravity waves (Levich 1962). Our discussion thus provides, at least in part, an answer to Plutarch's question of the calming effect of oil on the sea.

10.6   Cellular Convection Induced by Surface Tension Gradients

In the preceding section, we have examined a variety of steady thermocapillary and diffusocapillary flows. Not all such flows are stable and in fact surface tension variations at an interface can be sufficient to cause an instability. We consider here the cellular patterns that arise with liquid layers where one boundary is a free surface along which there is a variation in surface tension. It is well known that an unstable buoyancy driven cellular convective motion can result when a density gradient is parallel to but opposite in direction to a body force, such as gravity. An example of this type of instability was discussed in Section 5.5 in connection with density gradient centrifugation.

In a series of beautiful experiments at the turn of the 20th century, Bénard (1900) observed that hexagonal convection cells formed within thin films of molten spermaceti about 0.5–1 mm deep that were heated from below, with the cell spacing somewhat more than three times the liquid depth. These cells are now referred to as *Bénard cells* and a plan photograph of the cells from one of Bénard's original photographs is shown in Fig. 10.6.1. The liquid film is molten spermaceti on a flat surface, which is heated from below by steam. The upper surface of the film is in contact with the air. Although Bénard initially assumed that surface tension at the free surface of the film was an important factor in the cell formation, this idea was abandoned for some time as the result of the work of Rayleigh (1916) who analyzed the buoyancy driven natural convection of a layer of fluid heated from below. He found that if hexagonal cells formed, the ratio of the spacing to cell depth almost exactly equaled that measured by Bénard, an agreement which we now know to have been fortuitous.

Rayleigh showed that if the cells are to form, then the vertical adverse temperature gradient must be sufficiently large that a particular dimensionless parameter proportional to the magnitude of the gradient exceed a critical value. We now term this parameter the *Rayleigh number*

$$\mathrm{Ra} = \frac{g\gamma\beta h^4}{\nu\alpha} \qquad (10.6.1)$$

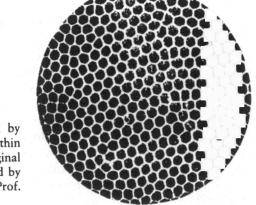

Figure 10.6.1 Plan photograph taken by Henri Bénard of hexagonal cells in a thin film of molten spermaceti from his original experiments on convection cells induced by surface tension gradients. [Courtesy of Prof. Simon Ostrach. From Bénard 1900.]

Here, β is the magnitude of the vertical temperature gradient across the liquid layer of height h and γ is used to denote the thermal expansion coefficient of the fluid (Eq. 3.2.17) to avoid confusion with the thermal diffusivity α.

It was the experimental work of Block (1956) which put to rest the confusion surrounding the interpretation of Bénard's experiments, and which demonstrated conclusively that Bénard's results were not a consequence of buoyancy but were surface tension induced. Among other things, he showed that cellular convection took place for Rayleigh numbers more than an order of magnitude smaller than required by the Rayleigh theory. Most importantly if the cells are buoyancy induced, then if the thin film is cooled from below the density gradient and gravity will be in the same direction and the film will be stably stratified. In such an experiment Block observed Bénard cells. He also produced Bénard cells and then removed them by covering the surface with a monolayer of surfactant. These together with other experiments led him to conclude that for thin films of thicknesses less than 1 mm, variations in surface tension due to temperature variations were the cause of Bénard cell formation and not buoyancy as postulated by Rayleigh. It is now generally agreed that for films smaller than about a few millimeters, surface tension is the controlling force, while for larger thicknesses buoyancy is the controlling force and there the Rayleigh mechanism delimits the stable and unstable regimes.

The phenomenon of surface tension induced Bénard cells is commonly observed in the drying of paint with the appearance of what is usually called an "orange peel" pattern. The cause of the orange peel or Bénard cells is the surface tension gradient induced along the paint film by the rapid cooling effect at the free surface associated with the evaporation of the volatile solvents in the paint. Again, the orange peel effect is independent of whether the free surface of the paint layer is topside or underside, the latter case being convectively stable.

The mechanism of Bénard cell formation, also termed the *Marangoni instability*, was first elucidated and demonstrated theoretically by Pearson (1958) who, unaware of Block's experimental work, showed that if there was an adverse temperature gradient of sufficient magnitude across a thin liquid film with a free surface that such a layer could become unstable and lead to cellular convection. Following Pearson, the instability mechanism is illustrated in Fig. 10.6.2. There a small disturbance is assumed to cause the film of initially

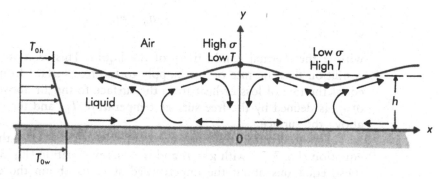

Figure 10.6.2 Instability mechanism for Bénard cell formation induced by a surface tension gradient.

uniform thickness to be heated locally at a point on the surface. This results in a decreased surface tension and a surface tension gradient that leads to an induced motion tangential to the surface away from the point of local heating. From mass conservation, this motion in turn induces a motion of the bulk phase toward the surface. The upwelling liquid coming from the heated region is warmer than the liquid/gas interface. The motion is thus reinforced creating cellular convection patterns and will be maintained if the convection overcomes viscous shear and heat diffusivity. We would also note at this point that the same model, with a simple redefinition of the parameters, also applies to surface tension gradients induced by concentration variations.

Pearson's theoretical treatment was based on a linear stability analysis of the type described in Section 10.4 in connection with jet stability to small disturbances and paralleled Rayleigh's analysis for buoyancy driven instability. He assumed an infinite homogeneous liquid film of uniform thickness h whose lower surface is in contact with a rigid heat conductor at a fixed temperature and whose upper surface is free. Gravity is neglected (Ra = 0) and a linear temperature distribution across the film is assumed, with the high temperature at the lower surface. The surface tension is a function of temperature alone, and the rate of heat loss from the free surface is also a function of temperature only.

The details of the analysis are somewhat lengthy and we therefore aim the presentation at deriving the dimensionless parameters that define the film stability, following the approach of Pearson. To do this, we write down the linearized equations and boundary conditions for the velocity and temperature disturbances recognizing that the problem is one of coupled flow and heat transfer.

The undisturbed state is one in which the free surface is flat, the fluid is static, and the heat transfer is purely by conduction. The unperturbed temperature distribution in the liquid is given by

$$T = T_{0w} - \beta y \qquad (10.6.2)$$

where T_{0w} is the unperturbed steady-state temperature of the bottom surface of the film at the plane $y = 0$. The unperturbed rate of heat loss per unit area (heat flux) from the upper free surface at the plane $y = h$, is

$$q_0 = k\beta \qquad (10.6.3)$$

with k the thermal conductivity of the liquid. This relation derives from the condition that the rate of heat supply to the free surface from the liquid must equal the rate of loss of heat from the surface to the air above. The magnitude of q_0 is defined by the free surface temperature T_{0h} and the cooling by the air above the surface.

The governing equations are continuity (Eq. 3.5.1), the Navier–Stokes equation (Eq. 3.5.2 with $\mathbf{g} = 0$) and the energy equation (Eq. 3.5.4). Linearizing these equations about the unperturbed state to obtain the equations for the disturbed field leads to

$$\left(\frac{\partial}{\partial t} - \nu\nabla^2\right)\nabla^2 v' = 0 \tag{10.6.4}$$

$$\left(\frac{\partial}{\partial t} - \alpha\nabla\right)\nabla^2 T' = 0 \tag{10.6.5}$$

where v' is the perturbation velocity in the y-direction and T' the perturbation temperature from the initial state defined by $T' = T - T_0$.

The boundary conditions on the velocity from the no-slip condition and continuity are

$$v' = \frac{\partial v'}{\partial y} = 0 \quad \text{at} \quad y = 0 \tag{10.6.6}$$

At the free surface, which is assumed to be nondeformable, corresponding to the limit of an appropriately defined capillary number $\text{Ca} \to 0$,

$$v' = 0 \quad \text{at} \quad y = h \tag{10.6.7}$$

The last boundary condition on the velocity comes from the balance between the change in surface tension due to temperature variations along the surface with the tractive force induced at the free surface. Here, as in Section 10.5, the rate of change of surface tension is taken to be linear with temperature

$$\sigma = \sigma_0 - \sigma_T T'_h \tag{10.6.8a}$$

where σ_0 is the unperturbed surface tension tension at the free surface and σ_T is the surface tension gradient evaluated at the unperturbed free surface temperature

$$\sigma_T = -\left(\frac{\partial\sigma}{\partial T}\right)_{T=T_{0h}} \tag{10.6.8b}$$

With the aid of continuity, the boundary condition is then written

$$\mu\frac{\partial^2 v'}{\partial y^2} = \sigma_T\nabla_1^2 T' \quad \text{at} \quad y = h \tag{10.6.9a}$$

where

$$\nabla_1^2 = \frac{\partial^2}{\partial x^2} + \frac{\partial^2}{\partial z^2} \tag{10.6.9b}$$

The critical parameter, although not the only one, governing the instability of a thin liquid film due to temperature induced surface tension gradients follows from the inhomogeneous boundary condition Eq. (10.6.9), the homogeneous boundary conditions introducing no parameters. With h the characteristic length scale, $v \sim \alpha/h$, and $T' \sim \beta h$ the boundary condition is seen to introduce the dimensionless parameter

$$\text{Ma} = \frac{\sigma_T \beta h^2}{\mu \alpha} \tag{10.6.10}$$

This parameter is termed the *Marangoni number*. As discussed below, if Ma exceeds a critical value, an unstable convective flow will develop. The Marangoni number can also be interpreted as a thermal Peclet number (Eq. 3.5.16) if the characteristic velocity for the surface tension driven viscous flow is taken to be that of Eq. (10.5.5). We emphasize that this velocity is not a given parameter but rather a derived quantity. Expressing this velocity in terms of the imposed uniform temperature gradient β, with the aid of continuity, we arrive at Eq. (10.6.10). Interpreted as a Peclet number, the Marangoni number is a measure of the heat transport by convection due to surface tension gradients to the bulk heat transport by conduction.

We next consider the boundary conditions on the disturbance temperature. At the lower surface, in the case where the conductivity of the rigid plate is large compared to the liquid, it is

$$T' = 0 \quad \text{at} \quad y = 0 \tag{10.6.11a}$$

which corresponds to a fixed temperature at the surface. Alternatively if the plate is of low conductivity compared to the liquid

$$\frac{\partial T'}{\partial y} = 0 \quad \text{at} \quad y = 0 \tag{10.6.11b}$$

which corresponds to a fixed heat flux surface.

At the free surface, from Eq. (10.6.3)

$$-k \frac{\partial T'}{\partial y} = Q T' \quad \text{at} \quad y = h \tag{10.6.12}$$

where the heat flux from the free surface is defined by

$$q = q_0 + Q T'_h \tag{10.6.13}$$

Here, Q is the surface heat transfer coefficient, that is, the rate of change with respect to temperature of the heat flux from the free surface to the air. Again, with h the characteristic scale, the boundary condition Eq. (10.6.12) introduces another dimensionless parameter called the *Biot number*

$$\text{Bi} = \frac{hQ}{k} = \frac{h/k}{1/Q} \tag{10.6.14}$$

The right hand term of the equation shows the Biot number to characterize the ratio of the thermal resistance of the liquid layer to the thermal resistance of the external environment.

In the next step of the analysis, the forms for the perturbation velocity and temperature are taken to be satisfied by normal modes

$$v' = -\frac{\alpha}{h}\, f\!\left(\frac{y}{h}\right) F\!\left(\frac{x}{h},\frac{z}{h}\right) e^{pt\alpha/h^2} \tag{10.6.15a}$$

$$T' = \beta hg\!\left(\frac{y}{h}\right) F\!\left(\frac{x}{h},\frac{z}{h}\right) e^{pt\alpha/h^2} \tag{10.6.15b}$$

where p is the amplification factor. Introducing these solutions into the linearized disturbance equations (10.6.4) and (10.6.5), and separating variables

$$[p - \mathrm{Pr}(D^2 - \tilde{k}^2)](D^2 - \tilde{k}^2)f = 0 \tag{10.6.16a}$$

$$[p - (D^2 - \tilde{k}^2)]g = -f \tag{10.6.16b}$$

where $D \equiv d/dy$ and $\tilde{k}$ is a real horizontal wave number. Together with the boundary conditions, these linear equations define the eigenvalue problem. The solutions depend upon the Prandtl number appearing in Eq. (10.6.16a), the Marangoni and Biot numbers from the boundary conditions, and the wave number. However, Pearson only sought those solutions for neutral stability ($p = 0$), corresponding to the onset of convection, so that the instability is time independent as a consequence of which the results are independent of Pr.

From Pearson's solutions for Bi = 0, with the fixed temperature boundary condition (Eq. 10.6.11a), the critical Marangoni number $\mathrm{Ma}_c \approx 80$, while for the fixed heat flux boundary condition (Eq. 10.6.11b) $\mathrm{Ma}_c \approx 48$. The critical Marangoni numbers increase with increasing values of the Biot number. Thus increasing the thermal resistance of the film is stabilizing as might have been anticipated on physical grounds. For a survey of extensions of Pearson's treatment to finite capillary and Rayleigh numbers, the reader is referred to the review of Davis (1987).

The analysis with disturbance quantities of the form of Eqs. (10.6.15) indicates a periodic structure in the x, z plane but the shape of the cells associated with the solution is not specified and higher order nonlinear theory is required to define a particular cellular structure. Palm (1960) has shown that in the parallel Rayleigh problem for steady buoyancy driven convection of a liquid film heated from below, the cells approach a hexagonal form as a consequence of the variation of the kinematic viscosity with temperature.

References

ADAMSON, A.W. 1982. *Physical Chemistry of Surfaces*, 4th edn. New York: Wiley.

BÉNARD, H. 1900. Les tourbillons cellulaires dans une nappe liquide. Deuxiéme partie: Procédés mécaniques et optiques d'examen lois numériques des phénomènes. *Rev. Gén. Sci.* 11, 1309–1328.

BLOCK, M.J. 1956. Surface tension as the cause of Bénard cells and surface deformation in a liquid film. *Nature* 178, 650–651.

BOGY, D.B. 1979. Drop formation in a circular liquid jet. *Ann. Rev. Fluid Mech.* 11, 207–228.

BRETHERTON, F.P. 1961. The motion of long bubbles in tubes. *J. Fluid Mech.* 10, 166–188.

DAVIES, J.T. & RIDEAL, E.K. 1963. *Interfacial Phenomena*, 2nd edn. New York: Academic.

DAVIS, S.H. 1987. Thermocapillary instabilities. *Ann. Rev. Fluid Mech.* 19, 403–435.

DE GENNES, P.G. 1985. Wetting: statics and dynamics. *Rev. Mod. Phys.* 57, 827–863.

DRAZIN, P.G. & REID, W.H. 1981. *Hydrodynamic Stability*. Cambridge: Cambridge Univ. Press.

DUSSAN V., E.B. 1979. On the spreading of liquids on solid surfaces: Static and dynamic contact lines. *Ann. Rev. Fluid Mech.* 11, 371–400.

GIBBS, J.W. 1906. *The Collected Works of J. Willard Gibbs. vol. I, Thermodynamics.* New York: Longmans, Green. (Reprinted 1961, New York: Dover.)

GOEDDE, E.F. & YUEN, M.C. 1970. Experiments on liquid jet instability. *J. Fluid Mech.* 40, 495–511.

GREENSPAN, H.P. 1977. On the dynamics of cell cleavage. *J. Theor. Biol.* 65, 79–99.

HÅRD, S., HAMNERIUS, Y. & NILSSON, O. 1976. Laser heterodyne apparatus for measurements of liquid surface properties—theory and experiment. *J. Appl. Phys.* 47, 2433–2442.

HERBOLZHEIMER, E. 1988. The effect of surfactant on the motion of bubbles in a capillary. Exxon Research and Engg. Co., Annandale, N.J. To be published.

HIEMENZ, P.C. 1986. *Principles of Colloid and Surface Chemistry*, 2nd edn. New York: Marcel Dekker.

LANDAU, L.(D.) & LEVICH, B.(V.G.) 1942. Dragging of a liquid by a moving plate. *Acta Physicochim. URSS* 17, 42–54.

LANDAU, L.D. & LIFSHITZ, E.M. 1987. *Fluid Mechanics*, 2nd edn. Oxford: Pergamon Press.

LANGLOIS, W.E. 1985. Buoyancy-driven flows in crystal-growth melts. *Ann. Rev. Fluid Mech.* 17, 191–215.

LEVICH, V.G. 1962. *Physicochemical Hydrodynamics*. Englewood Cliffs, N.J.: Prentice-Hall.

LIN, C.C. 1955. *The Theory of Hydrodynamic Stability*. Cambridge: Cambridge Univ. Press.

LIGHTHILL, J. 1978. *Waves in Fluids*. Cambridge: Cambridge Univ. Press.

MILLER, C.A. & NEOGI, P. 1985. *Interfacial Phenomena*. New York: Marcel Dekker.

NEWMAN, J.S. 1991. *Electrochemical Systems*, 2nd edn. Englewood Cliffs, N.J.: Prentice-Hall.

OSTRACH, S. 1980. Natural convection with combined driving forces. *PhysicoChem. Hydrodynamics* 1, 233–247.

OSTRACH, S. 1983. Fluid mechanics in crystal growth—The 1982 Freeman Scholar Lecture. *J. Fluids Engg. (Trans. ASME)* 105, 5–20.

PALM, E. 1960. On the tendency towards hexagonal cells in steady convection. *J. Fluid Mech.* 8, 183–192.

PEARSON, J.R.A. 1958. On convection cells induced by surface tension. *J. Fluid Mech.* 4, 489–500.

PIMPUTKAR, S.M. & OSTRACH, S. 1980. Transient thermocapillary flow in thin liquid layers. *Phys. Fluids* 23, 1281–1285.

QIAN S-X., SNOW, J.B., TZENG, H-M. & CHANG, R.K. 1986. Lasing droplets: Highlighting the liquid-air interface by laser emission, *Science* 231, 486–488.

RAYLEIGH, LORD 1894. *The Theory of Sound, vol. II*, 2nd edn. London: Macmillan. (Reprinted 1945, New York: Dover.)

RAYLEIGH, LORD 1916. On convection currents in a horizontal layer of fluid, when the higher temperature is on the under side. *Philos. Mag.* 32, 529–546.

ROWLINSON, J.S. & WIDOM, B. 1982. *Molecular Theory of Capillarity*. Oxford: Clarendon.

RUSCHAK, K.J. 1976. Limiting flow in a pre-metered coating device. *Chem. Eng. Sci.* **31**, 1057–1060.

RUSCHAK, K.J. 1985. Coating flows. *Ann. Rev. Fluid Mech.* **17**, 65–89.

RUSCHAK, K.J. 1987. Flow of a thin liquid layer due to ambient disturbances. *AIChE J.* **33**, 801–807.

SCHWARTZ, L.W., PRINCEN, H.M. & KESS, A.D. 1986. On the motion of bubbles in capillary tubes. *J. Fluid Mech.* **172**, 259–275.

SPIERS, R.P., SUBBARAMAN, C.V. & WILKINSON, W.L. 1974. Free coating of a Newtonian liquid onto a vertical surface. *Chem. Eng. Sci.* **29**, 389–396.

SZEKELY, J. 1979. *Fluid Flow Phenomena in Metals Processing.* New York: Academic.

TAYLOR, G.I. 1961. Deposition of a viscous fluid on the wall of a tube. *J. Fluid Mech.* **10**, 161–165.

TAYLOR, G.I. 1963. Cavitation of a viscous fluid in narrow passages. *J. Fluid Mech.* **16**, 595–619.

WHITE, D.A. & TALLMADGE, J.A. 1965. Theory of drag out of liquids on flat plates. *Chem. Eng. Sci.* **20**, 33–37.

WILSON, S.D.R. 1982. The drag-out problem in film coating theory. *J. Engg. Math.* **16**, 209–221.

XU, J-J. & DAVIS, S.H. 1983. Liquid bridges with thermocapillarity. *Phys. Fluids* **26**, 2880–2886.

XU, J-J. & DAVIS, S.H. 1984. Convective thermocapillary instabilities in liquid bridges. *Phys. Fluids* **27**, 1102–1107.

YIH, C-S. 1968. Fluid motion induced by surface tension variation. *Phys. Fluids* **11**, 477–480.

Problems

10.1 A small quantity of a substance such as a fatty acid is added to the surface of a liquid such as water so that it spreads out to form a monomolecular film on the liquid, thereby lowering its surface tension.

a. If the surface excess concentration of the substance in the surface film is proportional to its bulk concentration, show from the Gibbs equation that the surface tension decrease of the liquid $\Delta\sigma = RT/\bar{A}$, where $\bar{A}$ is the area per mole of the substance in the surface film.

b. The decrease in the liquid surface tension with the bulk substance concentration is measured and the slope of the measured surface tension versus the natural logarithm of the bulk substance concentration at 20°C is found to be $-10\,\mathrm{mN\,m^{-1}}$. What is the surface excess concentration? What is the area of surface occupied by a molecule of the adsorbed material?

10.2 A volume V of a liquid of density ρ and surface tension σ is contained between two parallel, concentric circular disks that are oriented horizontally and separated from each other by a distance a. A thin circular capillary tube of radius a is connected to the upper disk at its center and oriented vertically. The liquid level in the tube measured upward from the center of the space between the disks ($a/2$) to the meniscus is H (the meniscus height itself is taken to be small). The distance the liquid extends radially outward between the disks, measured from the centerline of the capillary to the meniscus formed between the disks, is held to a radius R, such that $R \gg a$ and $H \gg a$. The static contact angle is denoted by θ and the system is open to the atmosphere.

a. Assuming $\sigma / \rho g a^2 \gg 1$, obtain an implicit relation for R and an implicit relation for H as a function of V, a, σ, ρ, and g.

b. If the liquid is water, what are the numerical values of H, R, and V when $H = R = 10a$. Is any assumption made satisfied?

10.3 A semi-infinite solid slab is dipped vertically into a large pool of liquid open to the atmosphere. As a result of surface tension, the liquid rises above the pool level to meet the slab on its face at the contact angle $\theta_0 < \pi/2$. The geometry is two-dimensional, and a rectangular Cartesian coordinate system is chosen with origin at the intersection of the slab face and the undisturbed pool surface. The coordinate y is measured vertically upward from the undisturbed pool level, and the coordinate x is measured into the raised liquid.

a. If $\delta(x)$ is the vertical height of the liquid-air interface (at the pool level $\delta(x) = 0$), show that $\delta(x)$ is governed by the differential equation

$$\frac{\delta''}{(1 + \delta'^2)^{3/2}} = \frac{\rho g}{\sigma} \delta$$

b. Assume that $(1 + \delta'^2)^{3/2} \approx 1$ and solve for $\delta(x)$, using the result of part a.

c. Suppose that the criterion for the assumption in part b is that $(1 + \delta'^2)^{3/2} \leq 1.05$. Determine the minimum contact angle at the surface $\theta_{0\min}$, for which this requirement is met.

d. In the case that $\theta_0 < \theta_{0\min}$, at a distance $x = x_1$ the tangent to the liquid-air interface will have an angle $\theta = \theta_1 = \theta_{0\min}$. Let $z = x - x_1$ and obtain an expression for $\delta(z)$ valid for $z \geq 0$. Section 10.3 states that the differential equation governing the interface geometry is invariant to a shift in origin. Does this result suggest the truth of the statement and why?

10.4 a. If gravity cannot be neglected, to what would the Landau-Levich differential equation governing the film thickness in dip coating (Eq. 10.3.9) be modified? What does the parameter measure that appears in this modified equation?

b. If the film thickness ratio $\eta = \delta/\delta_f$ is close to 1, both the Landau-Levich differential equation and the differential equation accounting for gravity can be linearized. Carry this out and show that under an appropriate transformation the linearized equation with gravity included has the same form as the linearized Landau-Levich equation. Show that this transformation depends only on the capillary number, if the effect of gravity is small but finite, and that the transformation is a measure of the first-order effect of gravity in decreasing the asymptotic film thickness δ_f.

10.5 It was shown from linear stability theory that a circular jet will become unstable to axisymmetric disturbances whose wavelength $\lambda = 2\pi/k$ is greater than the circumference of the undisturbed jet $2\pi a$. Suppose that as a consequence of small axisymmetric disturbances the radius r of the jet surface is given by

$$r = a_0 + \alpha_0 \cos kz$$

where a_0 is a dimension of the order of the undisturbed jet radius a, z is the cylindrical coordinate along the symmetry axis, $\alpha_0 \cos kz \ll a_0$, and $k > 0$.

a. Show that $a_0 \approx a[1 - \frac{1}{4}(\alpha_0/a)^2]$.

b. Show for the small deformation considered that the surface area of the deformed jet is $z[2\pi a_0 + \frac{1}{2}\pi(k\alpha_0)^2 a]$, and justify the conclusion that the jet will be unstable to axisymmetric disturbances whose wavelength is greater than the undisturbed jet circumference.

10.6 Consider an air bubble moving in a capillary filled with a viscous liquid (Fig. 10.5.2). The capillary number is small. With surfactant in the liquid and with its diffusivity neglected, a surface tension gradient is set up by convection which is expressed by Eq. (10.5.31).

a. If mass transfer due to diffusion of the surfactant along the surface and between the bubble surface and the liquid is no longer negligible, what will happen to the pressure drop required to drive the bubble and why?

b. Derive an estimate for how the gradient in surface tension would be modified if the velocity at the bubble surface in the lubrication film region u_s were not zero (when the bubble is viewed as stationary) but still small compared with the bubble velocity U.

c. Diffusion leads to surfactant being adsorbed from the liquid at the nose of the bubble and desorbed from the rear cap of the bubble to the liquid. A finite value of u_s also results. Consider diffusion to the nose of the bubble, and from Eq. (10.5.22) show with $u_s \ll U$ an order-of-magnitude estimate for u_s is given by

$$\frac{u_s}{U} \sim \frac{Ca^{2/3}l^2}{Pe}\left(\frac{dc}{d\Gamma}\right)^2\left(\frac{\sigma}{\Gamma}\frac{d\Gamma}{d\sigma}\right)^2$$

where $Ca = \mu U/\sigma$
$Pe = Ua/D$
$D =$ dissolved surfactant diffusivity
$l =$ bubble length

The surface concentration Γ, like the surface tension σ, is an appropriate characteristic value. In deriving the result, take the diffusion boundary layer thickness on the nose of the bubble to be given in order of magnitude by $\delta \sim a/(u_s a/D)^{1/2}$.

10.7 Show that the Rayleigh number (Eq. 10.6.1) can, like the Marangoni number, be interpreted as a thermal Peclet number.

Appendix A
SI Units and Physical Constants

SI Base Units

Of the seven dimensionally independent base units, the candela is not used in the book.

Quantity	Name	Symbol
Mass	kilogram	kg
Length	meter	m
Time	second	s
Temperature	kelvin	K
Amount of substance	mole	mol
Electric current	ampere	A
Luminous intensity	candela	cd

SI Derived Units

Listed are some named derived units relevant to the material in the book.

Quantity	Name	Symbol	Definition
Force	newton	N	$kg\, m\, s^{-2}$
Pressure	pascal	Pa	$N\, m^{-2} = kg\, m^{-1}\, s^{-2}$
Energy	joule	J	$N\, m = kg\, m^2\, s^{-2}$
Power	watt	W	$J\, s^{-1} = kg\, m^2\, s^{-3}$
Electric charge	coulomb	C	$A\, s$
Electric potential difference	volt	V	$J\, C^{-1} = kg\, m^2\, s^{-3}\, A^{-1}$
Electric resistance	ohm	Ω	$V\, A^{-1} = kg\, m^2\, s^{-3}\, A^{-2}$

363

Electric conductance	siemens	S	$A\,V^{-1} = kg^{-1}\,m^{-2}\,s^3\,A^2$
Electric capacitance	farad	F	$C\,V^{-1} = kg^{-1}\,m^{-2}\,s^4\,A^2$
Frequency	hertz	Hz	s^{-1}

SI Prefixes

Listed are prefixes to designate submultiples to 10^{-12} and multiples to 10^{12} of a base unit.

Factor	Prefix	Symbol	Factor	Prefix	Symbol
10^{-1}	deci	d	10	deca	da
10^{-2}	centi	c	10^2	hecto	h
10^{-3}	milli	m	10^3	kilo	k
10^{-6}	micro	μ	10^6	mega	M
10^{-9}	nano	n	10^9	giga	G
10^{-12}	pico	p	10^{12}	tera	T

Physical Constants

Quantity	Symbol	Value	SI Units
Avogadro number	N_A	6.022×10^{23}	mol^{-1}
Boltzmann constant	k	1.381×10^{-23}	$J\,K^{-1}$
Elementary charge	e	1.602×10^{-19}	C
Faraday constant	F	9.648×10^4	$C\,mol^{-1}$
Gas constant	R	8.314	$J\,K^{-1}\,mol^{-1}$
Permittivity of vacuum	ϵ_0	8.854×10^{-12}	$C\,V^{-1}\,m^{-1}$
Standard acceleration of gravity	g	9.807	$m\,s^{-2}$
Standard atmosphere	p_0	1.013×10^5	Pa
Zero of Celsius scale	T_0	273.15	K

Appendix B
Symbols

All standard mathematical and chemical symbols are taken to have their usual meaning. Both Cartesian tensor and boldface vector notation have been employed in the book. In the following list only the boldface form is given for vector quantities to avoid confusion with the use of the subscripts i and j in Cartesian tensor notation and the use of a subscript i or j to denote a species.

No attempt has been made to define every symbol and, in particular, symbols used only locally and that are not referred to again are generally not included. An abbreviated verbal definition and, where appropriate, an equation number or symbolic definition are given. Illustrative SI base or derived units are given for dimensional quantities.

The book encompasses several subjects, so in trying to use standard notation wherever possible, repetition of symbols becomes unavoidable. The most common usage has generally been adopted within the limitation of maintaining self-consistency and avoiding repetition where it might be confusing.

Symbol	Definition	SI Units
a	Activity	—
a	Amplitude of water wave	m
a	Long semiaxis of prolate spheroid or short semiaxis of oblate spheroid	m
a	Radius of cylinder, also cylindrical capillary, pipe, jet, particle, and collector	m
a	Radius of sphere, also spherical particle, collector, and drop	m
a	Radius of nose of bubble in capillary	m
A	Area, also column cross-sectional area and porous medium surface area	m^2

A	Hamaker constant	J
A_{cyl}	Parameter in stream function for cylinder, Eq. (8.3.24b)	—
A_{sph}	Parameter in stream function for sphere, Eq. (8.5.4b)	—
b	Long semiaxis of oblate spheroid or short semiaxis of prolate spheroid	m
b	Radius of spherical cell in cell model	m
Bi	Biot number, Eq. (10.6.14)	—
Bo	Bond number, Eq. (10.2.1)	—
c	Molar density or concentration	mol m^{-3}
c	Reduced ion concentration, Eq. (3.4.11)	mol m^{-3}
c	Wave speed	m s^{-1}
c_a	Reduced ion concentration at anode	mol m^{-3}
c_c	Reduced ion concentration at cathode	mol m^{-3}
c_g	Solute gelling concentration	mol m^{-3}
c_m	Mixing-cup concentration, Eq. (4.4.29), Problem 4.3	mol m^{-3}
c_p	Specific heat at constant pressure	J kg^{-1} K^{-1}
c_{sat}	Saturated solution concentration	mol m^{-3}
c_w	Solute concentration at membrane	mol m^{-3}
c_0	Electrolyte concentration far from charged surface	mol m^{-3}
c_0	Reduced ion concentration, initial	mol m^{-3}
c_0	Solute concentration, initial, at channel inlet, at tube axis	mol m^{-3}
c^*	Dimensionless solute concentration, Eq. (4.3.4)	—
c^*	Dimensionless solute concentration defect, Eq. (4.4.17)	—
$\bar{c}$	Average solute concentration over tube cross section, Eqs. (4.6.2), (4.6.19)	mol m^{-3}
$\bar{c}$	Mean molecular speed	m s^{-1}
$\bar{c}_0$	Average solute concentration, initial	mol m^{-3}
Δc_w	Solute concentration difference across membrane	mol m^{-3}
C	Total equivalent concentration in solution, Eq. (6.3.8)	equiv m^{-3}*
C_{diff}	Mass transfer coefficient, Eq. (4.5.8)	—
C_f	Skin friction coefficient, Eq. (4.5.7)	—
$\bar{C}$	Total equivalent concentration in resin, Eq. (6.3.9)	equiv m^{-3}*
Ca	Capillary number, Eq. (10.2.8)	—
d	Diameter of sphere, also spherical particle, drop	m
d_e	Equivalent or hydraulic diameter, Eq. (4.7.8)	m
D	Diffusivity or diffusion coefficient, translational, effective Eq. (3.4.15)	m^2 s^{-1}

---

* Not SI unit.

D_{eff}	Dispersion coefficient, Taylor Eq. (4.6.27), Taylor-Aris Eq. (4.6.35)	$m^2 s^{-1}$
D_{rot}	Rotational diffusion coefficient	s^{-1}
D_{12}	Binary diffusion coefficient	$m^2 s^{-1}$
D_{12}	Brownian diffusion coefficient describing relative motion of two particles, Eq. (8.2.2)	$m^2 s^{-1}$
Da	Damköhler number, Eqs. (4.1.17), (4.4.11)	—
e	Specific internal energy	$J kg^{-1}$
E	Electric field	$V m^{-1}$
E_{cyl}	Cylindrical collector efficiency, Eq. (8.3.17)	—
E_{sph}	Spherical collector efficiency, Eq. (8.3.14)	—
E_x	Electric field, x component, parallel to direction of electrophoretic motion	$V m^{-1}$
$\mathscr{E}$	Equilibrium potential	V
$\mathscr{E}°$	Standard electrode potential	V
f	Dimensionless variable defining concentration in diffusion layer, Eq. (4.4.17)	—
f_{ij}	Translational friction tensor, Eq. (5.1.3a)	$kg s^{-1}$
$\bar{f}$	Mean translational friction coefficient, Eq. (5.1.10b)	$kg s^{-1}$
F	Faraday constant, $N_A e$	$C mol^{-1}$
F_{Ad}	Attractive London force along particle and collector line of centers	N
F_n	Net external radial force on particle driven to or from collector, Eq. (8.4.3)	N
F_{St}	Net external radial hydrodynamic force on particle driven to or from collector	N
F_x	Force component in x direction	N
F	Frictional force exerted by body in translational motion on fluid or by fluid on body	N
g	Standard acceleration of gravity	$m s^{-2}$
g_{eff}	Effective gravity in wave motion, Eq. (10.4.10)	$m s^{-2}$
g	Gravitational acceleration	$m s^{-2}$
G	Hindered settling factor, Eq. (5.4.17)	—
G	Negative of pressure gradient	$Pa m^{-1}$
G	Force on particle characterizing high-frequency molecular motions	N
h	Gap distance along particle and collector line of centers	m
h	Half-width of channel, electrodialysis cell, electrophoresis cell	m
h	Local height of liquid layer driven by thermocapillarity	m

h	Meniscus height	m
h	Sediment layer height	m
h	Spacing between parallel flat plates, charged flat plates, and flat electrodes in electrolytic cell	m
h	Specific enthalpy	$J\,kg^{-1}$
h_0	Gap distance along line of centers of identical spheres	m
H	Height of suspension layer	m
H	Instantaneous height of free surface of liquid in capillary tube	m
H_0	Equilibrium height of free surface of liquid in capillary tube	m
i_{lim}	Limiting current density	$A\,m^{-2}$
i	Current density, Eq. (2.5.8)	$A\,m^{-2}$
i_s	Surface current density	$A\,m^{-1}$
I	Total current	A
I_{cyl}	Mass flow rate of particles to cylindrical collector per unit length of cylinder	$kg\,s^{-1}\,m^{-1}$
I_{sph}	Mass flow rate of particles to spherical collector	$kg\,s^{-1}$
j	Mass flux	$kg\,m^{-2}\,s^{-1}$
j^*	Molar flux	$mol\,m^{-2}\,s^{-1}$
$\bar{J}$	Average solute mass flux over tube cross section, Eq. (4.6.25)	$kg\,m^{-2}\,s^{-1}$
J	Mass flux with respect to mass average velocity, Eq. (2.4.10)	$kg\,m^{-2}\,s^{-1}$
J^*	Molar flux with respect to molar average velocity, Eq. (2.4.11)	$mol\,m^{-2}\,s^{-1}$
k	Boltzmann constant, R/N_A	$J\,K^{-1}$
k	Crowding factor	—
k	Permeability of porous medium	m^2
k	Rate constant, νth-order homogeneous reaction	$(mol\,m^{-3})^{1-\nu}\,s^{-1}$
k	Thermal conductivity or thermal conduction coefficient	$W\,m^{-1}\,K^{-1}$
k	Wave number, Eqs. (10.4.2b), (10.4.28)	m^{-1}
k	Wetting coefficient, Eq. (10.1.9)	—
k_g	Gel permeability	m^2
k'	Rate constant, νth-order heterogeneous reaction	$mol^{1-\nu}\,m^{3\nu-2}\,s^{-1}$
l	Bubble lubrication film length	m
l	Mean free path, lattice spacing, or distance between particle collisions	m
l	Pan length	m
L	Characteristic length	m
L	Gel or resin packed bed column length	m
L_D	Development length for concentration profile in channel	m
L_U	Development length for velocity profile in channel	m
m	Consistency index, Eq. (9.1.8)	$Pa\,s^n$

m	Mass of a substance, molecule, or particle	kg
M	Molar mass	$kg\,mol^{-1}$
M	Molecular weight	—
$\bar{M}$	Mean molar mass, Eq. (2.4.5)	$kg\,mol^{-1}$
Ma	Marangoni number, Eq. (10.6.10)	—
n	Number of moles of a substance	mol
n	Particle number density or concentration	m^{-3}
n_0	Initial number of moles of solute	mol
$\mathbf{n}$	Unit normal vector	—
N_A	Avogadro number	mol^{-1}
N_{Ad}^{cyl}	Adhesion group for cylinder, Eq. (8.4.17)	—
N_{Ad}^{sph}	Adhesion group for sphere, Eq. (8.5.18)	—
p	Pressure	Pa
p	Ratio of spheroid semiaxes, a/b	—
p_a	Atmospheric pressure	Pa
p_e	Excess pressure over undisturbed value due to wave motion	Pa
Δp	Pressure difference across curved surface due to surface tension	Pa
Δp	Pressure difference across membrane	Pa
P	Probability of displacement	—
Pe	Peclet number, Eqs. (3.5.16), (3.5.17), (4.4.12), (4.6.8), (5.5.22), (9.2.3), (9.2.9), also ratio of Brownian diffusion time to convection time	—
Pe_D	Diffusion Peclet number, Eq. (3.5.17)	—
Pe_T	Thermal Peclet number, Eq. (3.5.16)	—
Pr	Prandtl number, Eq. (3.5.20)	—
q	Charge	C
q_s	Surface charge density	$C\,m^{-2}$
$\mathbf{q}$	Heat flux	$W\,m^{-2}$
Q	Heat transfer coefficient	$W\,m^{-2}\,K^{-1}$
Q	Volume flow rate	$m^3\,s^{-1}$
$\tilde{Q}$	Volume flow rate per unit width	$m^2\,s^{-1}$
r	Radial coordinate, plane polar, cylindrical or spherical polar	m
r_b	Radial distance of bottom of sector-shaped cell from axis of rotation	m
r_i	Mass rate of production of species i per unit volume	$kg\,m^{-3}\,s^{-1}$
r_m	Radial distance of meniscus from axis of rotation in sector-shaped cell	m
r_*	Radial distance of shock interface from axis of rotation in centrifuge	m
$\mathbf{r}$	Particle displacement vector	m
R	Gas constant	$J\,mol^{-1}\,K^{-1}$
R	Radius of curvature of static meniscus at apparent tangency point	m
$\bar{R}$	Mean translation coefficient, Eq. (5.1.10a)	m
R_F	Flory radius, Eq. (9.2.2)	m
R_i	Molar rate of production of species i per unit volume	$mol\,m^{-3}\,s^{-1}$

R_{ij}	Translation tensor, Eq. (5.1.3a)	m
R_s	Solute rejection coefficient of membrane	—
$R_{1,2,3}$	Translation coefficients for translation of a spheroid parallel to its semiaxes, Eqs. (5.1.6), (5.1.7), also any body parallel to its principal axes, Eq. (5.1.10a)	m
R_i'	Molar rate of production of species i per unit area of reaction surface	$mol\,m^{-2}\,s^{-1}$
Ra	Rayleigh number, Eq. (10.6.1)	—
Re	Reynolds number, Eq. (3.5.14)	—
Re_x	Reynolds number, Ux/ν	—
s	Sedimentation coefficient, Eq. (5.5.2)	s
s	Specific entropy	$J\,K^{-1}\,kg^{-1}$
s_0	Sedimentation coefficient, infinitely dilute mixture	s
S	Specific area, Eq. (4.7.11)	m^{-1}
S	Spreading coefficient, Eq. (10.1.10)	$N\,m^{-1}$
Sc	Schmidt number, Eq. (3.5.21)	—
St	Strouhal number, Eqs. (3.5.12), (5.5.25)	—
t	Time	s
T	Temperature	K
T'	Perturbation temperature	K
u	Fluid velocity, x component, parallel to channel walls, pan surface or plate surface	$m\,s^{-1}$
u	Speed of plane kinematic wave	$m\,s^{-1}$
u_{bot}	Speed of kinematic shock moving up from container bottom	$m\,s^{-1}$
u_E	True electrophoretic velocity in electrophoresis cell	$m\,s^{-1}$
u_{eff}	Speed with which a point with ionic fraction x_B in solution moves	$m\,s^{-1}$
u_{EO}	Electroosmotic velocity in electrophoresis cell	$m\,s^{-1}$
u_{ex}	Speed of ion exchange zone front	$m\,s^{-1}$
u_L	Liquid velocity in electrophoresis cell	$m\,s^{-1}$
u_{max}	Maximum fluid velocity at center of circular or straight channel with fully developed velocity profile, Eq. (4.2.14)	$m\,s^{-1}$
u_{max}	Velocity at free surface of falling liquid film or liquid film driven by thermocapillarity	$m\,s^{-1}$
u_{obs}	Observed particle velocity in electrophoresis cell	$m\,s^{-1}$
u_r	Radial component of fluid velocity in spherical polar coordinates	$m\,s^{-1}$
u_{top}	Speed of kinematic shock moving down from container top	$m\,s^{-1}$
u_θ	Polar component of fluid velocity in spherical polar coordinates	$m\,s^{-1}$

$\bar{u}$	Average longitudinal fluid velocity in channel, Eq. (4.4.6a)	$m\,s^{-1}$
u'	Longitudinal fluid velocity in channel with respect to moving axis x'	$m\,s^{-1}$
u'	Perturbation fluid velocity, x component	$m\,s^{-1}$
$\mathbf{u}$	Disturbance fluid velocity due to wave motion	$m\,s^{-1}$
$\mathbf{u}$	Fluid velocity, mass average velocity, Eq. (2.4.6)	$m\,s^{-1}$
$\mathbf{u}_s$	Velocity of surface active material	$m\,s^{-1}$
$\mathbf{u}^*$	Molar average velocity, Eq. (2.4.7)	$m\,s^{-1}$
U	Bubble interface speed	$m\,s^{-1}$
U	Characteristic speed	$m\,s^{-1}$
U	Electroosmotic velocity past a plane surface	$m\,s^{-1}$
U	Electrophoretic particle speed	$m\,s^{-1}$
U	Mean fluid velocity in tube	$m\,s^{-1}$
U	Particle fall speed, hindered settling speed	$m\,s^{-1}$
U	Superficial velocity in porous medium equal to uniform velocity upstream of medium, Eq. (4.7.7)	$m\,s^{-1}$
U	Uniform free stream flow velocity	$m\,s^{-1}$
U	Uniform speed of plate in Couette flow	$m\,s^{-1}$
U	Uniform velocity of fluid at channel inlet, equal to average longitudinal velocity in channel with no fluid removal or addition	$m\,s^{-1}$
U	Withdrawal speed of plate in dip coating	$m\,s^{-1}$
U_e	Mean interstitial or effective pore velocity in porous medium, Eqs. (4.7.6), (4.7.13), (6.3.13)	$m\,s^{-1}$
U_p	Cross-sectional average particle velocity	$m\,s^{-1}$
U_r	Radial drift velocity of particles in centrifugation	$m\,s^{-1}$
U_0	Particle fall speed in infinitely dilute suspension	$m\,s^{-1}$
$\mathbf{U}$	Translational velocity of body or particle	$m\,s^{-1}$
v	Fluid velocity, y component, normal to channel walls or plate surface	$m\,s^{-1}$
v	Mobility of particle in solution	$mol\,s\,kg^{-1}$
v	Specific volume	$m^3\,kg^{-1}$
v_{ij}	Mobility tensor, Eq. (5.1.3b)	$s\,kg^{-1}$
v_w	Permeation velocity of solution through membrane	$m\,s^{-1}$
$\bar{v}$	Mean mobility, Eq. (5.1.10c)	$s\,kg^{-1}$
$\bar{v}$	Solute particle partial specific volume	$m^3\,kg^{-1}$
v'	Perturbation fluid velocity, y component	$m\,s^{-1}$
V	Average speed of liquid ahead of bubble	$m\,s^{-1}$

V	Applied voltage	V
V	Volume of solution or mixture, also particles, column, and porous medium	m^3
V_A	London attractive energy between two molecules or particles	J
V_c	Coarse particle volume	m^3
V_f	Fine particle volume	m^3
V_l	Carrier liquid volume	m^3
V_A^{pl}	London attractive energy per unit area between two infinite flat plates	$J\,m^{-2}$
V_A^{sph}	London attractive energy between two identical spheres	J
V_e	Elution volume, Eq. (4.7.3)	m^3
V_{in}	Internal volume of gel pores accessible to solvent	m^3
V_R	Repulsive potential energy between two charged particles	J
V_R^{pl}	Repulsive potential energy per unit area between two infinite flat plates of same charge	$J\,m^{-2}$
V_R^{sph}	Repulsive potential energy between two identical spheres of same charge	J
V_{void}	Volume of voids in porous medium	m^3
V^*	Dimensionless applied voltage, Eq. (6.2.11)	—
w	Fluid velocity, z component	$m\,s^{-1}$
w'	Perturbation fluid velocity, z component	$m\,s^{-1}$
x	Boundary layer coordinate along surface in streamwise direction	m
x	Cartesian coordinate in direction of wave motion on free surface	m
x	Cartesian coordinate parallel to direction of motion of spherical particle and translating with it	m
x	Cartesian or cylindrical coordinate in direction of flow or direction of motion	m
x	Interface coordinate in direction of motion of plane kinematic wave	m
x_A	Equivalent ionic fraction of A in solution, Eq. (6.3.4)	—
x_B	Equivalent ionic fraction of B in solution, $1 - x_A$	—
x_i	Mole fraction of species i in solution, Eq. (2.4.4)	—
$\bar{x}_A$	Equivalent ionic fraction of A in ion exchange resin, Eq. (6.3.5)	—
$\bar{x}_B$	Equivalent ionic fraction of B in ion exchange resin, $1 - \bar{x}_A$	—
x'	Cylindrical coordinate in direction of flow and translating with mean flow speed, Eq. (4.6.15)	m

$\mathbf{x}$	Spatial coordinate vector, rectangular Cartesian coordinate system x, y, z	m
y	Boundary layer coordinate normal to surface	m
y	Cartesian coordinate normal to direction of motion of spherical particle and translating with it	m
y	Cartesian coordinate normal to surface or direction of flow	m
z	Cartesian coordinate normal to direction of wave motion on free surface or thermocapillary motion in pan	m
z	Cartesian coordinate transverse to direction of motion of spherical particle and translating with it	m
z	Charge number	—
z	Concentration boundary layer transformed ζ coordinate, Eq. (4.5.23)	—
z	Cylindrical coordinate along symmetry axis in stability analysis of liquid jet	m
α	Fraction of column cross-sectional area available to solute	—
α	Thermal diffusivity	$\mathrm{m^2\,s^{-1}}$
β	Amplification factor in jet instability	$\mathrm{s^{-1}}$
β	Particle diameter to minimum separation between particles, Eq. (9.3.9)	—
β	Temperature gradient	$\mathrm{K\,m^{-1}}$
β_{ii}	Collision frequency per unit volume of monodisperse particles	$\mathrm{s^{-1}\,m^{-3}}$
β_{12}	Collision frequency per unit volume of a_2 particles with all a_1 test particles	$\mathrm{s^{-1}\,m^{-3}}$
γ	Dimensionless function of wall potential, Eq. (8.1.12b)	—
$\dot{\gamma}$	Shear rate	$\mathrm{s^{-1}}$
Γ	Polarization parameter, Eq. (4.4.30)	—
Γ	Surface excess concentration	$\mathrm{mol\,m^{-2}}$
δ	Local film thickness formed in dip coating and by bubble in capillary	m
δ_D	Diffusion or concentration boundary layer thickness	m
δ_f	Constant limiting film thickness formed in dip coating and by bubble in capillary	m
δ_g	Gel layer thickness	m
δ_U	Viscous or velocity boundary layer thickness	m
Δ	Initial width of solute slug	m
Δ_c	Capillary length, Eq. (10.2.2)	m
ϵ	Permittivity	$\mathrm{C\,V^{-1}\,m^{-1}}$
ε	Void fraction or porosity of porous medium, Eq. (4.7.9)	—
ε_{ij}	Rate-of-strain tensor, Eq. (2.2.15)	$\mathrm{s^{-1}}$
ε_{kk}	Dilatation, Eq. (2.2.17)	$\mathrm{s^{-1}}$
ζ	Concentration boundary layer transformed y coordinate, Eq. (4.5.14)	m

ζ	Radial displacement of disturbed jet surface, Eq. (10.4.25)	m
ζ	Vertical displacement of disturbed free surface, Eq. (10.4.1)	m
ζ	Zeta potential	V
η	Apparent or effective viscosity of suspension	Pa s
η	Concentration or viscous boundary layer transformed y coordinate, Eqs. (4.3.5), (4.4.18), (4.5.11), (6.2.9), (8.3.7)	—
η	Dimensionless local film thickness, Eq. (10.3.8)	—
η_r	Relative viscosity, η/μ	—
η_{cr}	Coarse relative viscosity, Eq. (9.4.4)	—
η_{fr}	Fine relative viscosity, Eq. (9.4.4)	—
η_{nr}	Net relative viscosity, bimodal suspensions, Eq. (9.4.4)	—
η^0	Zero-shear-rate viscosity	Pa s
η^∞	Infinite-shear-rate viscosity	Pa s
η_r^∞	High shear relative viscosity	—
$[\eta]_\rho$	Intrinsic viscosity, polymer rheology, Eq. (9.2.4)	$m^3\,kg^{-1}$
$[\eta]_\phi$	Intrinsic viscosity, suspension rheology, Eq. (9.3.1a)	—
$[\eta]_\rho^0$	Intrinsic viscosity scaled to zero-shear-rate viscosity	—
θ	Fraction of surface adsorption sites occupied	—
θ	Polar angle or in cylindrical coordinates azimuthal angle about symmetry axis	—
θ	Static contact angle	—
λ	Filter coefficient, Eq. (8.5.13)	m^{-1}
λ	Particle-to-tube radius	—
λ	Wavelength, Eq. (10.4.2c)	m
λ_D	Debye length, Eq. (6.4.5)	m
λ^*	Debye length ratio, λ_D/a	—
Λ	Molar conductivity, Eq. (2.5.11)	$S\,m^2\,mol^{-1}$
μ	Chemical potential	$J\,mol^{-1}$
μ	Viscosity or viscosity coefficient	Pa s
μ_{in}	Internal viscosity of fluid drop	Pa s
μ_p	Bingham plastic viscosity	Pa s
ν	Kinematic viscosity	$m^2\,s^{-1}$
ν	Number of ions formed if solute dissociates	—
ν	Overall order of reaction	—
ν_i	Order of reaction with respect to species i or stoichiometric coefficient	—
ξ	Concentration boundary layer transformed x coordinate, Eqs. (4.4.16), (6.2.10)	—
ξ	Coordinate in flow direction in ion exchange column	m
ξ	Reduced radial variable, Eq. (5.5.20b)	—
ξ	Reduced x coordinate, Eq. (10.3.8)	—
π	Osmotic pressure	Pa

π	Reduced density variable, Eq. (5.5.20a)	—
$\Delta\pi$	Osmotic pressure difference across membrane	Pa
ρ	Mass density or concentration	kg m^{-3}
ρ	Solute or suspended particle density or concentration	kg m^{-3}
ρ_E	Electric charge density	C m^{-3}
ρ_{fl}	Fluid solvent density	kg m^{-3}
ρ_{in}	Solute concentration in gel	kg m^{-3}
ρ_m	Maximum particle concentration in sedimented layer	kg m^{-3}
ρ_0	Dilute suspension particle concentration	kg m^{-3}
ρ_0	Initial solute concentration	kg m^{-3}
$\bar{\rho}$	Mean density averaged over tube cross section, Eq. (4.6.26)	kg m^{-3}
$\bar{\rho}$	Mean density averaged over number of species	kg m^{-3}
ρ^*	Dimensionless particle concentration, Eq. (8.3.6)	—
σ	Electrical conductivity of solution, Eq. (2.5.10)	S m^{-1}
σ	Partition coefficient, Eq. (4.7.1)	—
σ	Surface tension	N m^{-1}
σ_b	Bulk electrical conductivity of fluid	S m^{-1}
σ_s	Surface electrical conductance	S
σ_{sg}	Solid-gas surface tension	N m^{-1}
σ_{sl}	Solid-liquid surface tension	N m^{-1}
σ_T	Negative of surface tension gradient	$\text{N m}^{-1}\text{K}^{-1}$
σ'_s	Mean value of conductivity of double layer shell around spherical particle	S m^{-1}
σ	Reference stresses	N m^{-2}
τ	Characteristic time	s
τ	Flocculation time	s
τ	Reduced time variable, Eq. (5.5.20b)	—
τ	Shear stress	N m^{-2}
τ	Time to reach equilibrium	s
τ	Wave period, Eq. (10.4.2a)	s
τ_{ij}	Stress tensor	N m^{-2}
τ_0	Yield stress	N m^{-2}
$\boldsymbol{\tau}$	Stress dyadic or tensor	N m^{-2}
ϕ	Disturbance velocity potential	$\text{m}^2\,\text{s}^{-1}$
ϕ	Electrostatic or electric potential	V
ϕ	Volume fraction, particles in suspension, polymer in fluid, grains in porous medium	—
ϕ_c	Coarse particle volume fraction, Eq. (9.4.3)	—
ϕ_f	Electric potential in bulk fluid	V
ϕ_{ff}	Fine filler volume fraction, Eq. (9.4.2)	—
ϕ_m	Maximum packing fraction, fluidity limit	—
ϕ_t	Total solids volume fraction, Eq. (9.4.1)	—
ϕ_w	Wall potential	V
$\Delta\phi$	Potential drop in fluid	V

Φ	Axial component of electric potential in circular capillary, Eq. (6.5.9)	V
Φ	Radial component of disturbance velocity potential, Eq. (10.4.28)	$m^2\,s^{-1}$
Φ	Viscous dissipation function, Eq. (3.2.6)	$W\,m^{-3}$
ψ	Electric potential in double layer, Eq. (7.2.8)	V
ψ	Radial component of electric potential in circular capillary, Eq. (6.5.9)	V
ψ	Stream function for flow past cylinder	$m^2\,s^{-1}$
ψ^*	Dimensionless radial component of electric potential in circular capillary, $zF\psi/RT$	—
Ψ	Stream function for flow past sphere	$m^3\,s^{-1}$
Ψ_{lim}	Stream function for limiting particle trajectory in flow past sphere	$m^3\,s^{-1}$
ω	Angular speed of rotation or of centrifuge	s^{-1}
ω	Radian frequency of wave motion, Eq. (10.4.2a)	s^{-1}
ω	Wave frequency, Eq. (10.4.23)	s^{-1}
ω_i	Mass fraction of species i in solution, Eq. (2.4.3)	—
ω_{12}	Total collision frequency of a_2 particles with a_1 test particle	s^{-1}
ω	Angular velocity	s^{-1}
Ω_{ij}	Rotation tensor, Eq. (5.1.16)	m^3

Subscripts

a	Parallel to a axis of prolate spheroid
A	With reference to exchangeable ion A
b	Parallel to b axis of prolate spheroid
B	With reference to exchangeable ion B
i	With reference to ith species
j	With reference to jth species
lim	Limiting value, current density, flux, flow velocity
m	Conditions at midplane between charged plates
max	Corresponding to maximum amplification factor
min	Corresponding to minimum wave speed
p	Particle
s	Surface layer
stat	Stationary level in electrophoresis cell
w	Value at wall or surface
0	Charge neutral condition far from charged surface
0	Infinitely dilute suspension
0	Particle-free value in Couette flow
0	Reference or unperturbed state, standard conditions, free stream value, far from surface, value at entrance to channel
1,2	Conditions above and below downward-moving kinematic discontinuity, respectively

1,2   Conditions at hot side wall temperature T_1 and cold side wall temperature T_2, respectively

1,2   Conditions ahead of and behind bubble in capillary, respectively

1,2   With reference to spherical particles of radius a_1 and a_2, respectively, in mixture of both particles

+   With reference to positively charged ion

−   With reference to negatively charged ion

Superscripts

‾   With reference to ions or ionic concentrations in ion exchange resin phase

′   Perturbation component

*   Reduced dimensionless variable

Mathematical

D/Dt Material derivative, Eq. (2.2.3)

∇_s Surface gradient

$\langle \ \rangle$ Time average value

Author Index

379

Subject Index

Page numbers followed by "t" indicate tabular material.